AF556756

PEST MANAGEMENT IN CITRUS

ABOUT THE BOOK

Citrus is a highly prized fruit crop which is grown between 40° North and 40° South latitude in more than 125 countries with an estimated production of 93,749 million tonnes. The production figure would be much higher but for the fact that pests and diseases inflict enormous losses every year to the citrus crop the world over. This book aims at bringing together the scattered information about major pests and diseases as well as highlights strategies for their control.

The book comprises of chapters contributed by eminent experts in their respective fields and deal intensively and extensively with pests and diseases caused by insects, nematodes, fungi, viruses, bacteria, mollicutes and viroids. These chapters present a comprehensive picture of diseases and pests with particular focus on symptoms, transmission, host range, production of diagnostic reagents, detection technology, taxonomy, ecology, epidemiology and effective management practices of the concerned pests or diseases. At the end of each chapter there is an exhaustive bibliography that includes the latest references on important aspects and salient features of biotic stresses of citrus-which enhances the utility of the book.

This immensely useful book is designed to update the knowledge of undergraduate and postgraduate students of Economic Entomology, Mycology, Nematology and Plant Protection. It will also provide ample guidance to all persons engaged in National Agriculture Research System (NARS), professors and teachers of Agriculture Universities as well as progressive farmers, orchardists and plant protection workers.

The book will certainly be a valuable reference tool for scientists, professors, teachers, and students as well as for progressive farmers.

ABOUT THE AUTHORS

Dr. Krishna Prasad Srivastava, an eminent entomologist was born on April 1, 1938 at Varanasi. He started his research career from Indian Agricultural Research Institute, New Delhi as an entomologist from 1963. He served IARI in various capacities and retired from active service in 1998 from the most prestigious position of Principal Scientist. He worked under PL-480 from 1965 to 1977 and completed his Ph.D. degree during this period. His work on insect genetics and host-plant resistance specially in sorghum is an excellent piece of research work and has been recognised world over. He visited advanced laboratories in United States of America. Dr. Srivastava is not only an excellent researcher but also a good teacher and taught advanced courses in entomology and Plant Virology (virus-vector).

Dr. Srivastava is a Fellow of Plant protection Association of India (PPAI), Entomological Society of India, Indian Phytopathological Society and Indian Society of Genetics and Plant Breeding. He is life member of many professional academic societies and also served as Vice-President of PPAI. He is in the editorial Board of several national journals of reputes. He published over 150 research papers and authored two books on plant protection and also published scientific dictionaries on agriculture, botany and entomology. He is utilising his post-supernuation time fruitfully in expressing his ideas in various ways.

Dr. Y.S. Ahlawat, Professor of Virology at the Indian Agricultural Research Institute, New Delhi was born on October 10, 1943 in Kandera village of Meerut district. He did his M.Sc. and Ph.D. degrees from Agra University in Botany with specialisation in Plant Pathology. His major interest has been in Plant Virology where he made outstanding contribution specially in the area of ultrastructure and characterisation of viruses infecting citrus. He investigated several new viral diseases infecting a variety of crops. Indian citrus ringspot and citrus mosaic badnavirus are his very important discoveries and are first records in citrus.

The research contributions made by Dr. Ahlawat have been recognised Nationally and Internationally. He received Dr. H.D. Chapman award in 1991 from Indian society of Citriculture and again he was given the most prestigious international James M. and Adeline Research award in 1998 by the International Organisation of Citrus Virologists, Riverside, California on his outstanding contribution to citriculture.

Dr. Ahlawat is Fellow of Indian Phytopathological Society (IPS). He had been the Jt. Secretary (1989-91) and President, IPS in 1995. He served as Secretary General of the Indian Society of Citriculture from 1995-97 and presently is the Vice-President of this Society. He is member of several national and international scientific organisations and visited large number of advanced laboratories in USA, Brazil, Australia, China, France and Japan. He organised national and international scientific meetings. His students are serving in many prestigious organisations in the country.

PEST MANAGEMENT IN CITRUS

Edited by

K.P. SRIVASTAVA

Y.S. AHLAWAT

Forwarded by

S.P. GHOSH

RESEARCH PERIODICALS & BOOK PUBLISHING HOUSE

USA UK INDIA TAIWAN

ISBN 0-9656038-4-9

First Published 1999

Published by
RESEARCH PERIODICALS & BOOK PUBLISHING HOUSE
P.O. Box 720728, Houston, Texas 77272 USA
Tele. : (713) 779-2999, Fax : (713) 779-2992
Tollfree (800) 521-0061, E mail : rpbs@rpbs.com.

Printed by
RAINBOW PROCESSORS & PRINTERS
New Delhi-110059

Dedicated to

Dr. Dhamo K. Butani
(01.02.1923–13.12.1998)

Our
Friend, Philosopher & Guide

Tel. : Off. 91-11-3382534
Res : 91-11-6443196
Fax : 91-11-3382534/3387293
Telex : 031-62249 ICAR-IN
E-Mail : spg@icardelhi.nic.in

DR. S.P. GHOSH
Dy. Director General
(Horticulture)

भारतीय कृषि अनुसंधान परिषद्
कृषि भवन, डा० राजेन्द्र प्रसाद रोड, नई दिल्ली - 110001
INDIAN COUNCIL OF AGRICULTURAL RESEARCH
Krishi Bhawan, Dr. Rajender Prasad Road, New Delhi - 110001

FOREWORD

World citrus production totalled 87.9 million tons during 1994, covering approximately 23% of all fruits. Disease and insect pests posed serious threats to citrus industry in the past and plant protection is most important component in citriculture. The impact of growing disease-free citrus has been experimentally demonstrated through out the world. However, none of the potential improved practices can be fully realised so far, particularly in majority of the developing countries. Pathogen-detection technology and control of major pests and disease prevalent in different countries are of utmost importance.

The book entitled, "PEST MANAGEMENT IN CITRUS" edited by K.P. Srivastava and Y.S. Ahlawat is aimed to describe various methods available for pathogen detection which will certainly help in developing pest and disease-free plantations of citrus in developing countries. The authors of various chapters included in the book are specialists in their fields.

The book contains four major chapters on Introduction, Citrus Entomology, Major pathogenic diseases of citrus and Parasitic nematodes of citrus. The book is intended to aid plant protection scientists and other workers involved in citrus production worldwide. It will serve as guide to identification and understanding of citrus pests including diseases that afflict citrus trees, nursery plants, fruits and post harvest problems. The description of each pest and disease includes its geographical distribution, symptoms, transmission, host range, production of diagnostic reagents, detection technology, taxonomy, ecology, epidemiology, effective management practices and relative commercial importance. Illustrations have been given with a view to differentiate various pests and diseases. Brief accounts have been given of the life cycles of the biotic agents and epidemiology stressing features that can help in disease prevention and control.

The book is designed to update the knowledge of under graduate and post graduate student of economic entomology, mycology, virology, nematology and plant protection. It will also be useful to all who are related to the science of citriculture and will also be a useful tool in developing PRA in national Plant Quarantine System.

S. P. Ghosh

(S.P. GHOSH)

CONTENTS

List of Figures

List of Illustrations (Black & White)

List of Illustrations (Coloured)

PROLOGUE

CITRUS : Pest Problems are always there in each and every citrus orchard since evergreen trees are cultivated on diverse soil types even at sea level and at a height of over 2000 meters above sea level. The soil types for citrus cultivation range from loose sand with very low organic matter to loams, heavy clays and soils rich in organic contents. The soil pH range from low to high (4-9 pH) to even alkaline soils. The Irrigation facilities may also vary from irrigated desert to soil exposed to heavy rainfall. The citrus crop is a complex combination to various rootstocks and scions, cultivated in different ecological regions using different horticultural practices. Enormous amount of losses caused by number of insects, mites, diseases starting from seedling stage, to storage, which is an major constraint in production.

Orchardists till the end of last century used to accept these losses as natural inevitable calamities. But with the advance in science and technology, ways and means have been evolved and advocated to safeguard the orchard against the ravages of such relentless enemies. The losses can be substantially avoided or reduced by efficient pest management; for this, it is essential to identify the key pests of that particular area and to have sound knowledge of these insect pests, diseases — their habit, and habitat, time of occurrence, life-cycle etc.

As there are several hundred insect species reported damaging the citrus trees, some or the other species are always found in each and every orchard all the year round. Diseases are caused by fungi, bacteria, viruses, viroids, phytoplasma and other agents. Fungal diseases have been attributed to be associated with decline in certain parts of India. Bacterial diseases specially canker is responsible in reducing the quality and yield of produce. During last five decades virus diseases have emerged as main factor restraining the vigour and productivity of citrus in all the major citrus producing regions of the world. Plant parasitic nematodes, the microscopic worms, are very widespread in the developed as well as the developing countries. It is neither practicable nor economical to keep on spraying the trees throughout the year. So only feasible alternative is to 'learn to live with the pests'. Manage the pest population in such a way that no pest goes beyond the limit of economic-injury-level. Budwood certification programmes can achieve the target of producing healthy citrus crop.

Citrus orchards are contiguous both in space as well as time — as these are usually grown over extensive continuous areas and maintain for several decades. Ecologically such conditions are very conducive for rapid multiplication of insect pests and diseases. Since the food material is available to insects throughout the year the carryover of the pest cannot be easily avoided. In the last four decades a number of virus diseases have appeared in most region of the world where citrus is grown. Virus diseases of citrus are comparatively more important due to their perpetuation through inadvertent vegetative propagation which also help in wide spread of the disease in different geographic regions. Use of resistant varieties has limited scope as these are difficult to develop quickly and can be planted only at the time of stating a new orchard. Similarly, cultural practices, such as, crop rotation, change of sowing dates, use of healthy seed materials or seed treated with pesticides cannot be practiced in the established orchards. Hence utmost care, vigilance, surveillance and cooperative endeavour are necessary to keep the orchards free of pest problems.

During the last half a century more than one thousand scientific papers and articles have appeared in various journals, magazines and technical reports in India and abroad on insect pests and diseases of citrus orchards. Besides these are chapters in various books, notable among these being; *Insects of citrus and other subtropical fruits* by H.J. Quayle (1938), *Handbook of Economic Entomology for South India* by T.V. Ramakrishna Ayyar (1943, reprinted 1963); *Subtropical fruit pests* by Walter Ebling; *Important Fruit Pests of North-west India*, by Hem Singh Pruthi and

H.N. Batra (1960); *Text book of Agricultural Entomology*, by Hem Singh Pruthi (1969); *Insect Pests of crops*, by S. Pradhan (1969, revised 1983); *Insects and Mites of Crops in India* by M.R.G.K. Nair (1975, revised 1983), *Insect pests of citrus fruits* by D.U. Garg (1978); *Insects and Fruits* by Dhamo K. Butani (1979); *Citrus decline in India, causes and control* by K.L. Chadha, N.S. Randhawa, O.S. Bindra, J.S. Chohan and L.C. Knorr (1970), *Agricultural Insect Pests of the Tropics and their Control*, by Dennis S. Hill (1975, revised 1983), *Virus and Mycoplasma diseases of plants in India*, by S.P. Raychaudhuri and T.K. Nairiani (IInd edition, 1989), *Diseases, Pests and Weeds in Tropical Crops*, edited by J. Kranz, H. Schmutterer and W. Koch (1977); and *Virus and Mycoplasma Diseases of Fruits crops in India* by V.V. Chenulu and Y.S. Ahlawat (1993), *Problems of Citurs diseases in India* by S.P. Raychaudhuri and Y.S. Ahlawat (1982).

Morphology, nature and extent of damage and bionomics of almost all the major insect pests, diseases and most of the minor ones, have been studied by various workers and control measures suggested. But all this information is laying scattered in various journals and magazines. Moreover, it is not unlikely than a insect and disease recorded today as a minor pest of little economic importance, may become major pest tomorrow. So with this aim in view, the available information on both major and minor pests of citrus trees have been collected and presented herein along with their recommended control measures. The appendices include lists of insects, mites injurious to citrus trees. Besides, an exhaustive bibliography has been appended. It is high time that budwood certification and cross protection programmes are taken up in India as the technology for these programmes is fast developing in India and elsewhere. A rupee spent in such management strategies will multiply thousand fold.

ACKNOWLEDGEMENT

The authors wish to place on record their sincere thanks to Dr. S.P. Ghosh, Deputy Director General, Indian Council of Agricultural Research, New Delhi for kindly writing the foreword and various suggestions for finalising this book. The authors are extremely grateful to late Dr. Dhamo K. Butani who expired while the book was in its final stage for his kind help in various ways.

We also appreciate the efforts made by Smt. Sangeeta. Namrata and Shweta for line drawing and proof reading. The authors are also thankful Mrs. Kusum Srivastava and Mrs. Urmila Ahlawat for their constant cooperation that made possible for this treatise to see the light of the day. Thanks are also due to Mr. Arvind Jain for his keen interest in publishing this book in its present form.

Last but not the least the authors are thankful to Prof. Ram Badan Singh, Director and Prof. Anupam Varma, Dean, Indian Agricultural research Institute, New Delhi for encouragement and help in various ways.

K.P. Srivastava
Y.S. Ahlawat

1

INTRODUCTION

K.P. Srivastava and Dhamo K. Butani

Citrus — a World famous fruit, belongs to family Rutaceæ (Sapindales : Dicotyledons). The family comprises of 150 genera and about 1600 species of plants found predominantly in tropical, sub-tropical and warm temperate regions of the World, especially in arid zones. The Rutaceæ is further divided into seven subfamilies consisting of 11 tribes (Engler, 1931). Three of these subfamilies, namely Aurantioideæ (33 genera), Rutoidæ (30 genera) and Toddalioideæ (25 ganera) are found in South-east Asia. The remaining four subfamilies are spread over in the New World, Australia and other parts of Asia. Aurantioideæ — the orange subfamily, classed two tribes, Clauseneæ and Citreæ, classified into six subtribes containing 215 species and 65 varieties of plants (Swingle and Reece, 1967). 'Citrus Fruit Trees Group' comes under Citrinaæ which includes the genus *Citrus* and its close relatives, namely *Fortunella, Eremocitrus, Clymenia, Poncirus* and *Microcitrus*. Of these only *Fortunella* (kumquats), *Poncirus* (trifoliate orange) and *Citrus* (eight species) are commercially important (Chapot, 1975). The important species of *Citrus* include, *C.reticulata* Blanco (mandarins), *C.grandis* (Linnæus) Osbeck (shaddo or pummelo), *C.paradisi* Macf. (grape fruit), *C.aurantium* Linnæus (sour orange), *C.sinensis* (Linnaeus) Osbeck, (sweet orange), *C.aurantifolia* (Christon) Swing (acid lime), *C.limon* (Linnaeus) Burm. (lemon) and *C.medica* Linnæus (citron) albeit, the exact number of species of *Citrus* has not been agreed upon, as new data emerging from chemistry and molecular biology are rapidly changing the understanding of the taxonomy of *Citrus* species and their relatives (Jones, 1990).

Citrus is grown in belt between 40° North and 40° South

latitude in more than 125 countries with an estimated production of 93.749 million tonnes (FAO, 1996). The perennial evergreen trees are cultivated on diverse soil types even at sea level and at a height of over 2000 meters above sea level (Ducharme, 1969). The soil types of citrus cultivation range from loose sand with very low organic matter to loams, heavy clays and soil rich in organic matter. The soil pH range from low to high to even alkaline soils. The irrigation facilities may also vary from irrigated deserts to soil exposed to heavy rainfall. Citrus orchard is a complex combination of various rootstocks and scions cultivated in different ecological regions using different horticultural practices.

True citrus fruits, Kumquats, trifoliate orange, mandarin, sour orange, sweet orange, lime, citron, lemon and allied types including natural hybrids are native to a very large Asiatic area extending from the Himalayan foot hills of North-eastern India to North Central China and the Philippines in the East, Myanmar, Thailand, Indonesia and New Caledonia in the South-east. Grape fruit (*C.paradisi*) originated from Barbados (West Indies) sometime before 1790 as a mutant or possibly a hybrid of species introduced from the Far East. The two main species of *Fortunella* are native of Southern China are genus *Poncirus* to Central and Northern China.

Citrus, a highly prized fruit crop having significant importance in the fruit economy of our country is the second largest fruit industry in India with respect to area under cultivation and third largest with respect to production — after mango and banana — ranks sixth among top citrus producing countries of the World. The major citrus producing countries are, USA, Spain, India, Italy, Japan, Argentina, Mexico, Brazil, Morocco, Algeria, Greece, South Africa, Australia, Israel, Egypt and Jamica. In India, the estimated area under cultivation is 0.454 million hectares a production of 3.8 million tonnes, which constitutes 4.8 per cent of the total World production (Anonymous, 1981). The oranges occupy the first place among the various *Citrus* species grown in India. The average yields in India is about 7.51 tonnes per hectare which is much lower than 20 to 25 tonnes per hectare in Brazil, USA, China, Spain and Mexico (Singh, 1997).

Ebeling (1959) has listed 875 species of insects and mites known to feed on citrus trees in different parts of the World; this number by now must have swelled to over one thousand species. In India, however, only about 250 species of insects have been reported feeding on various *Citrus* species. Of these those that cause heavy losses regularly, included lemon butterflies. Citrus psylla, whiteflies., coccids (mealybugs, and scale insects) and citrus leaf miners. In neglected orchards, bark eating caterpillars, stem boring beetles, leaf eating caterpillars, fruit sucking moths and fruitflies have also been recorded as major pests. Besides, termites (especially in sandy soils), aphids, thrips, leaf eating beetle and weevils, fruit sucking bugs, blossom midge, flower moth and wasps are also often found causing some negligible to minor damage. Among the non-insect pests, mention may be made of mites and nematodes that cause regular loss.

Lemon butterflies (*Papilio* species) are the most destructive pests of citrus seedlings. These are widely distributed from Taiwan to Saudi Arabia including Indian subcontinent. The caterpillars feed voraciously on leaves, defoliating the entire seedling or the young tree. These are found on almost all *Citrus* species, though malta (*C.sinensis*) is their preferred host. In India *Papilio demoleus demoleus* Linnæus is the most commonly found species. Besides, *P.helenus* Linnæus, *P.helenus daksha* Linnæus, *P. machaon asiatica* Menestries, *P.menmon* Linnæus, *P.polymnestor* Cramer, *P.polytes alphenor* Cramer, *P.polytes pammon* Linnæus, *P. polytes polytes* Linnæus, *P.porinda* Moore and *P.rumtanzovia* Eschscholz have also been recorded (Butani, 1979).

Epidemic of citrus blackfly, *Aleurocanthus woglumi* on Nagpur mandarin during last 2-3 decades had put the citrus industry of Central India in the doldrums. Due to sap sucking by nymphs and adults coupled with copious secretion of honeydew by these insects, results in rapid growth of sooty mould (*Capnodium* species), locally known as *Kolshi*. The fungus covers the entire leaf surface with a thin superficial black coating, which interferes with the photosynthetic activity of the plant with the result growth of the tree and its fruit bearing capacity is adversely affected. Besides, this species, *Dialeurodes citri* also cause substantial loss especially in Punjab and elsewhere. Citrus psylla, *Diaphorina citri* Kuwayama is another homopteran serious pest especially of young flushes of citrus trees. Due to desapping by enormous number of almost stationary nymphs, the plants are devitalised their growth is arrested and fruiting capacity adversely affected. The insect is active vector of phytoplasma-like organism that causes the dreaded 'Greening disease' which in turn contributes decline of citrus orchard through direct and indirect carrier of the disease (Ahlawat, 1997).

Citrus bark eating caterpillars, *Inderbela quadrinotata* Walker and *T.tetraonis* Moore are the important pests of citrus trees quite serious in neglected orchards. Tender shoots when attacked wilt and dry up. The pest being nocturnal in habit, is seldom noticed.

Citrus trunk borer, *Monohanmus versteegi* Nitzoma is serious pest of mandarin orange especially in North-eastern region and Sikkim. In Assam 15 to 60 per cent orchards are infested with this trunk borer. The pest is also responsible for citrus decline in West Bengal. The grubs bore horizontally into the trunk near the tree base and make tunnel near the pith and the tree eventually dies, especially the young ones.

Citrus leafminer, *Phyllocnistis citrella* Stainton a native of tropical and subtropical Asia, attacks all species of *Citrus* and also some related species of family Rutaceæ. The leafminers are usually found on abaxial surface of leaves but may also be seen on adaxial surface, causing the typical injury, characterised by zigzag galleries coupled with epidermis appearing as a silvery film. The finally results in necrotic tissue, leaf curling and often abscission of the infested tissue. Leafminer larvae per leaf and the number of days of mining were observed to be positively correlated with visual estimates of leaf damage and negatively correlated with net photosynthesis (Schaffer *et al.*, 1997). Nine citrus related species and 349 root stocks were evaluated for resistance to the leafminer. Two citroids, *Murraya kœnigii*

(Linnæus) Sprengel and *Glycosmis-pentaphylla* (Retzius) Correa, exhibited antibiotic effects against first instar larvae and may provide a source of resistant gene to breeders or biotechnologists (Jacas *et al.*, 1997).

Thrips (Thysanoptera) have also been observed as pest of citrus trees though of minor importance. Citrus thrips, *Scirtothrips citri* (Moultan) infests the small citrus fruitlets soon after petal fall, causing scarring on the fruit rind.

Aphids are another important pest of citrus trees. Brown citrus aphid *Toxoptera citricida* (Krirkady) and citrus tristeza virus which have combined and created a most serious problem all over the World. Disastrous epidemic of this virus and aphid combination have been reported from Argentina, Brazil, Colombia and Peru by Rocha-Pena *et al.*, 1995).

The brown citrus aphid is believed to have originated from China. Until 1900, it was confined to South-east Asia, Australia, New Zealand, the Pacific Islands, South Africa and South America but thereafter it spread widely and got established in several, countries especially in Central America and many Islands in the Caribbean basin.

Tsai (1998) studies the development and reproduction of this aphid on eight host plants-rough lemon (*Citrus jambhiri* Lush), sour orange (*C.aurantium*), box orange (*Severinia buxifolia* (Poir) Tenene, calamondin X *Citrofortunella microcarpa* (Bunge) Winjnands, Lime berry (*Triphasia trifolia* (Burm f) P. Wilson and orange jassamenia (*Murraya paniculata* (Linnæus) Jack. The average number of nymphs produced by a female were 17.7 to 58.8 The mean population generation time ranged between 9.7 and 12.2 days.

Ants also play an important role in the dynamics of citrus production and successful pest management. Some ant species disturb natural enemies as they forage in the trees for honeydew produced by homopteran insects. Other species such as leaf cutting ants and fire ants cause foliar damage or sting and bite the workers so severely, that it results in interference with the production of harvest. About 40 species of ants have been reported on citrus from 30 countries (Haney, 1988).

Mites — not insects but arachnids belonging to phylum Arthropoda — have also been recorded feeding and desapping the citrus trees. Dhooria and Butani (1984) reported seven species of mites as pests of citrus, from India, namely, *Eutetranychus orientalis* (Klein), *Panonychus citri* (Mc Gregor), *Schizotetranchus hindustanicus* (Hirst), *Brevipalous californicus* (Banks), *Polyphagotarsonemus latus* (Banks), *Aceria sheldoni* (Ewing), and *P.oleivora* (Ashmead). Vacante *et al.* (1988) listed 164 species of mites belonging to 26 different families from Mediterranean area alone.

The citrus red mite, *Phyllocoptrula oleivora* (Ashmead) infests fruits, leaves and young twigs of all citrus species in most humid regions of the World. Due to their short life-cycle and high reproductive rate mite population can often reach phenomenal level in just a few weeks during warm Summer. Bergh and Mc Coy (1997) studied the aerial dispersal of citrus red mite and could capture the mites upto 135 meters from an infested grove. Phillips and Walker (1997) reported increase in flower and young fruit abscission caused by eriophyid mite, *Aceria sheldoni* and reported abscission rates of both flowers and fruits increased significantly with increasing the mite population causing distortion and further observed decrease in auxin activity and other biochemical changes in the axillary buds infested with this mite.

Enormous losses are caused by number of diseases starting from seedling stage to storage, which is a major constraint in citrus production. Diseases are caused by fungi, bacteria, viruses, viroid, phytoplasma etc. Various fungal diseases attack almost all citrus species during all the stages of their growth. The influence of these diseases is highly destructive. Bacterial diseases are responsible for reducing the quality and yield of citrus fruits. Some bacterial diseases often cause extensive losses to citrus orchards.

Fungal diseases have been attributed to be associated with citrus decline in certain parts of India. The fungi that cause diseases in different commercial species of citrus are primarily the different species of *Phytophthora, Diplodia, Colletotricum, Fusarium, Macrophomina, Curvularia, Geotrichum, Oospera, Rhizoctenia* and *Alternaria.* Among these *Phytophthora* species play the major role in citrus decline especially in ill-drained soils. *Colletotricum gloeosporioides, D.natelensis, C.tuberculata* and *Fusarium* species have been reported to be associated with citrus dieback affected trees in India. However, the actual role of these fungi in decline complex needs to be further ascertained.

In Punjab gummosis of citrus (*Phytophthora* spp.) is one of the major problems in raising a successful citrus orchard. The infested trees show symptoms of foot rotting with profused gumming, typical trunk girdling with pale green foliage, defoliation and twigs dieback leading to gradual death of trees. Thus this problem is causing a serious threat to the future of citriculture. The prevalence of the gummosis or foot rot diseases in citrus orchards in Punjab varies between 8.3 and 45.3 per cent. (Thind and Sharma, 1996).

Other fungal diseases of citrus reported from India include, pink disease caused by *Botryobasidium salmonicolor* (Berk and Brown) Venkatarayan; powdery mildew caused by *Odium tingitanum* Corda; scab caused by *Elsinœ fawcelli* and stem rot or twig blight caused by *Diplodia natalensis* Evans, *Fusarium solani* f.*auranticola, Curvularia tuberculata* Jain and *Hendersonia toruloidea* Natt. (Lele and Butani, 1975).

Citrus canker caused by *Xanthomonas exenopodis citri* (Hasse) Dowson is the most important bacterial disease of citrus in India and elsewhere. The heavy infection of this pathogen particularly on acid lime is dreadful. The infection of this bacterium can kill the tree, if not properly controlled. Recently a xylem limited bacterium *Xylella fastidiosa* has been found to infect citrus trees in Brazil, causing a serious disease referred as variegated chlorosis.

During last five decades, virus disease have emerged as main factor restraining the vigour and productivity of citrus in all the citrus producing regions of the World. Tristeza (four strains) and greening transmitted by *Diaphorina citri* are the most important while leaf curl transmitted by whitefly, *Bemisia tabaci* Gennadius though of regular occurrence is of lesser importance.

Three viruses, Citrus tristeza closterovirus (CTV), citrus ringspot capillovirus (CRVS) and citrus mosaic badnavirus (CMBV); *Liberobactor asiatium* causing greening disease and the viroids (exocortis and yellow vein) have been identified as the major pathogens of the decline syndrome. Studies on these pathogens revealed the presence of strains in CTV and greening bacterium and biotypes in their vectors. Such informations are useful for the management of these diseases by cross protection technology and identification of suitable areas for nurseries.

In India, the yellow corky vein and the leaf yellow mid vein are characterised with similar symptoms as yellowing of the vein followed by corking of the vein on the underside of the leaves. Yellowing does not extend down the petioles into branches; corked leaves become crinkled and curled. Affected trees give not only poor yield but poor quality of fruits as well. Azad (1993) compared yield of two acid lime varieties — *kagzi* and *goal nebu* and reported that the two varieties suffered a weight loss of 60.4% and 89.7% respectively in the yield as compared to healthy trees.

Tristeza disease of citrus, caused by citrus tristeza virus (CTV) — a closterovirus, occurs in almost all citrus producing areas of the World. CTV is a phloem limited virus transmitted semi-persistently by aphid-feeding and also by grafting infected budwood. Severe epidemics of tristeza occurred in Argentina and Brazil during thirties when over 30 million citrus trees were killed (Bar-Joseph *et al.*, 1989; Costa, 1956) and in Spain during 1960s and Venezuela during eighties where 10.0 and 6.6 million trees respectively were killed (Rocha-Peno, 1995).

Four aphid species — cotton aphid (*Aphis gossypii* Glover), spirea aphid (*A.spiræcola*), black citrus aphid (*Toxoptera aurantii* Boyer de Fonscolombe) and brown citrus aphid (*T.citricida*) have been as associated with the natural movement of CTV. The most efficient vector of CTV all over the Worldwide is *T.citricida* which is 25 times more efficient in transmitting virus followed by *A.gossypii*. Gottwald *et al.* (1998) studied the spread pattern of CTV by monoclonal antibody probes via enzyme-linked immunosorbent assay (ELISA) in four citrus orchards in Northern Costa Rica and four orchards in Dominica Republic following the introduction of *T.citricida.* According to Gompertz non-linear model the beta-bionomical index of dispersion for various quadrat sizes suggested aggregations of CTV positive trees. Mehta *et al.* (1998) developed a rapid and simple reverse-transcription polymerase chain reaction (RT-PCR) method for detection of CTV in three aphid species. These include efficient CTV vector *T.citricida,* cotton aphid *A.gossypii* and non vector green peach aphid *Myzus persicæ* (Sulzer). A short procedure for nucleic acid extraction from single or groups of aphids was developed.

Citrus tristeza virus (CTV), a closterovirus has positive sense RNA genome encapsidated in flexuous particle about 2,000 nm long (Bar-Joseph and Lee, 1989). Virions contain two capsid proteins, a 25 kDa coat protein covering about 95% of the particle length, and a diverged 27 kDa coat protein (Pappu *et al.*, 1994; Sekiya *et al.*, 1991 and Febres *et al.*, 1996) that covers one end of the particle, forming a 'rattlesnake' structure.

Two complete genomic sequence from CTV isolates 'T-36 from Florida 1 Karaseu *et al.*, 1995 and Pappu *et al.*, 1997) and VT from Israel (Mawarsi *et al.*, 1996) have been reported recently and found to contain 19,296 and 19,226 nuclotides (nt) respectively, organized in 12 open reading frames (ORFs) potentially encoding at least 17 protein products. López, *et al.*, 1998 studied the molecular variability of the 5'- and 3'- terminal regions of CTV and concluded the presence of a conserved putative "Zinc finger" domain adjacent to a basic region in p^{23}, the predicted product of ORF, suggest that this protein might act as a regulatory factor during virus replication.

There are number of post harvest diseases reported on citrus fruits all over the World. Alternaria rot [*Alternaria alternata* (Fr.) Keissler, and *A.citri* Ell & Pierce], Anthracnose [*Glomorella cingulata* (Stonem) Spauld & v. Schrenk] conidial state [*Collectrotrichum gloesporiodes* (Penz) Sacc.], Aspergillus black mould rot (*Aspergillus niger* v. Tieghen) bacterial canker (*Xanthomonas campestris pv. citri* (Hasse) Dye], black pit (*Pseudomonas syringæ* pv. *syringæ* van Hall), black spot (*Guignardia citricarpa* Kiely), brown rot (*Phytophthora* spp.), cottony rot *Sclerotinia minor* Jagger, *S.sclerotiorum* (Lib.) de Bary, Fusarium rot (*Fusarium* spp.), greasy spot rind bloch (*Mycosphaerella citri* Whiteside), green mould rot (*Penicillium digitatum* Sacc.), blue mould rot (*P. italicum* Wehmer), grey mould rot [*Botryotinia fuckeliana* (de Bary) Whetzel, conidial state; *Batrytis cinerea* Pers.], Melanose (*Diaporthe citri Wolf.*) scab (*Elsinœ* and *Sphaceloma* spp.), Septoria spot (*Septoria* spp., sour rot (*Geotrichum candidum* Likk), stem-end rot (*Botryospharia ribis* Grossenb. & Duggar, conidial state *Dothiorella gregaria* Sacc., *Diaporthe citri* Wolf, conidial state *Phomopsis citri* Fawcett, *Physalospora rhodina* (Berk & Curt.) Cooke, conidial state, *Botryodiplodia theobromæ* Pat.) Trichogramma rot (*Trichoderma viride* Perb. ex S.F. Gray), Aspergillus rots (*Aspergillus niger* & *A. flavus* Link ex *A.niveus* and *A.varioclor*), ceratocystis rot (*Ceratocystis fimbriata* Ell, & Halsted), *Cercospora* spot [*Cercosphora angolenis* now called as *Phæoramularia angolonsis* (de Carvalho & O. Mendes) P.M. Kirk], charcoal rot (*Macrophomina phaseolina*), freckle (*Ascochyta pisi*) pink mould rot (*Trichotheciu roseum*), Pleospora rot [*Pleosphora herbarum* (Pers) Rabenh], Rhizophus rot (*Rhizopus stolonifer*), sooty blotch [*Gloeodes pomigena* (Schw.) and *Stomiopeltis citri*], sooty mould (*Capnodium cladosphorium* or *Meliola* Yeast rot [*Candida krusei* (Castell)]. Apart from diseases in general, grapefruits, kemons and limes are most susceptible; oranges least while mandarins are intermediate to chilling injury.

Fruits harvested early in the season may be 'degreened' by ethylene released at temperatures nearer to 20°C penicillium moulds can be established in unhealed wounds (Snoden, 1990). In India post harvest losses of orange varied from 6.0 and 30.7 per cent depending upon the mode of transport, transit time and season (Madan and Ullasa, 1993). Effectiveness of fungicide imazalil provided very effective control of citrus green mould (*Penicillium digitatum*) under heated aqueous solution (Similanich *et al.*, 1997). At Costa Rica epidemiology studies were conducted on greasy spot, caused by *Mycosphærella citri* Whiteside by Hidalgo *et al.* (1997). Wounds on citrus fruits are the primary infection for green and blue moulds of citrus caused by *Penicillium digitatum* Sacc. and *P.italicum* Wehmar respectively.*Pseudomonas syringæ* Stroaens (ESC-10 and ESC-11) provided effective biological control of blue moulds (Smilanick *et al.*, 1996).

Economic loss due to blue mold, *Lasiodiplodia* rot incidence varied from 1.5 to 2.1 per cent (Godava, 1994). Certain antagonistic micro-organisms effective against post harvest disease are reported by Pathak (1998), these include *Bacillus subtilis* for *Lasiodiplodia theobromæ, Trichoderma viridis* for green mold *Penicillium digitatum, Pseudomonas syringæ* and *P.cepacia* for blue mold *Penicillium* rot.

Plant parasitic nematodes — the microscopic thread-like worms are very widespread throughout the World. These parasites penetrate the roots, remove nutrients from plant system, hamper plant growth and make the plants vulnerable to attack by other organisms. Roots being the major target of their attack, nematodes remain underground, out of sight of cultivators as well as scientists (nematologists) and cause unhindered damage which is cumulative in nature. Due to these reasons these subterranean organism are known as 'hidden enemies of farmers' as these are responsible for causing 12 per cent annual yields loss $ (= US $ 77 billions) worldwide to the major life sustaining and economically important crops (Sasser and Fredhman, 1987).

The first record of a nematode *Heterodera radicicola* found associate with citrus roots was reported by Neal (1887) from Florida (USA). Since then many other parasitic forms of nematodes associated with citrus roots have been reported from several parts of the World. However, only six species have proved to be pathogenic; the two most economically important being *Tylenchulus semipenetranus* and *Radopholus similus*, the disease is called spreading decline. Effected plants show reduced vigour, small leaves, chlorosis, defoliation, dieback of twigs and small fruits. The yields may be reduced that maintenance of the orchards become quite uneconomical. Till 1969, 189 species of nematodes belonging to 39 genera were reported in association with citrus roots (Ducharme, 1969) and subsequently in less than 20 years the number of associated species increased to more than 200 within 44 genera worldwide (Anonymous, 1986). Some important new species described after 1990 include root knot nematode, *Meloidogyne jianyangensis* on mandarin oranges (Zang-Baojun *et al.*, 1990), *M.citri* on unshui in China (Shaosheng *et al.*, 1990), *Xiphinema karochensis* (Nasira *et al.*, 1992) and *Paralongidorus lemoni* on citrus (Nasira *et al.*, 1993), another species of burrowing nematode (*Radopholus citri*) in Indonesia (Machon and Bridge, 1996). In one of the surveys conducted around Nagpur, mandarin orchard, citrus nematode *T.semipenetrans* was observed to be the most prominent in 73.7% orchards. The population in soil ranged 8-2880 nematode prominent larvae per 250 cc soil and in roots from 2-900 nematodes per gram roots.

The strategy for the control of insect pests, mites, diseases and nematodes in citrus orchards is necessarily to be different form the other crops because of the nature of utilisation in form as fresh fruit jam, squash, marmalades and fruit juice. This invite for extra attention especially in case of chemical control. Pesticides that are higher toxic and are known to leave hazardous residues cannot be recommended off-hand. Proper care and precautions need be taken while recommending the use of pesticides. Thus reference has to be given to other known effective methods of control by which the population build-up of insects, mites, diseases and nematodes can be avoided. These observation led to seriousness on integrated pest management including use of resistant varieties, manipulation of agronomic practices and biological control measures to minimise the use of insecticides. For this, it is of paramount importance to know the habits, habitat, distribution, nature of damage, life table and seasonal occurrence of various pests. It is also essential to work out the economic injury level so that control operations are undertaken at suitable time. Varieties that are less damaged by pests may have to be identified and developed for commercial cultivation. It can be claimed that relevance of integrated control and pest management is required more in different citus species than in other crops.

REFERENCES

Ahlawat, Y.S., 1997 : Viruses, greening bacterium and viroids associated with citrus (*Citrus* species) decline in India. *Indian Journal of Agricultural Sciences*, **67** : 52-57, New Delhi.

Anonymous, 1986 : Plant parasitic nematodes of bananas, citrus, coffee, grapes and tobacco. Union Carbide Agricultural Products Company, N.C. 71p.

Anonymous, 1998 : Indian Horticulture Database, National Horticulture Board.

Azad, F., 1993 : Reduction in yield of acid limes unfected with citrus corky vein virus in Assam. p. 458. *In*: *Proceeding 12th Conference 10 CV, Riverside*, Califorinia.

Bar-Joseph, M. and R.E. Lee, 1989 : Citrus tristeza virus CMI/AAB Descriptions, *Plant Virus* : 353.

Bar-Joseph, M., R. Marcus and R.F. Lee, 1989 : The continuous challenge of citrus tristeeza virus control. *Annual Review Phytopathology*, 26 : 292-316, Palo Alto, California.

Bergh, J.C. and C.W. MC Coy, 1997 : Aerial dispersal of citrus rust mite (Acari: Eriophyidae) from Florida citrus growers. *Environmental Entomology*, **26** (2) : 256-457, Lanham, Maryland.

Butani, Dhamo K., 1979 : Insect pest of citrus and their control, *Pesticides*, **13** (4) : 27-33, Mumbai.

Cambra, M., D. Villaiba and P. Moreno, 1988 : Present, situation of the citrus tristeza virus in the Valencian community. p. 1-7. *In* : L.W. Timmer, S.M. Garnsey, L. Navarro (Eds.), : *Proceedings Conference International Citrus Virology* 10th, Riverside, California.

Chapot, H., 1975 : The Citrus plant. *In* : E. Hoefinger (ed.) *Citrus Technical Monograph 4*, CIBA-GEIGY Agrochemicals, Basele, Switzerland — 6-13.

Costa, A.S., 1956 : Present status of the tristeza disease if citrus in South America. *FAO, Plant Protection Bulletin*, **4** : 97-105, Rome.

Dhooria, M.S. And Dhamo K. Butani, 1784 : Citrus mites, *Eutetranychus orientalis* (Klein) and its control. *Pesticides*, **18** (9) : 35-38, Mumbai.

Ducharme, E.P., 1954 : Cause and nature of spreading decline of citrus. *Proceedings of the Florida State Horticultural Society*, **67** : 75-81, Deland.

Ducharme, E.P., 1969 : Nematode problem of citrus. p. 222-237. *In*: J.E. Peachy (ed.) *Nematodes of tropical Crops*, Commonwealth Agricultural Bureau, England.

Ebeling, Walter, 1959 : *Subtropical Fruit Pests*, University of California Press, Los Angeles, 436pp., Los Angeles.

Engler, A., 1931 : Rutaceae. p. 187-357. *In* : A. Engler and K. Pranld (Eds.) *Die naturilichem of lanzenfamilien*, band 19 A. Verlag von Wihelm engelmann Leipzig, Leipzig.

Fao Yearbook, 1996 : FAO Statistic series, Vol. 50, No. 135, Rome, Italy, 1997.

Febres, V.J., L. Ashoulin, M. Mawassi, A. Frank, M. Bar-Joseph, K.L. Manjunath, R.F. Lee and C.L. Niblett, 1996 : The p. 27 protein is present at one aend of citrus tristeza virus perticle. *Phytopathology*,

86: 1331-1335, St. Paul, Minnesota.

GODARA, S.L., 1994 : Studies on postharvest diseases of orange fruits. Ph.D. thesis Rajasthan Agricultural University, Bikaner.

GOTTWALD, T.R., S.M. GARNSEZ AND J. BORBON, 1988 : Increase and pattern of spread of citrus tristeza virus infections in Costa Rica and the Dominican Republic in the presence of the brown citrus aphic, *Toxoptera citricida, Phytopathology*, **88** : 621-636, St. Paul, Minnesota.

HANEY, P.B., 1988 : Identification, ecology and control of the ants in citrus. A World survey : p. 1227-1251. *In* : R. Goren and K. Mondel (eds.) *Proceedings of Sixth International citrus Congress* Tel Aviv (March 6-11, 1988).

HIDALGO, H., T.B. SUTTON AND F. ARAUZ, 1997 : Epidemiology and control of Citrus greasy spot on Valencia orange in the humid tropics of Costa Rica. *Plant Disease*, **81** : 1015-1022, St. Paul, Minnesota.

JACAS, J.A., A. GARRIDO, C. MARGAIX, J. FORNER, A. ALCAIDE AND J.A. PINA, 1997 : Screening of different citurs root stocks and citrus-related species for resistance to *Phyllocnistis citrella* (Lepidoptera: Gracillariidæ). *Crop Protection*, **16** (3) : 301-305, Guildford survey, U.K.

JONES, D.T., 1990 : A background for the utilisatioin of citrus genetic resources in South-East Asia, I-classification of the Aurantioideæ. p. 31-37. *In* : *Proceedings 4th International Asia Pacific Conference on Citrus Rehabilitation*. Chiang Mai, Thailand, UNDP-FAO Regional project.

KARASEV, A.V., V.P. BOYKO, S. GOWDA, O.V. NIKOLAEVA, M.E. HILF, E.V. KOONIN, C.L. NIBLETT, K. CLINE, D.J. GUMPF, R.F. LEE, S.M. GARNSEY, D.J. LEWANDOWSKI AND W.O. DAWSON, 1995 : Complete sequence of citrus tristeza virus RNA genome. *Viroogy*, **208** : 511-520, Baltimore, New York, N.Y,

LELE, V.C. AND DHAMO K. BUTANI, 1975 : Trends in plant disease control in India with Particular reference to fruit crops. *Pesticides Annual, 1975* : 75-96, Mumbai.

LÓPEZ, C., M.A. AYLLÓN, J. NAVAS-CASTILLO, J. GUERRI, P. MORENO AND R. FLORES, 1998 : Molecular variability of the 5'- and 3'- terminal regions of citrus tristeza virus RNA. *Phytopathology*, **88** : 685-91, St. Paul, Minnesota.

MACHON, J.E. AND J. BRIDGE, 1996 : *Radophelus citri* n.sp. (Tylanchida: Pratylechidæ) and its pathogenicity on citrus. *Fundamental and Applied Nematology*, **19** (2) : 127-133, Rue Linois, Paris.

MADAN, M.S. AND B.A. ULLASA, 1993 : Post harvest losses in fruits p.1795-1810. *In*: K.L. Chadha and O. Pareek (eds.), *Advances in Horticulture*, **4**, New Delhi.

MAWASSI, M., E. MIETKIEWSKA, R. GOFMAN, G. YANG AND M. BAR-JOSEPH, 1996 : Unusual sequence relationship between two isolates of citrus tristeza virus. *Journal of General Virology*, **77** : 2359-2369 London.

MEHTA, P., R.H. BRILANSKY AND S. GOWDA, 1997 : Reverse-Transcription polymeras chain reaction detection of citrus tristeza virus on aphids. *Plant Disease*, **81** : 1066-1069, St. Paul, Minnesota.

NASIRA, R.K. AND M.A. MAQBOOL, 1992 : Description of *Xiphinema karachiense* sp.n. and morphometric data on *Enchodelus macrodorus* (de Man 1880) Thome 1939 (Nematoda: Dorylaimida) from Pakistan.

Fundamental and Applied Nematology, **15** : 421-426, Rue Linois, Paris.

NEAL, J.C., 1989 : The root knot disease of the peach, orange and other plants in Florida, due to the work of Angullila. *Bulletin of Bureau of Entomology*, U.S. Department of Agriculture, No. 20, 31pp., Washington, D.C.

PAPPU, H.R., A.V. KARASEV, E.J. ANDERSON, S.S. PAPPU, M.E. HILF, V.J. FEBRES, R.M.G. ECKLOFF; M. MCCAFFERY, V. BOYKO, S. GOWDA, V.V. DOIJA, E.V. KOONIN, D.J. GUMPF, K.C. GLINE, S.M. GARNESEY, W.O. DAWSON, R.F. LEE AND C.L. NIBLETT, 1994 : Nuclotide sequence and organization of eight 3' open reading frame of the citrus tristeza closterovinus, *Virology*, **199** : 35-46, New York, N.Y.

PAPPU, H.R., S.S. PAPPU, C.L. NIBLETT, R.F. LEE AND E. CIVEROLO, 1997 : Comparative sequence analysis of the coat protein biologically distinct citrus tristeza closterovirus isolates. *Virus Genes*, **7** : 255-264.

PATHAK, V.N., 1988 : Postharvest fruit diseases – Pathosis and management. *Journal of Mycology and Plant Pathology*, **28** (2) : 87-113, Udaipur.

PHILIPS, P.A. AND G.P. WALKER, 1997 : Increase in flower and young fruits abscission caused by citrus bud mite (Acari: Eriophyidæ) feeding in the axillary buds of lemon. *Journal of Economic Entomology*, **90** (5) : 1273-1282, Lanham, Maryland.

RAE, D.J., D.M. WATSON, W.G. LIANG, B.L. TAN, M. LI, M.D. JUYANG, Y. DING, J.J. XIONG, D.P. DU, J. TANG AND G.A.C. BEATTIE, 1996 : Comparison of Petroleum spray oils Abaectin, Cartap and Methomyl for control of citrus leafminer (Lepidoptera Gracillariidæ) in Southern China, *Journal of Economic Entomology*, **89** (2) : 493-500, Lanhan, Maryland.

ROCHA-PENA, M.A. AND R.F. LEE, 1991 : Serological techniques for detection of citrus tristeza virus. *Journal Virology Methods*, **34** : 311-331, Amestradam.

ROCHA-PENA, M.A., R.F. LEE, R. LASTRA, C.L. NIBLETT, F.M. OCHOA-CORNOA, S.M. GARNSEY AND R.K. YOKOMI, 1995 : Citrus tristeza virus and its aphid vector *Toxoptera citricida*. Threats to citrus production in the Caribbean and Central and North America. *Plant Disease*, **79** : 437-445, St. Paul, Minnesota.

SASSER, J.N. AND D.W. ERECHMAN, 1987 : A world perspective on nematology. p. 7-14. *In* : *Vistas on Nematology*. J.A. Veech and D.W. Dickson (eds.) Hyattsville, Maryland.

SCHAFFER, B., J.E. PENA, A.M. COLLS AND A. HUNSBERGER, 1997 : Citrus leafminer (Lepidoptera: Gracillariidæ) in lime : Assessment of leaf damage and effects on photosynthesis. *Crop Production*, **16** (4) : 337-343, Washington D.C.

SEKIYA, M.E., S.D. LAWRENCE, M. MCCAFFERY AND K. CLINE, 1991 : Molecular cloning and nucleotide sequence of the coat protein gene of citrus triteza virus. *Journal General Virology*, **72** : 1013-1020, New York, N.Y.

SHAOSHENG, L.G., G. RIXIA, W. ZIMING, 1990 : *Meloidogyne citri* n.sp. (Meloidogynidæ) a new root knot nematoæ parasitizing citrus in China. *Journal of Fujian Agricultural College*, **19** : 305-314.

SINGH, H.P., 1996 : Citrus : p.433-444, *In*: R.S. Paroda and K.L. Chadda (Eds.), *50 years of Crop Science Research in India*. Indian Council of Agriculture Research, New Delhi.

SINGH, S., 1997 : Citrus decline-cause remedies and future strategies. In: *National Symposium Citriculture, NRC for citrus*, Nagpur, pp. 42-55.

SMILANICK, J.L., C.C. GOUIN-BEHE, D.A. MARGOSAN, C.T. BULL AND B.E. MACKEY, 1996 : Virulence on citrus *Pseudomonas syringæ* Straens that control post havest Green Mold of citrus fruit. *Plant Disease*, **80** : 1123-1128, St. Paul, Minnesota.

SMILANICK, J.L., I.F. MICHAEL, M.F. MANSOUR, B.E. MACKEY, D.A. MARGOSAN, D. FLORES AND C.E. WEIST, 1997 : Improved control of green mold of citrus with imazalil in warm water compared with its use in wax. *Plant Disease*, **81** : 1299-1304, St. Paul, Minnesota.

SNOWDON, A.L., 1990 : *A colour atlas of post harvest diseases and disorders of fruits and vegetables*, Volume **1**. General introduction and fruits, London, 302pp.

SWINGLE, W.T. AND P.C. REECE, 1967 : The botany of citrus and its wild relatives, p. 190-430. *In* : Reuther, L.D. Batchelor and H.J. Webber (eds.) *The citrus industry*, Vol. 1, University of California, Berkeley, Berkelay.

TSAI, JAMES, H., 1988 : Development, survivorship and reproduction of *Toxoptera citricida* (Kirkaldy) (Homoptera: Aphididæ) on eight host plants. *Environmental Entomology*, **27** (5) : 1190-1195, Lanham, Maryland.

THIND, S.K. AND J.N. SHARMA, 1996 : Incidence and control of citrus gummosis on kinnow mandarin. *Indian Journal of Horticulture*, **53** (2) : 118-120, New Delhi.

VACANTE, V., N. NUCIFORA AND T. GARZIA, 1988 : Citrus mites in the Mediterranean area. p.1325-1334. *In : Proceedings of the Sixth International Citrus Congress*. Tel Aviv, March 6-11, R. Goren and K. Mendel [eds.] Balaban Publishers, Philadelphia/Rehovot Margraf Scientific Book, D-6992, Weikershein.

YANG-BAOJUN, HU-KAIJI, CHEN-HUI, ZHU WEISHENG, 1990 : A new species of root-knot nematode *Meloidogyne jianyangensis* n.sp. parasitzing mandrin orange. *Acta Phytopathologica-Sinica* (China), **20**: 259-264, Beijing, China.

2

CITRUS ENTOMOLOGY

DHAMO K. BUTANI* AND K.P. SRIVASTAVA**

CITRUS ENTOMOLOGY is a biological science — branch of agriculture — that concerns study of insect pests vis-a-vis citrus cultivation. It includes not only the insects harmful to citrus trees (orchards) but also parasitoids, predators and pathogens of these pests as also other beneficial insects such as pollinators etc. that to the relentless service and help in cross-fertilisation, without which many of the citrus varieties would not bear fruit and may have vanished long ago.

Ebeling (1959) has listed as many as 875 species of insects and mites feeding on various species *of Citrus* trees around the World; the number must have by now swelled to above one thousand species. In India, however, only about 250 species of insects have been found feeding on various *Citrus* species (Pruthi and Mani, 1945; Wadhi and Batra, 1964; Nayer *et al.*, 1976; Butani, 1976 a & b; Hill, 1983; Hill and Hill, 1994). Of these the major pests that cause regular heavy losses in India, include, lemon butterflies, citrus psylla, whiteflies, coccids (mealy bugs and scale insects) and citrus leaf-miners. Those that cause serious damage but only occasionally or in certain pockets or neglected orchards are bark eating caterpillars, stem boring beetles, leaf eating caterpillars, fruit sucking moths and fruit flies. Besides, there termites, aphids, thrips, leaf eating beetles and weevils, fruit sucking bugs, blossom midge, flower moth and wasps that may be found often in citrus orchards causing usually negligible to minor damage. Among the non-insect pests, mention may be made of mites and nematodes.

To describe the various insect pests, the insects can be

* Senior Scientist (Retired) 58A/BB Janakpuri, New Delhi-110058.

** Principal Scientist (Retired), D-62, Durga Market, Naharpur, Sector 7, Rohini, Delhi-110085.

grouped in different categories; may be according to their systematic position or nature of damage or season and time of occurrence etc. In this compendium, the pests have been listed according to the plant parts attacked by these insects — roots to fruits; irrespective of their status — major, sporadic or minor.

ROOT INHABITING INSECTS

Termites

Termites or whiteants (Isoptera) are social insects that live in colonieis — termitarium (termitaria). Whiteant is a misnomer as these insects are neither white nor ant-like. These are cosmopolitan in distribution and feed on a vast range of hosts, devouring almost anything containing cellulose, starch or sugar. Termitarium — the nest of termites — may be underground or above ground, and is inhabitated by a royal pair (king and queen) and several thousands of sterile individuals comprising of soldiers and workers and also members of complementary castes (7 to 8 per cent) consisting of immature stages of both sexes which subsequently grow into soldiers or workers as per need of the colony. There is a marked division of labour among the various groups (castes), though living under the same roof. The various castes are:

Queen — the mother of entire colony, lives in a special prepared cell, known as *Royal Chamber*, situated 0.6 to 1.5 metres below the ground surface. After the nuptial flight, she mates, shed her wings and start a new colony. She gradually grows in size by distending her abdomen, till she attends a length of 40 to 80 mm. The only function of the queen, once the colony is established, is to lay eggs and she is capable of laying some millions of eggs a piece during her lifetime.

King — reproductive caste — winged male, bigger in size than soldiers and workers having two pairs of equal wings. After the nuptial flight, shed wings and keeps constant company to the queen; its sole function is to fertilise the queen and help her initially to start a fresh colony.

Soldiers — most specialised members of termite colony; consists of about 3 per cent of total population in a termitarium. These are mainly concerned with the defence of the colony and are easily recognisable due to the enormous size of their mandibles and chitinised head.

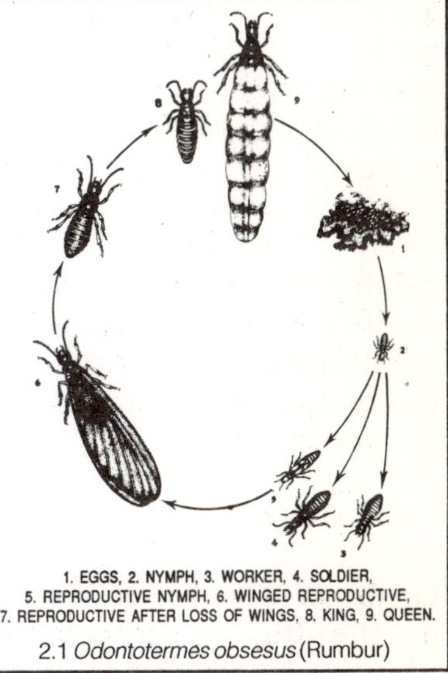

1. EGGS, 2. NYMPH, 3. WORKER, 4. SOLDIER, 5. REPRODUCTIVE NYMPH, 6. WINGED REPRODUCTIVE, 7. REPRODUCTIVE AFTER LOSS OF WINGS, 8. KING, 9. QUEEN.

2.1 *Odontotermes obsesus* (Rumbur)

Workers — labour caste; constitute over 90 per cent population of the colony. These are immature females, soft-bodied, greyish-white in colour, blind and wingless. Except reproduction and defence, all other work is done by this caste. They feed and tend the queen, forage for food, rear fungus gardens in a special chamber, where the queen lays her eggs.

Termites are active all the year round with slight slackness during monsoons; their activity is accelerated during years of drought. According to Khanna *et al.* (1956), texture of soil and amount of moisture present in the soil are the two main factors responsible for termite damage. This insects prefer sandy soil and cause more damage in nurseries and young orchards as their feeding on roots of saplings often kill the plants.

There are about 2000 described species of termites round the World. Of these only one species, *Odontotermes obesus* (Rambur) (Termitidæ : Isoptera), has been reported attacking citrus trees in India, that too as minor pest. Besides, citrus, other fruit trees reported as its hosts from India include, apple, grapevine, guava, mango, sapota and pomegranate. External morphology of this species has been studied in detail by Kushwaha — soldier (1955, 1956a); alate and workers (1956b; 1959a); chaetotaxy (1959b). Attempts have also been made by many workers to study life-cycle of termites in general, but without much success. Beeson (1941) observed that a single queen can lay upto 30000 eggs per day whereas Arora and Gilotra (1960) recorded this species to lay 273 eggs every 15 minutes. Longevity of the queen is estimated to be 6 to 9 years.

To control the termites spray dip treatment or soil treatment with imidacloprid @ 175 to 245 of a.i/ha, or chlorpyrifos 0.5 to 1.0 kg a.i/ha or gamma HCH 0.5 to 1.0 kg a.i/ha is recommended.

Chafer Beetles

Grubs of (cock)chafer beetles (Scarabæoidea : Coleoptera) are the other subterranean pests. These feed mainly on roots of various host plants. Butani (1977, 1979) has listed two rutelids and eight melolonthids species feeding on roots of citrus trees, specially in the nurseries. These include, *Adoretosoma citricola* Ohaus, *Anomala tenella* (Blanchard), *Aserica nilgiriensis* Sharp, *Autoserica insanabilis* Brenske, *Ectinophoplia nitidventris* Arrow. *Lachnosterna* (*Holotrichia*) *consanguinea* Blanchard, H. *repetita* Sharp, *H. rufoflava* Brenske, *Schizonycha repetita* (Fabricius), and *S. ruficollis* (Fabricius). Adults of these species are polyphagous leaf defoliators. In case of citrus, these grubs/beetles are not of any significant importance.

Eggs are laid during June-July in the moist soil, 50 to 150 mm deep. In Rajasthan, egg, grub, pupal and total life-cycle durations last for 7 to 12 days, 8 to 21 weeks, 2 to 3 weeks and 11 to 16 weeks respectively (*L. consanguinea*, Rai *et al.*, 1969). Adults emerge usually during November, remain inactive during the entire Winter and Spring seasons and become active with the onset of monsoon, when these come out in large number (at night), feed on foliage and again hide during day down in the soil or under big clods.

No control measures are adopted against these pests on citrus. However, raking the soil around the trees and mixing 5% gamma HCH (lindane) or dust is quite useful in checking the pest population.

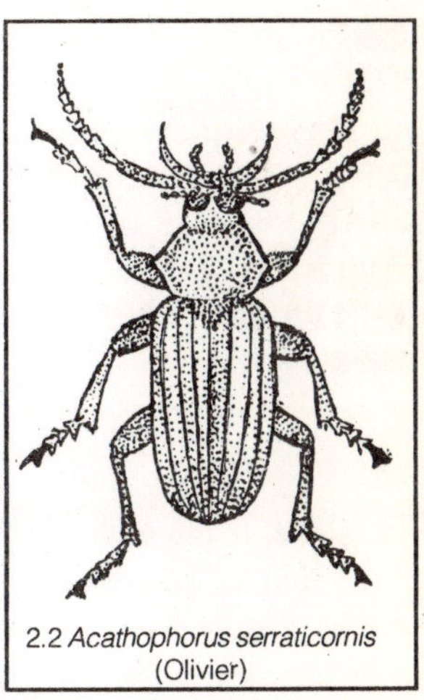
2.2 *Acathophorus serraticornis* (Olivier)

Root Boring Beetles

Acanthophorus serraticornis (Olivier) (Cerambycidae: Coleoptera) — the giant root borer — is a common pest of mango that has been reported as a minor pest of citrus trees from Tamil Nadu by Sathiyanandam *et al.* (1971). Khader and Balasubramanian (1977) also observed the grubs feeding on acid-lime (*Citrus aurantifolia* and Sathgudi orange (*C. reticulata*) roots at the depth of 300 to 600 mm in the soil. The affected trees loose vigour, the leaves droop down and ultimately dry away.

Full grown grubs are about 160 mm long and 80 mm in girth; white in colour, with black head; body smooth, without any hair and devoid of legs. Adults are 50 to 90 mm long beetles, reddish-tawny in colour with head and pronotum black; antennæ 12-jointed; eyes large, pronotum unevenly convex; elytra coriaceous and dull but sparsely punctured and glossy near the base (Gahan, 1906). The grub duration, on an average, is about a year but it may extend upto four years (Garg, 1978). Pupal period is about one month.

Dorysthenus rostratus (Fabicius) (Cerambycidae : Coleoptera) is another root boring beetle recorded on citrus trees. Earlier reported as a minor pest of apple, jujube, java plum (*jamun*) and mango. Khader and Balasubramanian (1977) reported the grubs of this beetle boring roots of citrus trees in Tamil Nadu.

Pale yellow eggs laid singly in the soil at the depth of about five mm and about 20 to 30 mm apart. On emergence, the creamy white grubs bore into the roots, move downwards and pupate therein. Grubs are found upto depth of about 300 mm in the soil. Adult beetles are brownish black in colour; elytra hard and closely covering the entire abdomen.

Belionota prasina Thunberg (=*pyrotis* Illiger, *scutellaris* Weber) — a mango buprestid, has also been found boring into the stems of guava trees in Maharashtra (Nair, 1975) and citrus. The damage is usually confined to bast and sapwood wherein these grubs make winding shallow, broad galleries.

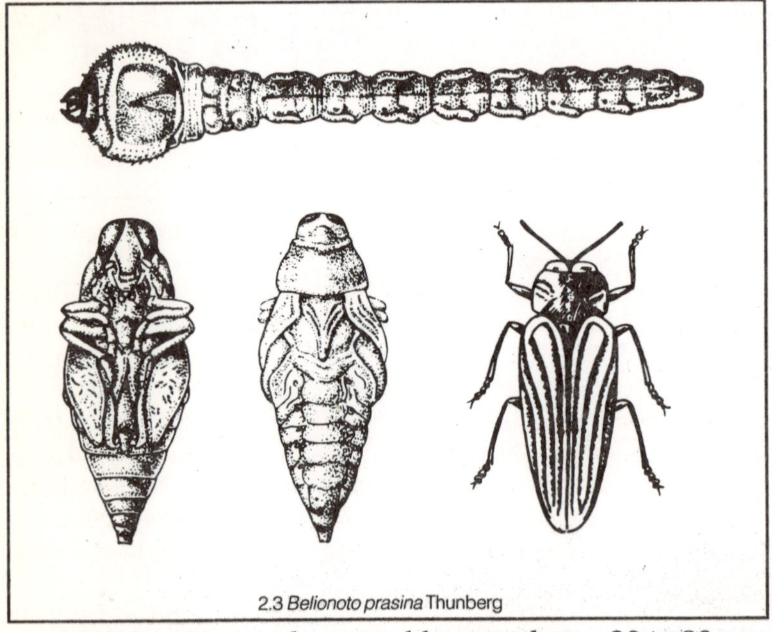

2.3 *Belionoto prasina* Thunberg

Adult beetles are elongate oblong in shape, 20 to 30 mm long and have bronze coloured head that is bright green at the vertex; prothorax is olive-green with coppery reflection and the legs are coppery-green. Elytra are rounded at shoulders and dark green in colour.

To control these root borers, deep ploughing around the trees is suggested, it will kill the grubs by exposing the same to sunshine and other vegeries of nature. Dusting 5% gamma HCH dust around the root zone @ half a kilogram per tree will prevent the females from ovipositing.

Large Brown Cricket

Brachytrypes protentosus (Lichtensyein) (Gryllidae : Orthoptera) is a major pest of jute in India and minor pest of various vegetables and fruits including citrus and mango. Eggs are usually laid during the end of monsoon, when the soil is still wet. Females make burrows in moist soil and deposit 40 to 50 eggs in each burrow. The eggs hatch in September and the freshly formed nymphs make fresh burrows and holes in the soil. The appearance of freshly excavated soil near the small holes on soil surface indicate the presence of this pest in the soil. Nymphs and adults feed on the seedlings during night and hide during day.

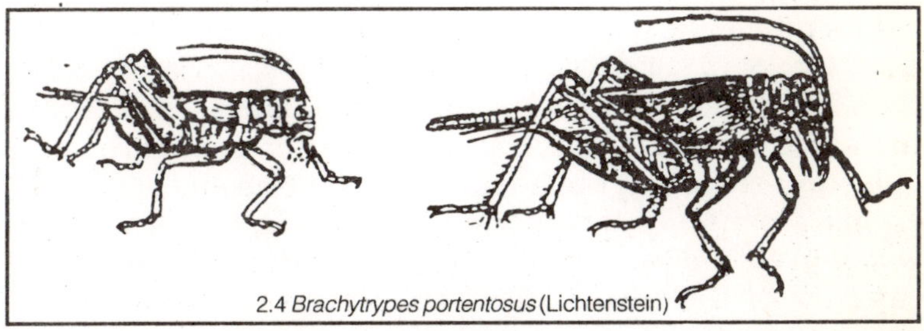

2.4 *Brachytrypes portentosus* (Lichtenstein)

Eggs are cylindrical in shape and white in colour. Nymphs and adults are greyish-brown dorsally and pale yellow ventrally. Both look alike in appearance except that the last two instars of nymphs have wing-pads while the adults have fully developed wings. Adults are 40 to 60 mm long and have long antennæ; forewings (tegmina) are thickened and leathery while hind wings are membranous and lie folded in fan-like fashion, below the forewings. Males stridulate by friction of similarly modified tegmina and the auditory organs are located on the foretibia. Females have a long and pointed (needle-like) ovipositor while the males possess a pair of anal stylets. Incubation, nymphal period and adult longevity last for 2 to 4, 28 to 32 and 16 to 20 weeks respectively (Sen, 1921) and there is only one generation in a year.

No separate control measures are adopted against this cricket on citrus trees. Dusting the seedlings with 5 to 10 per cent gamma HCH gives good protection against this pest as also most of the foliage feeders.

TRUNK BORERS

Bark Boring Caterpillars

The bark eating caterpillars are abnoxious pests of various fruit, forest and ornamental trees found throughout Indian subcontinent, including Bangaladesh, Myanmar, Pakistan and Sri Lanka. The species found damaging citrus trees include, *Indarbela* (*Arbela*)* *quadrinotata* (Walker) and *I.* (*A. tetraonis* (Moore). These occasionally appear in large number, causing severe damage which is more common in neglected and unclean orchards. Older trees are more prone to their attack than the young ones. Eggs are laid on the bark of the trees during May-June. Freshly hatched caterpillars nibble the bark of the trunk or main branches of the trees for couple of days, making irregular galleries thereon, then bore inside the trunk or the main branches, feed within the same till end of December, thereafter pass through a quiescent stage and pupate in April. This feeding of the caterpillars causes interruption in translocation of cell sap which in turn adversely affects the growth and fruit setting capacity of the tree. Large webby masses comprising of chewed wooden particles and faecal matter are conspicuously seen hanging or plastered on tree trunks or the main branches specially near the forking place, and covering the entry holes.

Eggs are spherical in shape and dirty white in colour. Full grown caterpillars are 50 to 60 mm in length having dirty brown body with dark brown head. Pupae 16 to 20 mm long, stout, reddish-brown in colour with two rows of spines on each abdominal segment arranged transversely on anterior and posterior margins. Adults fuscous; forewings pale rufous with numerous dark rufous bands; hind wings fuscous. Wing-spread 35 to 38 mm (males) and 40 to 50 mm (females) (Butani, 1977). A single female lays about 2000 eggs in about 120 clusters, each containing 15 to 20 eggs, on the bark of the trunk or main branch of the tree

* As the generic name *Arbela* is preoccupied, Fletcher (1922) proposed the new name *Indarbela* and placed, *quadrinotata* Walker, *tetraonis* Moore, dea Swinhoe and *theivora* Hampson under *Indarbela*.

sometime during April to June. These hatch in 8 to 10 days. Caterpillar duration lasts till next April while the pupal period varies between 21 and 31 days. Longevity of moths is about 3 days and there is only one generation in a year.

To prevent the occurrence of these borers, avoid overcrowding of the trees and keep the orchards clean. In case of infestation, clean the affected parts by removing all the webbings etc., and insert in the entry holes a swab of cotton-wool soaked in a good fumigant, like chlorosol, carbon disulphide or petrol and seal the same with mud. Half a litre of fumigant is sufficient for 500 holes (Sontakey, 1945). Khurana and Gupta (1972) suggested injecting 0.013% dichlorvos (DDVP), 0.05% trichlorfon or 0.05% endosulfan. This method though effective is combursome to carry out.

Anoplophora (*Monohammus, Monochamus*) *versteegi* (Ritsema) (Lamiidæ : Cerambycoidea) — orange trunk borer — has been reported as a major pest of citrus from Assam (Cherian and Sundaram, 1941), Sikkim (Dutta, 1966; Hayes, 1966), West Bengal as also Bangladesh. It is the predominent species responsible for citrus decline on the Darjeeling hills in West Bengal where it damages 40 to 60 per cent of orange trees (Nayar *et al.*, 1979), Besides *Citrus* spp., it also attacks drumsticks (*Moringa oleifera*) and mango trees.

Eggs are laid on the tree trunks and main branches, just above the soil level, upto a maximum height of 2.5 metres. Freshly hatched grubs feed beneath the bark and later bore inside the wood and tunnel through the pith taking a circuitous course. The entire grub duration as also the pupal period are passed within the tree trunk. Before pupation, the grubs tunnel at right angles to the vertical gallery and make an exit hole which is plugged with frass. The adults come out of these holes during the coming Summer. Adult beetles are sluggish. Initially they feed on leaves during day, devouring the entire leaf lamina along the midrib, keeping the margin intact;

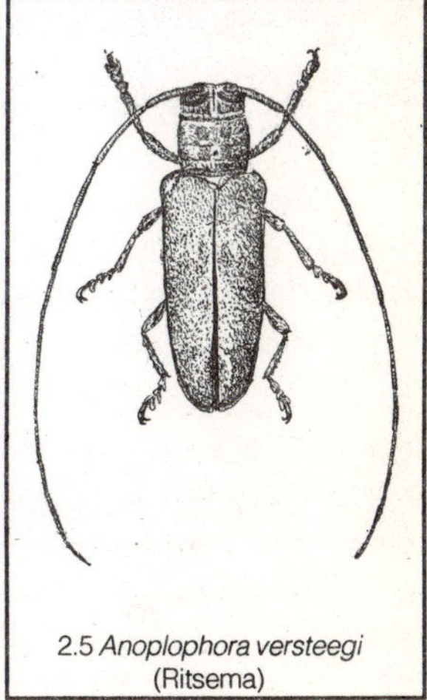

2.5 *Anoplophora versteegi* (Ritsema)

after a few days, they start feeding on bark of twigs and branches. The affected branches gradually dry up and the leaves on these branches soon wither away. Appearance of resinous exudation and sawdust-like powder on tree trunks indicate the presence of this pest in that tree.

To control this beetle, Chowdhury and Majid (1954) have suggested smearing the tree trunks, about one metre from ground level with a suspension of 6.5% lindane (wettable powder) in water (1 : 100). Butani and Jotwani (1984) recommended inserting in each entry hole, cotton-wool soaked in any fumigant like carbon bisulphide, carbon tetrachloride, chloroform or even petrol and sealing the treated hole with mud. An effective method no doubt, but very labourous and may be uneconomical.

Stem Boring Beetles

Three buprestid beetles and ten longihorn beetles have been reported as pests of citrus from various parts of India. These include. *Agrilus grisator* Kerremans, *A. mediocris* Kerremans *Anoplophora versteegi* (Ritsema) *Belionata prasina* Thunberg, *Aesopida malasiaca* Thomson, *Blepephæus succincter* (Chevrolat), *Chelidonium cinctum* (Guerin-Meneville), *Chloridolum alcmene* Thomson, *Demonax balyi* (Pascoe), *Gnatholea eburifera* (Thomson), *Oberea mangalorensis* Fœrster, *Stherias grisator (Fabricius)*, *Stromatium barbatum* (Fabricius) and *Xoanodera trigona* (Pascoe). Most of these are of minor importance but some do occur regularly in certain pockets, causing appreciable loss.

Chelidonium cinctum (Guerin-Meneville) (Cerambycidae : Coleoptera) — commonly called lime tree borer is a common pest of lime trees (*Citrus limon*) throughout India (Murthi, 1922) but destructive only in South India (Kannan, 1926; Cherian, 1942; Ramachandran, 1954). Freshly hatched grubs bore and feed in the superficial layer and make spiral (cork-screw-shaped) tunnels round the twigs, resulting in breaking of the bark from the woody portions. Later, the grubs enter the pith region and cut off the sap flow, causing gradual wilting of the affected twigs. As soon as the infested twigs die, the grubs move downwards, enter the thicker branches below and finally ramify in the main stem and feed on woody portion. Enroute very small holes are also made probably for ventilation purpose and it is the appearance of these minute holes that indicate presence of the pest inside. Besides, gummy exudates from the entrance of the tunnel, withering of twigs and accumulation of chewed fibrous material at the base of the tree are other typical symptoms of attack.

Eggs are flattened and yellowish in colour. Full grown grubs are 35 to 40 mm long. Adults measure 25 to 30 mm in length and are dull metallic-green to dark violet in colour having a yellow patch across the middle of elytra. Adults emerge from pupae in April-May but remain within pupal chambers till the onset of heavy monsoon (June-July). Mating and egg laying has been observed during June-July. Eggs are laid singly in the angles of twigs or thorns and are covered by a resinous fluid secreted by the female. A female lays 30 to 50 eggs in her life time. Incubation period is 11 to 12 days. Grub development takes 28 to 36 weeks while pupal period on an average, is 3 weeks. Pupation takes place inside the tunnels excavated by the grubs.

To control the pest activity, cut and burn the affected twigs. Prasad (1992) suggested two sprayings, at 2 to 3 weeks' interval, with 0.05% methyl-parathion or fenthion or injecting into the bored holes 0.05% to 0.1% dichlorvos with petrol and sealing the treated holes (with mud).

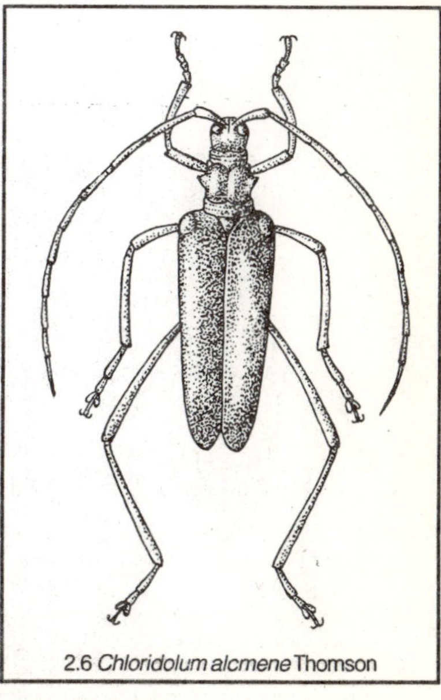
2.6 *Chloridolum alcmene* Thomson

Chloridolum alemene Thomson, — orange tree-borer — a minor pest of citrus, found throughout the Indian subcontinent including Myanmer. In India, it is more abundant in Coorg, Nilgiris, Assam and Andaman Islands, specially on orange (*Citrus aurantinum*). Described originally by Thomson (1865), it has also been listed by Gahan (1906). Adult beetle is metallic-green with antennæ and legs are blue becoming darker towards their extremities. Prothorax with closely rugulose punctured area is covered with black pubescence. Head is densely rugulose, punctate posteriorly, transversely straited at sides and sparsely punctate. It measures 20 to 36 mm in length and 5 to 8 mm in width.

Demonax balyi (Pascoe), a minor pest of citrus trees has been reported from South India. Adult beetle is reddish testaceous above, elytra clothed with yellowish tawny pubescene; each having three distinct tomentose black spots between middle and the apex. Antennæ and legs testaceous. Antennæ are longer than the body length in males but little shorter than the body length in females. Head sparsely clothed with yellowish pubescence. Prothorax marked with a narrow white band at the base on each side, Body length 8 to 11 mm and breadth two to three mm.

Stromatium barbatum (Fabricius) is another minor pest of citrus. Gahan (1906) mentions India, Sri Lanka, Myanmar, Mauritius, Bourbon and Madagascar as its habitat. Adult beetle is reddish-brown to brownish-black in colour, fully covered with tiny pubescence. Head above and at sides, and entire prothorax are very densely and rather coarsely punctured. Elytra are also coarsely and very densely punctured, each with two dorsal costae and a short sutural tooth at the apex. Body length 12 to 29 mm and breadth 4 to 7.5 mm.

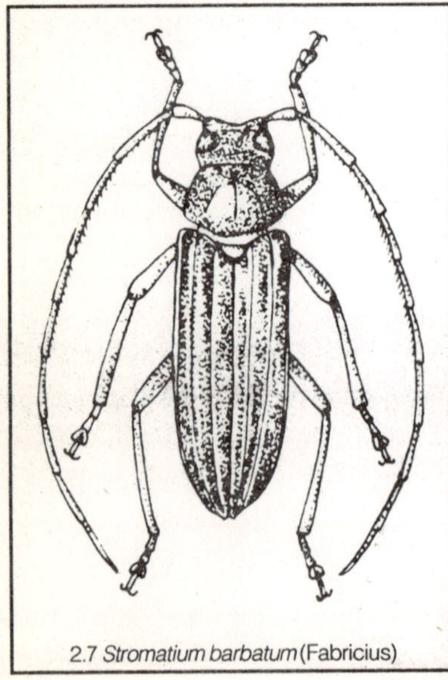
2.7 *Stromatium barbatum* (Fabricius)

Sthenias grisator Fabricius girdler beetle — is a major pest of grapevines all over India. It also attacks other fruit trees including citrus, cashewnut, jack-fruit, mulberry, mango and also some ornamentals. The adult beetles, ring or girdle the young green branches of the host trees. The girdled branch soon dries up; the females cut transverse slits under the bark of girdled branch and thrust in between the bast and sapwood, one to four eggs. On hatching the grubs tunnel into the wood. Pupation takes place in an elongate chamber inside the tunnel and beetles emerge out by circular exit-holes.

Eggs are elongate-oval in shape, about 4 × 1 mm in size, whitish in colour and enclosed in white parchment-like coverings. Freshly hatched grubs are 2 to 3 mm long, having dark brown head, globular throax and cylindrical abdomen, sparsely covered with hair. Full grown grubs are 11 to 14 mm long. Adults are medium-sized, about 24 mm long, stout beetles, greyish-brown in colour with white and brown irregular markings, resembling the bark of tree; elytra also have an elliptical greyish median spot and an eye-shaped patch.

FOLIAGE BORERS

Grasshopper

Poecilocerus pitcus (Fabricius) (Acriddidae : Orthoptera) — *Ak* grasshopper — has been reported from India, Pakistan and some North African countries. In India, it is widespread throughout the plains. As the name suggests, it is primary a leaf defoliator of madar (*Ak, Calotropis gigantea*) but has been found feeding on banana, citrus, castor, fig, grapevine, guava, jujube, melons, peach, papaya, various vegetables (okra etc.), ornamentals and the wild vegetation. The pest appears in April and continues its activity till onset of Win-

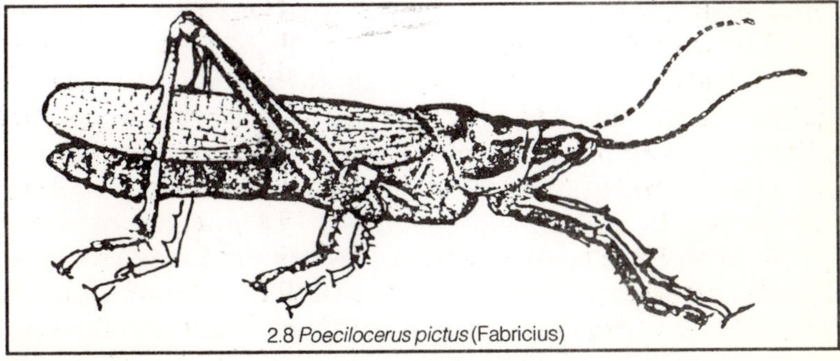

2.8 *Poecilocerus pictus* (Fabricius)

ter, maximum being during July-August. Both nymphs and adults feed on leaves and in case of severe infestation, which is rather rare in case of citrus, even the bark of the trees is not spread.

Eggs are elongate, curved and yellowish-orange in colour. Nymphs are yellowish-white in colour having orange and black stripes and dots all over their bodies. Adults are stout, yellowish with broad bluish-green stripes on head and thorax; antennae bluish-black with yellowish rings; abdomen yellowish with transverse blue-black bands; forewings bluish-green with yellow veins and reticulations; hind wings hyaline (Pruthi and Nigam, 1939).

Mating lasts for five to seven hours and pre-oviposition period is three to four weeks. The female thrusts its abdomen into the moist soil and lays 150 to 180 eggs at a depth of 150 to 200 mm, depending upon the texture of the soil. There are two generations in a year — a short one with eggs hatching in about one month (June-July) and nymphal period two months while the long generation eggs are laid during September to November, these over-winter and hatch around March or early April and become adults in another 10 to 12 weeks.

Generally no control measures are adapted against these hoppers on citrus trees. Verma *et al.* (1970) found spraying on castor trees with 0.03% lindane or 0.1% malathion to be effective.

Leaf Eating Caterpillars

The most common citrus pests found all over the World are the lemon butterflies (*Papilio* spp., Papilionidæ) or Swallow-tail butterflies. Most of these have beautifully patterned wings and are found on the wings throughout the year in gardens and orchards, visiting various flowers but causing practically no damage. These serve as children's delight, who are often seen running after these butterflies and catching the same. In Indian subcontinent, 14 species of these papilionids have been recorded, feeding on citrus leaves — *Papilio demoleus demoleus* Linnæus, *P. helenus helenus* Linnæus, *P. h. daksha* Moore, *P. machaon machaon* Linnæus,

B-2.1 Copulating adults of lemon butterfly *Papilio demoleus demoleus* (Linnaeus)

B-2.2 Eggs and caterpillars of *P. demoleus demoleus* (Linnaeus)

B-2.3 *Papilio polytes* Linnaeus

B-2.4 *P. machaon* Linnaeus

P. m. asiatica Menestries, *P. menmon menmon* Linnæus *P. polyctor polyctor* Boisduval, *P. polymnestor polymnestor* Cramer, *P. polytes polytes* Linnæus, P. *porinda* Moore, *P. protenor protenor* Cramer, and *P. rumanzovia* Eschscholz, *P. sikhimensis, Papilio demadecus* Esper, a closest relative of *P. demoleus demoleus* occurs on *Citrus* spp. in Africa, but has not yet been intercepted from India.

On all the species of *Papilio* reported so far, only *P.*

C-1 Caterpillars of *Papilio demoleus demoleus* (Linnaeus), (brown in colour)

C-2. Caterpillars of *Papilio demoleus demoleus* (Linnaeus) (Green in colour)

C-3. *P. machaon* Linnaeus

demoleus demoleus is the major pest of *Citrus* trees and is cosmopolitan in distribution; all others are of minor or no importance as these are either sporadic in occurrence or confined to certain pockets, causing little or no damage.

Papilio demoleus demoleus was originally described by Linnæus (1758). It is widely distributed from North Australia to Saudi Arabia including Iran, Pakistan, India, Sri Lanka, Bangladesh, Myanmar, China, Taiwan, South-east Asia, Indonesia and Philippines. It attacks almost all the *Citrus* spp. though prefers *C. grandis* and *C. sinensis*. In addition to *Citrus* spp., the caterpillars have been found feeding of leaves of bael (*Aegle marmelo*), jujube (*Zizyphus* spp.), wood apple (*Feronia* species) besides some ornamental and medicinal plants (Butani, 1979). Satinwood tree (*Chloroxylon swietenia*) is another potential host in Penninsular India.

The matings of males and females, after their nuptial flights, are rather complicated; following which the females are seen hovering over the host plants, laying one or two eggs at a time, mostly on the leaves but occasionally on leaf stalks as well. A single female lays 75 to 120 eggs in 2 to 5 days in Summer and upto 180 eggs in 3 to 6 days during Winter. On hatching the young caterpillars first eat the empty egg-shells and then start feeding voraciously on leaves by biting and gnawing the leaf lamina from edges inwards. The caterpillars are mostly found on dorsal surface of the leaves, in an exposed condition. Odour, taste and age of the tree influence the selection of food (Srivastava, 1955). Adults suck juices from flowers but do not cause any economic loss. The attack is found throughout the year, though rare in Winter. The attack is more pronounced in nurseries and young plantations where the seedlings and young trees are often completely defoliated.

Eggs are pale yellow in colour, round in shape; smooth and tough in texture and relatively big in size, 1.0 to 1.25 mm in diameter. Young caterpillars are tuberculate, blackish-brown in colour having milky-white conspicuous markings and are often mistaken as bird's excreta. Full grown caterpillars are 28 to 35 mm long, smooth and velvety (without tubercles), brilliant rich green in colour with dusky brown oblique bands on lateral abdominal segments, not meeting on the dorsum. When these caterpillars are disturbed, they throw out a bifid, Y-shaped osmaterium from behind the head which also emit strong odour. Pupae measure 25 to 52 mm in length and are of three different colours — parrot green, dark-grey or dull yellowish-brown (dry straw coloured). This colour variation is not associated with the sex of adult butterfly (Ghosh, 1914). Adult butterflies are conspicuously artistically designed; body ventrally yellow; wings ornamented with yellow and black markings, hind wings without tail-like prolongation and having base and dorsal margin specked with yellow scales, a broad yellow medial band, a post-discal series of yellow spots and a terminal series of yellow spots; beyond the cell a brick-red oval spot near upper margin and a blue spot near lower margin. Wing-spread varies between 80 and 100 mm.

Incubation, caterpillar, prepupal and pupal periods last for 3 to 8, 11 to 40, 2 to 3 and 8 to 36 days respectively. A single life-cycle is completed in 3 to 6 weeks during Summer but takes 13 to 15 weeks in Winter. Hibernation is in pupal stage. There are four to five overlapping generations in a year in Northern India and five to six in South India where there is hibernation.

Papilio helenus helenus was also described by Linnæus (1758). It is common in Assam and South India. The pest has also been reported from Myanmar, Malaysia and Thailand. Its preferred hosts are lime (*C. aurantifolia*) and orange (*C. sinensis*). The caterpillars are yellowish-green and have oblique bands on lateral sides which meet dorsally. Adults have black antennæ, head, thorax and abdomen with rich velvety brownish-black wings, hind wings have three white spots forming a conspicuous upper discal patch; this patch is more prolonged anally in females than in males. The butterflies are bigger in size than *P. demoleus demoleus*; wing-span being 110 to 120 mm.

Papilio helenus daksha is found throughout South India, specially on lime (*C. aurantifolia*) and orange (*C. sinnensis*). Similar in appearance to *P.h. helenus* except that its upper discal white patch is comparatively much larger and the wing-spread is 106 to 140 mm.

Papilio machaon machaon, a minor pest of various *Citrus* spp., was originally described by Linnæus (1758).

Papilio machaon asiatica Menestriaes is common in Kashmir and Northern Pakistan. These butterflies are smaller in size having dull black wings. Hind wings exhibit creamy yellow area and wing-span is 75 to 95 mm.

Papilio memnon memnon Linnæus — Oriental lemon-butterflies are widely distributed from South Japan to North India (more common in Assam) as also in Myanmar, Borneo and Sri Lanka. Caterpillars are dark velvety green in colour with bluish tinge and having two lateral oblique bands that meet dorsally. Male butterflies have forewings that are deep indigo dorsally and pale black ventrally; hind wings are opaque and dull black. Female butterflies are polymorphic and exhibit three distinct forms; two forms have tailless hind wings while the third form which is very common has tailed hind wings; these hind wings are dusky black in colour with five to seven white or yellowish discal patches. Wing spread is 120 to 160 mm.

Papilio polyctor polyctor Boisduval is found at the foot of Himalayan range from Pakistan to Myanmar through Kashmir hills, but is not so common as *P. helenus helenus* and *P. polytes polytes*. These butterflies are more abundant during rainy season. Caterpillars are green with four yellowish-grey oblique stripes on abdomen. Butterflies have wings, dull black dorsally and chocolate-brown ventrally; forewings have broad submarginal golden green band and hind wings a prominent bluish-green patch. Wing spread is 90 to 120 mm. Sexual dimorphism exists in the adults.

Papilio polymnestor polymnestor Cramer is a rare species reported from Bengal, Sikkim, Maharashtra and Karnataka. Caterpillars have rich velvety dark texture and are grass-green coloured dorsally with a yellowish shade above segments II to V and are covered with minute erect hair. Butterflies have dark brown body and antennæ; forewings having a pale blue discal band that is obsolescent anteriorly; hind wings with distal area pale blue with five irregular small red patches at the base. Wing spread is 140 to 160 mm.

Papilio polytes polytes Linnæus is common species in plains of India as also in South-West China and Indonesia. Caterpillars are rich glaueous green dorsally and slightly yellowish laterally. There is sexual dimorphism in the butterflies. Male butterflies are black with forewings having white marginal spots and hind wings with a white post dorsal band. Female butterflies are polymorphic, exhibiting three distinct variable forms — one resembling the male butterflies; the second having white dorsal patches on hind wings; while the third has red discal patches on hind wings. Wing spread varies between 80 and 100 mm.

Papilio protenor protenor Cramer is not a common species. It has been reported from Kashmir to Kumaon hills as also from Sikkim and Assam but only during Winter. Outside India, these butterflies are found in Myanmar, Southwest China and Thailand. Caterpillars are green with a conspicuous band on thorax and two oblique bands on lateral sides of abdomen which do not meet dorsally, butterflies have velvety indigo wings; forewings paler than hind wings and latter with a broad pale yellowish-white subcostal stripe. There is no swallow-tail projection of hind wings. Wing spread is 100 to 130 mm.

Papilio sikhimensis — a common species confined to Eastern Himalayas, Sikkim and Bhutan (Antram, 1986). The butterflies have black antennæ and cream coloured body with black dorsal streak, broad in males and narrow in females; ventrally the abdomen has black lines. Forewings with basal third, a transverse post discal band and preapical spot dull black arrorated with yellow scales. Hind wings have basal half creamy-yellow with the veins, dorsal margin and apex of cell black, a yellowish red spot at the tornal angle separated from a blue lunule above by means of a narrow black band; tail short and stout. Wing spread is 75 to 85 mm.

To check the carry over to lemon butterflies, remove and destroy all the potential host plants growing wild, in and

B-2.5 Dead larvae after spraying with entomogenus fungus, *Beauveria bassiana*

B-2.6 Dead larvae after spraying with *Bacillus thuriengiensis*

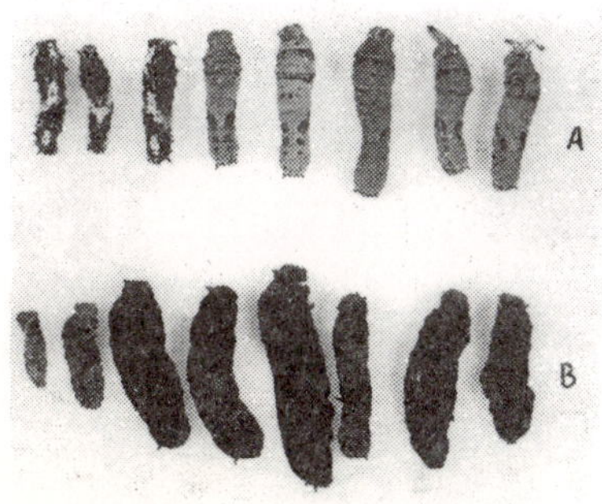

B-2.7 Lemon butter fly caterpillars treated with entomophagous nematode DD-136, (A) Untreated larvae (B) Treated larvae.

around citrus orchards. Hank-picking of various stages of pest and their destruction is very useful in mitigating the pest problem specially in nurseries and new orchards. In case of severe infestation, spray with 0.04% monocrotophos or quinalphos. Spraying with entomogenous fungus or *Bacillus thuringiensis* Berliner or nematode DD-136 strain, also gives very high mortality of these caterpillars.

In nature, eggs of *P. demoleus demoleus* are parasitised by *Trichogramma evanescens* (Westwood), *Pteromalus luzonensis* and *Telenomus* species (Pruthi and Mani, 1945; Prem Chand, 1995). The larval parasites recorded include, *Brachymera* species, *Charopes* species and *Erycia nymphalidaephaga* Baranov (Mishra and Pandey, 1965; Prem Chand, 1995) whereas *Brachymera* species and *Pterolus* species parasitise the pupae. The yellow wasp, *Polistes olivaceus* (Fabricius), preying mantis *Crebrator gemmatus* Stoll and some spiders prey upon the caterpillars (Atwal, 1964).

Other Leaf Eating Caterpillars

Besides, lemon butterflies, (*Papilio* spp.) there are a few other lepidopterous larvae often found feeding on leaves of various citrus species. These leaf defoliators include. *Archips micaceanus* (Walker) (= *Tortrix epicyrta* Meyrick, *T. isocyrta Meyrick, T. pensilis* Meyrick), *Parasa* (*Latoia*) *lepida* (Cramer), *Psorostica ziziphi* (Stainton), *Narosa propolia* Hampson, *Ulodemis trigrapha* Meyrick, *Chilades laius* (Cramer), *Freyeria putli* (Kollar), *Eumeta crameri* (Westwood), *Dasychira mendosa* (Hübner), *Jamides bochus* Cramer, *Tarucus theophrastus* (Fabricius), *Thiacidas postica* (Walker), *Macroglossum belis* (Linnæus), *Euproctis fraterna* Moore (Butani, 1979). All these are minor pests of citrus, some may cause sporadically severe damage.

Tarucus theophrastus (Fabricus) (Lycænidæ) commonly called ber (jujube) butterfly is widely distributed all over Indian subcontinent. Besides it has also been reported from UAR, Iran, Northern and Western Africa, Saudi Arabia and Baluchistan. It is a major pest of various species of jujube (*Zizyphus* spp.) and a minor pest of *Citrus* spp. Eggs are laid on tender leaves. On hatching the caterpillars feed on leaves and flower buds. The pest is more active curing May-June and again during September-October. Eggs are small, flat, round and pale green in colour. Full grown caterpillars are 11 to 15 mm long, fleshy and flattened with pale ochreous head and pale green body. Pupae are smooth and green in colour; head, thorax and wing pads speckled thickly with black. The butterflies have dark brown body and beautifully patterned bluish wings; wing spread is 22 to 31 mm (Bingham, 1907).

Thiacidas postica Walker (Notodontidæ) — a hairy caterpillar is commonly found in South India, specially during March to November thereafter the activity declines due to cold. Eggs are laid in batches glued together on ventral surface of leaves. A female lays 300 to 700 eggs; maximum during August-October. On hatching, the young caterpillars feed gregariously but later they disperse and feed voraciously defoliating the entire twig. Eggs are roundish 0.5 to 0.7 mm in diameter, chocolate coloured and beautifully sculptured with seven white raised longitudinal lines that start from the bottom and end in a small circle on the top. Full grown caterpillars are 20 to 30 mm long, dirty white to pale

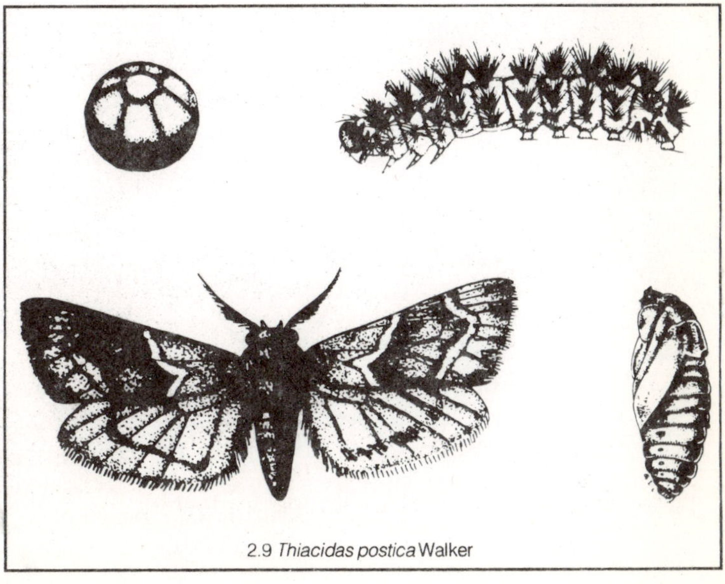

2.9 *Thiacidas postica* Walker

yellowish in colour and fully clothed with long fine setae (Gardner, 1941). Pupae are 15 to 18 mm long and brown coloured. Adult are medium-size moths, stout and dull colored; forewings greyish-brown with two black double lines, one antemedial and other postmedial; hind wings having an indistinct curved post-medial line. Wing expanse is 38-44 mm. Preoviposition and oviposition periods last for 2 and 4 to 5 days respectively; incubation period is 5 to 13 days; larval development takes 16 to 55 days; pre-pual period is 5 to 7 days but extends upto 135 days during Winter (maximum 208 days); pupal period is 7 to 39 days and adult longevity 2 to 9 days (Mehra and Shah, 1970). Total life-cycle occupies 38-124 days with five overlapping generations in a year.

Macroglossa belis (Linnæus) (Sphingidæ : Lepidoptera) — hawk moth, a minor pest of citrus trees is found all over India. Caterpillars are large, smooth having body with longitudinal oblique stripes; anterior segments retractile and a dorsal horn on VIII abdominal segment; pupate in soil in earthen cocoons. Moths are large, robust, diurnal and powerful fliers having long proboscis and antennæ hooked apically. forewings elongate, outer margin being very oblique, much larger than hind wings and heavily scaled.

Chilades laius (Cramer) (Lycænidæ : Lepidoptera) — a leaf defoliator that is common throughout Indian subcontinent and China, as a minor pest of lemon, lime, orange and pomelo. Fletcher (1919) mentions it destroying orange shoots in South India while Nayar *et al.* (1979) have reported that top shoots of oranges are damaged by these caterpillars. Caterpillars are onisciform with both ends tapering and the enlarged sides concealing the legs. Moths are moderate-sized with tail-like prolongations on hind wings.

Parasa (*Latoia*) *lepida* (Cramer) (Limacodidæ : Lepidoptera), commonly called castor slug caterpillar, occurs throughout India. Castor and coconut are its preferred hosts but it is also found on banana, citrus, country almond, fig. mango, pomegranate, wood-apple etc. Eggs are laid in clusters on ventral surface of leaves. On hatching, the caterpillers, move gregariously but sluggishly devouring all the leaves completely, leaving behind only the midribs.

Eggs are flat-ellipitical in shape and shinning yellow-

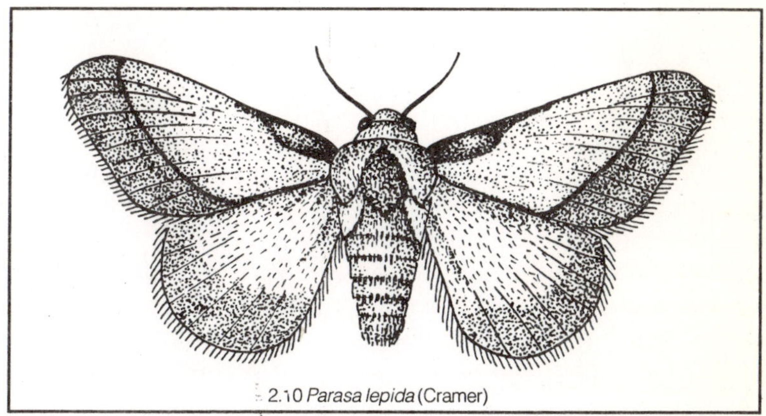

2.10 *Parasa lepida* (Cramer)

ish-green in colour. A single female lays 20 to 30 eggs. Full grown caterpillars are 25 to 30 mm long, fleshy, soft-bodied, apple-green in colour, hairy with greenish-blue stripes and two yellowish-green lateral stripes (one on each side). Caterpillars also have a series of tufts of spines on their body and these spines are highly irritant to touch. Ventrally, the caterpillars are flat and fleshy, therefore move very slowly, like slugs. Population takes place in shell-like compact, elliptical, chocolate-brown cocoons, attached to tree trunk. Moths are short, stout, forewings predominantly green, fringed with big brown patches.

Incubation, caterpillar, pupal and total life-cycle durations are on an average, one, six, three and ten weeks respectively (Ananthakrishnan and Abraham, 1955). Females pupae and adults live a little longer than male pupae and adults.

Normally no control measures are warranted against this pest on citrus trees. If and when these caterpillars are found feeding gregariously on the leaves, clip off and destroy such leaves along with the caterpillars. Vevai (1971a) suggests dusting with 1.3% lindane or 2% fenitrothion dust. Pruthi and Batra (1960) observed *Phycita dentilinella* Hampson preying upon this insect at Coimbatore.

Narosa proplia Hampson (Limacodidae), is another slung caterpillar reported from West Bengal feeding on citrus leaves (Nath, 1970).

Dasychira mendosa (Hubner) (Lymentriidae) popularly known as jute hairy caterpillar is a major pest of jute that also attacks apple, banana, citrus, falsa, grapevine, Java plum (jamun), jujube, mango, peach, plum, pomegranate, quince, strawberry etc. Eggs are spherical in shape 0.8 mm in diameter and creamy-pale-yellow in colour. Full grown caterpillars are 30 to 38 mm long, greyish-yellow with red stripes on prothorax and paired lateral tufts of greyish-white plumose hair on each segment. Pupae are creamy-white and measure 20 to 25 mm in length. Moths have pale yellow body with irregularly patterned brownish forewings. Wingspread is 26 to 30 mm (male) and 36 to 40 mm (females). Egg, larval, pre-pupal and pupal periods occupy 8 to 12, 20 to 38, 2 to 3 and 5 to 7 days respectively. One life-cycle is completed in 40 to 58 days and there are five to six generations in a year.

Dasychira mendosa fusiformis (Walker) is another species of hairy caterpiller reported as a minor pest of citrus trees.

Archips micaceana (Walker) (= *Cacœcia micaceana*) Walker, *C. epicyrta* Meyrick; *Tortrix epicyrta* Meyrick, *T. isocyrta* Meyrick, *T. pensilis* Meyrick) (Torticidae : Lepidoptera) — peach leaf roller — is found all over Indian subcontinent and Indonesia attacking apple, apricot, citrus, grapevines, guava, java plum (jamun), litchi, peach and pear. The caterpillars web up adjoining flower heads and feed within on corollas. In case of severe infestation, even the ripe fruits are webbed up and the caterpillars are found feeding on dried pulp and boring the seeds (Butani, 1979a). As a result of infestation, the fruit setting capacity of the trees is reduced and the fruits attacked loose their market value. Pupation takes place in silken cocoons on the leaves.

Incubation, caterpillar and pupal periods occupy on an average 8, 30 and 10 days respectively with three generations in a year; overwintering is in caterpillar stage (Butani, 1979 b).

Eumeta (Clania) erameri (Westwood) (Psychidæ) Lepidoptera) — bagworm or caseworm — is a polyphagous pest, recorded as serious pest of grapevine from Karnataka (Veeresh and Rajagopal, 1977). It is a minor pest of various

Citrus species as also of pomegranate and tamarind. Caterpillars bind together pieces of twigs, grasses, wood particles etc. by means of silken strands and construct a small compact bag-like covering over their body — only the head and part of thorax are seen protruding outside the bag for feeding purpose (nibbling leaf lamina). Pupation and fertilisation of female takes place within the case or bag. Females are apterous and devoid of antennæ, mouthparts and legs (apodous). Males are winged and have bipectinate antennæ and short mouthparts.

Tarucus theophrastus (Fabricius) (Lycaenidæ : Lepidoptera), commonly called ber (jujube) butterfly is widely distributed all over the Indian subcontinent. It has also been reported from UAR, Iran, Northern and Western Africa, Saudi Arabia and Baluchistan. It is a major pest of various jujube (*Zizyphus* species) and a minor pest of *Citrus* spp. (Pruthi and Mani, 1945; Pruthi and Batra, 1960). Eggs are laid on tender leaves. On hatching the caterpillars feed on leaves and flower buds. The pest is more active during premonsoon (May-June) and post-monsoon (September-October) periods.

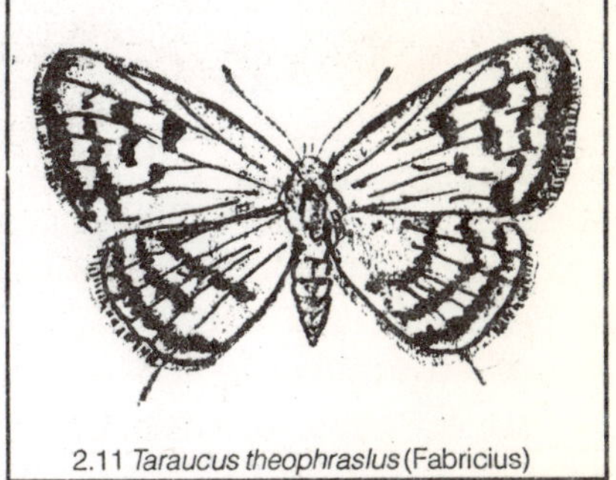

2.11 *Taraucus theophraslus* (Fabricius)

Eggs are small, flat, round in shape and pale green in colour. Full grown caterpillars are fleshy and flattened, having pale green body with pale ochreous head and measure 11 to 15 mm in length. Pupae are green in colour with head, thorax and wing-pads speckled thickly with black. The butterflies are dark brown to black in colour with beautifully patterned bluish wings; wing spread varies between 22 and 31 mm (Bingham, 1907).

In nature, the eggs are parasitised by *Telenomus transverscripes* Nixon and *T. otones* Nixon (Pruthi and Mani, 1945).

Euproctis fraterna (Moore) Lymentrridæ : Lepidoptera) — plum hairy caterpillar is also found all over the Indian subcontinent. It is a major pest of castor and plum and sporadic pest of apple, apricot, citrus, falsa (*Grewia subinaequalis*), grapevine, jujube, mango, mulberry, peach, pear, pomegranate, strawberry etc., causing occasionally severe damage (Butani, 1979b). Eggs are laid in clusters, covered with yellow hair, usually on ventral surface of leaves. A single female lays 150 to 300 eggs. On hatching the young caterpillars feed gregariously on leaf lamina, skeletonising the same completely, Later, these caterpillars disperse and start devouring the leaves. In case of severe infestation, the entire tree may be defoliated.

Eggs are flat, circular in shape and yellow in colour. Full grown caterpillars are 35 to 40 mm long having red head and dark brown body with white hair on the head and a tuft of long hair at the anal end. Pupation takes place in silken hairy cocoons within the leaf-folds. Adults are yellow coloured moths having pale transverse lines on forewings. Wing spread is 24 to 28 mm and 30 to 38 mm in case of males and females respectively. According to Teotia and Chaudhri (1966), egg, caterpillar and pupal periods on castor last for 4 to 10, 13 to 29 and 9 to 25 days respectively. There are three generations in a year. The caterpillars of winter brood feed upto December, hibernate for couple of months, resume feeding and pupate by and February.

Psorostica zizyphi (Stainton) (Oecophoridæ : Lepidoptera) — commonly called ber (jujube) leaf-roller, is widely distributed all over Indian subcontinent, including Pakistan and Sri Lanka. Citrus and jujube are its main hosts, besides aonla (*Emblica officinalis*). Eggs are laid along the midribs of leaves. On hatching, the tiny caterpillars mine and leaves; later the caterpillar roll up the leaves longitudinally and web together several apical leaves and feed within, first on epidermis and later on leaf

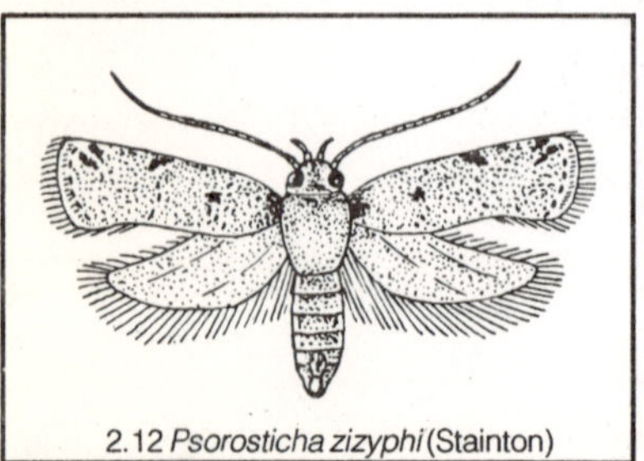

2.12 *Psorosticha zizyphi* (Stainton)

lamina. In case of older leaves only the chlorophyll is eaten away. Pupation takes place in transparent white cocoons spun within the folded leaves.

Full grown caterpillars are slender, about 15 mm long and yellowish-green in colour. Pupae are reddish-brown in colour and 6 to 8 mm long. Adults are small moths, brownish in colour with conspecuously rounded broad wings; wing spread is 16 to 20 mm. Incubation period is 3 to 5 days while caterpillar and pupal development takes 9 to 11 and 5 to 10 days respectively (Nair, 1975). Adult longevity is 15 to 16 days.

Leaf-rollers being minor pests, no separate chemical control measures are adapted against these on citrus. However, hand-picking and mechanical destruction of caterpillars is suggested.

Leaf Eating Weevils

Curculionid grey weevils *Myllocerus dentifer* Fabricius and *M. evasus* Marshall have been reported damaging citrus trees. Their grubs as usual live in the soil and feed on roots and other organic matter; only adults are seen nibbling the margins of citrus leaves. The damage caused however is of minor or no importance.

Adults of *Myllocerus dentifer* are 4 to 7 mm long, black in colour with greyish-brown or pale fawn scaling. Head had distinct shallow punctation beneath the scaling; rostrum is a little longer than head and prothorax is transverse. Elytra mottled with small dark spots along the striae; the striae being narrow and finely punctate. Adults of *M. evasus* are smaller in size than those of *M. dentifier*, being only three to four mm long. These weevils are also black in colour but with uniform pale green scaling. Head is narrow from back to front; antennæ are testaceous while prothorax is sub-conical having a broad base and very narrow apex. Elytra are parallel-sided and have very short recumbent setae.

Pellotrachelus pubes (Faust) (Curculionidæ : Coleoptera) is another minor pest of citrus, jujube and mango trees. The grubs of this weevil also feed on roots of the host trees, specially during Summer while the adults are found feeding on aerial parts, mostly on leaves but sometimes on flower-buds as well.

To control these weevils, spray with 0.03% monocrotophos or endosulfan (Butani, 1975). In case of seedlings or young trees, dusting may be done with 1.3% lindane dust. Deep summer ploughing around the trees expose grubs in the soil to direct sunlight which is fatal for the grubs, some grubs may even be picked up by their predators.

Citrus Psylla

Psyllids or jumping plant-lice (Psyllidæ : Homoptera) are small sucking bugs found all over the World on a large number of host plants — cultivated and wild. More than 30 species have been recorded infesting various fruit trees in India; of these only three have been reported on citrus trees — *Diaphorina citri* Kuwayama, *D. communis* Mathur and *Euphyllus minuta* (Crawford); first one being the most destructive pest. Other psyllids feeding and breeding on citrus trees include. *Psylla citrionga, P. citricola* and *Trioza crytreae* from China (Yang and Li, 1984), *P. murreyi* Mathur from Malaysia and *D. auberii* Hollis from Comoro Islands (Aubert, 1990).

Diaphorina citri Kuwayama commonly called citrus psylla, was originally described by Kuwayama (1907); later *Euphalerus citri* Crawford was also sunk as synonym of *D. citri*. The species is widely distributed throughout the tropical and subtropical Asia and the Far East. It has been recorded from India, Sri Lanka, Bangladesh, Myanmar, Malaysia, Indonesia, Philippines, Taiwan, Southern China etc. In India, it is a serious pest in Punjab, Delhi and Haryana and attacks almost all *Citrus* species (Butani, 1979).

Ferris (1923, 1925, 1926) has described in detail the external morphology of various nymphal stages, while Husain and Nath (1927) have given the description of egg, nymphal stages and adult. Eggs are almond-shaped and orange in colour. Full grown nymphs are flattened-circular

C-4. Nymphs of citrus psylla, *Diaphorina citri* Kuwayama

C-5. *Diaphorina citri* Kuwayama (adults)

in shape and yellowish-orange in colour. Adults are brownish, two to three mm long and are seen sitting on ventral surface of leaves with their heads almost touching the leaf surface and rest of the body raised up.

Eggs are laid singly or in clusters in the folds of half opened

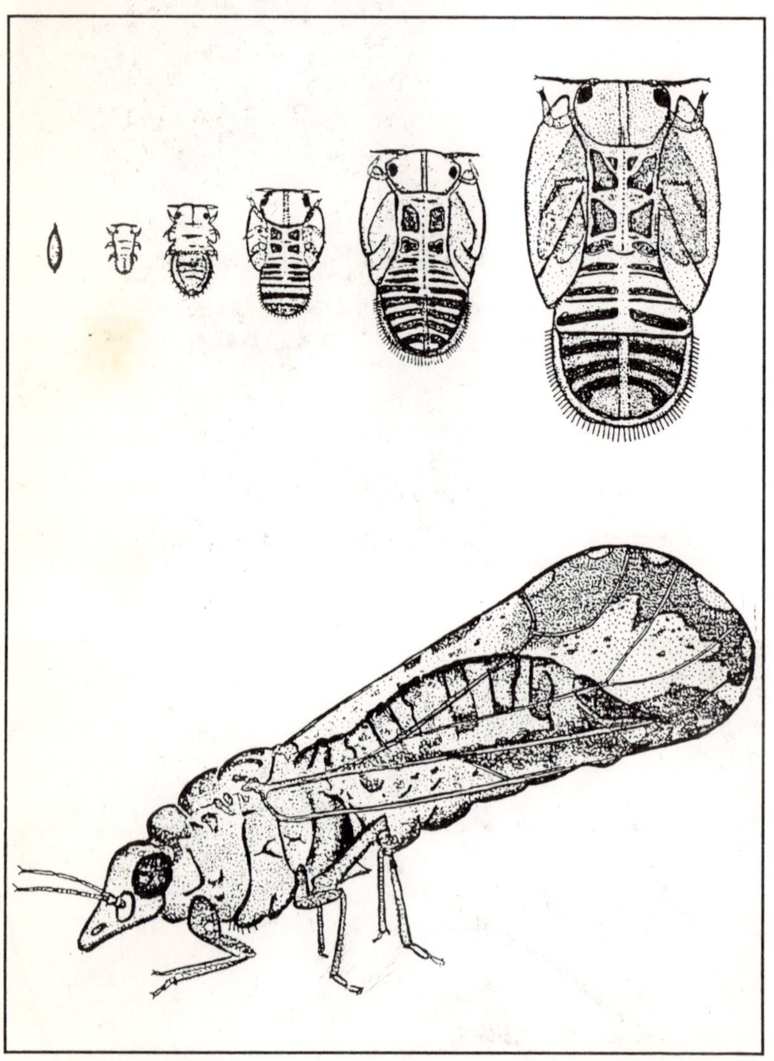

2.13 Differeeent stages of citrus psylla, *Diaphorina citri* Kuwayama (After Catling, 1970)

leaves or pushed in-between flower-buds and stems or petioles of leaves and axillary buds. A single female may lay as many as 800 eggs in about two months. The eggs are anchored in the tissues by means of short stalks. On hatching, the nymphs crowd on terminal shoots and buds and suck the cell sap therefrom; after two to three days, these nymphs disperse and move on to tender leaves and the older foliage. The affected plant parts gradually dry and die away. The insects also inject certain toxins along with their saliva, as a result, the adjacent branches that are not actually attacked also dry and die away. Besides, these bugs also give out copiuos quantity of honeydew that immediately attracts the fungal growth, covering the affected parts with a thin superficial black coating, which in turn hinders the photosynthetic activity of the trees, resulting in stunted growth and poor quality of fruit. In addition to all this damage, the psylla also act as vectors of citrus virus causing citrus decline (Bindra 1966; Capoor *et al.*, 1967). They also transmit greening disease (GD) or bacteria-like organism (BLO) that is transmitted by another species of psyllid *Trioza erytreae* (Ahlawat *et al.*, 1995).

Diaphorina citri survives a wide range of temperature from 45°C in arid climate to 7°C in subtropical wet regions (Ahlawat *et al.*, 1993). The pest is most active during Spring season, the population decreasing with the rise in temperature, but again increases rapidly soon after the heavy monsoon showers and continues till the mercury again falls down substantially (December-January)..

Adults mate soon after the emergence and the oviposition period varies between one and 12 days. Incubation period lasts for four to six days in Summer extending upto 22 days during Winter. Nymphal development takes 9 to 12 days in Summer extending upto 37 days during Winter (Mangat, 1966). A single life-cycle is completed in 15 days during Summer and 47 days in Winter; with nine to ten overlapping generations in a year. Overwintering is in adult stage and the overwintering adults may live as long as 189 days (Husain and Nath, 1927).

To control this pest, Bindra *et al.* (1970) reported the soil application with dimethoate @ 16 ml a.i. per tree to be effective. Sethi (1967) suggested spraying the trees with 0.05% malathion or 0.1% carbaryl while Atwal and Verma (1968) recommended 0.025% phosphamidon. Spraying with nicotine sulphate (1 : 600) or endosulfan is advocated by Butani (1973 d). Lakra *et al.* (1977) found spraying with systemic insecticides like monocrotophos, dimethoate and methyl-demeton, each at 0.037% to be effective while Prasad (1992) recommended 0.05% monocrotophos or phosphamidon.

In nature the nymphal pararite includes *Tetrastychus radiatus* Walker while the predators reported are *Coccinella septempunctata* (Linnæus), *C. rependa* (Thunberg), *Chilomenes sexmaculata* (Fabricius), *Brumus saturalis* (Fabricius) and *Chrysoperla* species (Prem Chand, 1995).

Whiteflies

Ceston in the 17th century, is said to have noticed the first whitefly but his account is inaccessible (Trehan and Butani, 1960). Reaumur (1736) and Linnæus (1758) considered whitefly to be a moth (Lepidoptera) and later the insect was briefly described by Roemer (1789). The hemipterous nature of whitefly was first recognised by Latreille (1796) who redescribed the family and conferred upon it the generic name *Aleyrodes*. Burmeister (1835) changed the name to *Aleurodes*. The family Aleurodidæ was established by Westwood (1840) with only one genus *Aleyrodes* under it. Signoret (1868) described 23 species of Aleyrodidæ; Maskell (1899) added three more species from India. Quintance and Baker (1913 & 1914) splitted Aleyrodidæ into 17 genera.

Today there are 125 genera (Mound and Halsey, 1978). that at include more than 1200 described species of whiteflies in the World, all grouped under one family Aleyrodidæ (Aleyrodoidea: Homoptera). The Aleyrodidæ is further divided into three subfamilies — Aleurodinæ, Aleurodicinæ and Udamaselinæ. Of these, the last one is based on a single specimen that too has probably been destroyed. Aleurodicinæ found mainly in South America are more primitive, much larger in size and have a less reduced wing venation than Aleurodinæ.

Peal (1903) in his *Monograph of Oriental Aleyrodidæ* has

listed only one species *Aleurodes eugeniae* var. *aurantii* Maskell on oranges from North Himalayas. Lefroy (1906) also repeated the same. Woglum (1913) visited India in 1910 and collected four species from various citrus trees — *Aleurocanthus citriperdus* (Quaintance and Baker). *A. woglumi* (Ashby), *Aleurolobus marlatti* (Quaintance) and *Dialeurodes citri* Ashmead. To this Fletcher (1917) added *Aleurocanthus spiniferous* Quaintance while Husain and Khan (1945) added *Aleurocanthus husaini* Corbett, *Aleurolobus citrifolii* Corbett, *Aleurotuberculatus murrayee* Singh and *Dialeurodes elongata* Dozier. Corbett (1935) published description of *Aleurocanthus punjabensis* Corbett which Husain and Khan (1945) considered to be that of *A. woglumi.* These authors have given a list of 27 species of whiteflies reported from various parts of the World, of which nine are from India (pre-partition) including *Bemisia giffardi* (Kotinsky). Pruthi and Mani (1945) have added to this list *Aleurocanthus spiniferous* (Quaintance). *Aleurothcixus howardi* (Quaintance), *Dialeurodes citri* (Riley & Howard) and *Dialeurodes citrifolii* (Morgan). Mound and Halsay (1978) has listed 66 species (Aleyrodinæ 49 and Aleyrodicinæ 17) on citrus trees around the World.

Whiteflies are not true flies but plant-lice a kin to aphids. Morphologically the whiteflies seem to be degenerate psyllids although in contrast to psyllids, the antennæ of whiteflies have fewer segments and forewings fewer veins. Moreover, the immature stages of whiteflies are always sessile. Ecologically whiteflies are equivalent of aphids — opportunist insects with transient populations. Whiteflies are small to tiny, sucking insects, mostly polyphagous, found all over the World. Adults have four broadly rounded opaque wings, venation highly reduced, and dusted with mealy was that is usually white in colour; body length one to three mm and wing expanse is 3 to 4 mm. Both nymphs and adults suck the vital juice from leaves, young shoots and even fruits. Nymphs also out copious quantity of a sweet sticky secretion — honeydew, on which black sooty mould fungi *Capnodium mangiferum* rapidly developed, covering the affected parts of the tree with a thin superficial black coating which in turn hinders the photosynthetic activity on the tree, giving the entire tree a sickly appearance and a severe attack may ultimately yield poor quality of fruits and reduced yield. In addition, these insects act as vectors by transmitting various virus diseases and the loss thus caused is often phenomenal; the affected trees seldom recover from the disease and the young ones may even gradually die away.

Whitefly egg has a short stalk that is inserted into the leaf tissue of the host plant by a female's ovipositor, usually on ventral surface of a leaf. First instar larvae are minute but have relatively long legs and antennae; these crawl actively in search of a succulent spot on the leaf on which they have hatched. Legs and antennæ become atrophied as second, third and fourth instar larvae are sessile. Adult develops within the fourth instar, that is why the fourth instar is known as pupal case. Most species produce abundant quantity of wax around the margins and on dorsal surface of their larvae. The adult usually emerges though a T-shaped slit in the dorsal surface of the pupal case. The white powdery wax that covers the body most species of whitefly is secreted from abdominal glands after the adult has emerged from its pupal case. Some species have dark spots on the wings and a few species are not white — *Aleurocanthus woglumi* Ashby, has black wings and little wax.

Aleurocanthus woglumi Ashby the notorious citrus blackfly, has been recorded from tropical Asia, East Africa, Central America. West Indies etc. (CIE map No. A-91). It is native to India and is primarily a pest of *Citrus* spp., though it has been reported feeding on more than 70 host plants — cultivated and wild, including avocado, grapevine, guava, mango, pear, plum, pomegranate, quince and sapota (Butani, 1979).

Eggs are laid in spiral pattern, which is produced by the female, rotating around her mouth parts, that are kept inserted in the leaf. A single female lays three to four such batches in her life time, each comprising of 35 to 50 eggs. Eggs are oval in shape with both sides rounded; creamy-white an colour when freshly laid, becoming brown and finally black. The eggs are attached to the leaf by a short pedicel situated near the posterior end of the egg. Freshly hatched larvae, called crawlers, are oval in shape, flattened and scale-like in appearance, dark brown to shiny black in colour and

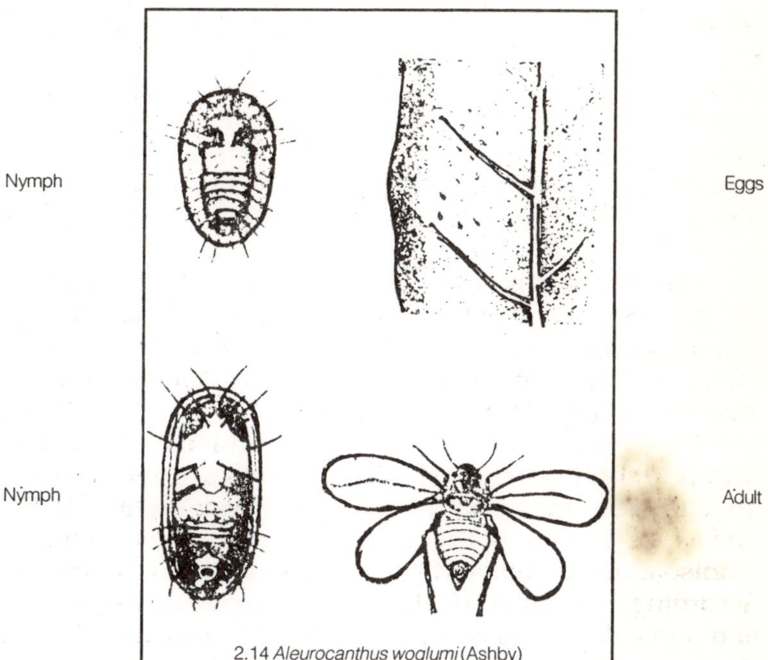

2.14 *Aleurocanthus woglumi* (Ashby)

conspicuously spiny, borded by a white fringe of wax. Full grown larvae are one to two mm long. Pupae are also about two mm long and black in colour. Adults on an average are one mm long males being 0.8 mm and females 1.2 mm. When freshly emerged, its head and thorax are bright red; eyes reddish-brown while antennæ and legs are whitish in colour (Dietz an Zetek, 1920). Within 24 hours of emergence, the adults become covered with a heavy pulverulence that gives them a slatey bluish look. The wings have black patches on white back-ground. Incubation takes 9-10 days while the nymphal development takes 45 to 115 days depending upon the climatic conditions (Pruthi and Mani, 1945). The insect completes, four to five generations in a year.

Aleurocanthus husaini Corbett, a closely related species to *A. woglumi* was originally described by Corbett (1939) from the specimen collected by Afzal Hussain from Kulu valley. It is a minor pest, reported damaging citrus trees in Himachal Pradesh and Punjab. Adults are about one mm in length having pale yellow body maculated with dark spots: the wings are also maculated and slate-grey in colour. Incubation period varies from 9 to 12 days in April, 7 to 9 days in August and 8-12 days in October. Nymphal period occupies between 26 and 52 days; adults emerge after 64 to 108 days of pupation and there are only two generations in a year (Husain and Khan, 1945; Pruthi and Mani, 1945).

Aleurocanthus spiniferus (Quaintance) — the spiny whitefly, has been reported from India. Myanmar, Japan, Malaysia, Indonesia, Thailand, Philippines, West Indies and West Africa (CIE map No. A-112). In addition to *Citrus* spp. it has also been recorded in India on grapevine, pear etc. Newly hatched nymphs are yellowish-green, gradually becoming brown and ultimately turn black. Adults are yellowish-orange to dark orange with head and thorax slightly darker and wings slate-grey. The wings have whitish spots and are reddish at the margins which are armed with hairs (spines). According to Kodama (1931), in Japan, incubation, nymphal and pupal durations occupy 11 to 27, 21 to 84 and 12 to 60 days respectively with usually four generations in a year; hibernation being in the last nymphal stage. In India, it is active throughout the year, without any hibernation period and has several generations in a year.

Dialeurodes citri (Ashmead) — the citrus whitefly — is one of the serious pest of citrus. It is of Indian origin distributed throughout Indian subcontinent, China, Taiwan, Japan, Philippines, Malaysia as also tropical America, specially Mexico as also Florida and California in USA (CIE map No. A-111). It is a polyphagous pest having a wide range of host plants, including, banana, cherry, persimmon and pomegranate.

Eggs are oval in shape, pale yellow in colour. Nymphs are also pale yellow with purple eyes, whereas pupae are subelliptical in shape; pale yellow in colour with orange tinge in the centre. Adults are 1.0 (male) to 1.5 (female) mm long having pale yellow body and dark red medially constricted eyes. Eggs are laid on ventral surface of leaves; fecundity of a female varies between 150-200 eggs (Patel and Talgeri, 1956).

Reinking (1921) counted as many as 2000 eggs on a single leaf. Incubation period is 10 to 20 days; nymphal and pupal periods occupy 25 to 71 and 114 to 159 days respectively (Husain and Khan, 1945). Longevity of males is 2 to 3 days while that of females varies between 7 and 10 days. In Punjab there are only two generations in a year (Atwal, 1962; Gupta and Joshi, 1973); while in Japan where the climate is much milder, there are normally three generations in a year. In northern India, adults are found in abundance during Spring and early Autumn while all stages of the insect are found during Summer (March to August) and only the pupal stage is present during Winter (October to February).

Dialeurolonga (*Dialeurodes*) *elongata* Dozier is another minor pest of citrus. It was first collected by A.W. Khan from Lyallpur (now Faislabad, Pakistan) and described by Dozier (1928). The adults are smaller in size than *D. citri* having pale yellow coloured body. Eggs are laid singly on ventral leaf surface of leaves. Incubation period is 6 to 9 days during April and 12 to 14 days during October-November. Larval and pupal periods last for 24 to 40 and 109 to 120 days during Summer but in Winter, these occupy 36 to 42 and 102 to 112 days respectively (Nair, 1975).

Aleurolobus marlatti (Quaintance) — an oriental species, collected first by C.L. Marlatt from Kummot (Japan) in 1901, described by Quaintance (1903) as *Alueurodes marlatti* and placed under *Aleurolobus* by Quaintance and Baker (1913). In India, it was collected by R.S. Woglum in 1910 from Lahore (now Pakistan). It is a minor pest of citrus trees. Eggs are laid singly on both sides of leaves; these hatch in 5 to 14 days; nymphal and pupal periods are 12 to 35 and 36 to 77 days respectively. During Winter, pupal period extends from December to last week of February; adults emerging early in March. All stages are found in nature during March through December but during January-February, only pupal stage is present (Hussain and Khan, 1945).

Aleurotuberculatus (*Aleurolobus*) *citrifolii* (Corbett) — a serious pest of *Citrus* spp. in Punjab (Pakistan) and a minor pest in India, was originally collected by A.W. Khan in October 1926 from Faislabad (Lyallpur, Pakistan) and described by Dozier (1928) as *Tetraleurodoides citriculus*. Corbett (1935) re-examined the same material and placed the specimen under *Aleurolobus* as a new species *citrifolii*. Mound and Halsey (1978) transferred it to *Aleurotuberculatus* as *A. citrifolii* (Corbett). This dark brown species is related to *Aleurotuberculatus psidii* (Singh). Eggs are laid singly on ventral surface of leaves. Incubation period is 10 to 12 days in Summer, 7 to 9 days during Autumn and 13 to 14 days in Winter. Nymphal development takes 33 to 61 days while pupal period is 39 to 104 days depending upon the climatic conditions (Husain and Khan, 1945).

Aleurotuberculatus murayae (Singh) — a rare species collected from Pusa (Bihar), it was described by Singh (1931) as *Aleurotrachelus murrayae* and transferred to *Aleurotuberculatus* by Takahashi (1932). Husain and Khan (1945) have given a detailed account of its pupal case; which is elongate-elliptical in shape and dark brown to black in colour.

Bemisia giffardi (Kotinsky) — A minor pest of citrus, common in Hawaii, Japan, China, Nepal, Thailand, Malaysia and Vietnam but not so common in India, where it has been confused with *Bemisia jasminum* David and Subramaniam. First recorded from Hawaii on citrus and described by Kotinsky (1907) as *Aleyrodes giffardi*; Quaintance and Baker (1914) placed it under *Bemisia*. Woglum (1913) and Takahashi (1942b) found it in India. Mound and Halsey (1978) regards *B. jasminum* as a synonym of *B. giffardi*.

Control of whiteflies is comparatively easier as these are soft-bodied, external pests.

Cultural : Avoid close planting of trees in an orchard and waterlogging (Butani, 1979). Avoidance of excess nitrogen fertilisation reduces whitefly density and the severity of honeydew damage to fruits (Deshmukh and Garg. 1984). Application of phosphatic fertilisers reduces the incidence of whiteflies specially on flooded and low lying areas.

Mechanical : Butani and Jotwani (1975) suggested prunning and immediate destruction of affected twigs in case of localised or small scale infestation. Destruction of alternate host plants from citrus orchards is also helpful in minimising the attach of these pests.

Chemical : Earlier workers suggested lime sulphur wash (Gupta and Haq, 1959; Atwal 1962, 1976; Gupta and Joshi, 1977). Atwal and Verma (1967) also recommended spraying with 0.05% malathion; Pal (1975) advocated 0.15% carbaryl while Butani and Jotwani (1975), Butani (1973, 1979) and Prasad (1992) have recommended 0.03% monocrotophos or phosphamidon or 0.05% quinalphos. Katole *et al.* (1993) observed fenvalerate to be effective against citrus black fly.

Microbial : The work on microbial control of whiteflies specially in India, is still in its infancy. Though Pruthi and Mani (1945) and Pruthi (1969) have emphasised spraying of entomogenous fungi on whiteflies, the successful cases reported so far include, *Aschersonia papillata* from Kumaon (Bose, 1953) and *A. aleyrodis* Webber from South India (Dhamaraju and Reddy, 1975). *Aschersonia aleyrodis* and *Aegerita webberi* Fawcell have been found effective in Florida (McCoy, 1978).

Biological : Most of the parasites and predators found in nature are more records as these have not been found efficacious enough to check the pest population. *A. woglumi* is parasitised by chalcids, *Encarsia merccti* Silvestri. *Eretmocerus serius* and *Propaltella divergens* Silvestri (Thompson, 1950; Prem Chand 1995) and a pyralid *Cryptoblabes gnidiella* (Milliere) (Thompson and Simmonds, 1964); while *Dialeurodes citri* is parasitised by *Aphelinus fuscipennis* (Howard), *Prospaltella* (= *Encarsia*) *lahorensis* (Howard) and *P. citrifila* Silvestri (Husain and Khan, 1945). Among the predators, Gandhi (1959) reported ladybird beetle, *Rodolia cardinalis* (Mulsant) preying on young crawlers. This beetle — an exotic predator — was actually imported from Australia in 1928 for control of citrus fluted scale (cottony cushion scale), *Icerya purchasi* Maskell and proved to be a great success (Butani, 1993). Other predators found in nature on eggs and nymphs of *D. citri* include, *Brumus suturalis* (Fabricius), *Chrysoperla* species, *Cryptognatha flavescens* (Motschulsky), *Scymnus punctatus* (Motschulsky) and *Varonia cardoni* (Weise) (Husain and Khan, 1945; Prem Chand, 1995). In addition, a black ant *Lasius* species preys on eggs and crawlers, whereas adult flies are preyed upon by a red spider mite *Amystus* species (Garg, 1978).

Aphids

Aphids or plant-lice (Aphidoidea : Homoptera) are polyphagous pests having a very wide range of host plants and are widely distributed all over the World. A large number of crops — cultivated or wild — harbour one to several species of aphids and citrus trees are no exception. About 25 aphid species have been reported on citrus around the World (Viggiani, 1988). Of these, the species found in India include, black citrus aphid *Toxoptera aurantii* (Boyer de Fonscolombe), brown citrus aphid *T. citricidus* (Kirkaldy) (= *T. tavaresi* Del Guercio; *Paratoxoptera argentinensis* Blanchard), green citrus aphid *Aphis citricola* van der Goot (= *spiraecola* Patch), mango aphid. *T. odinae* van der Goot, cotton or melon aphid *Aphis gossypii* Glover, (= *cucumeris* Forbes, *cucurbitti* Buckton), green apple aphid *A. pomi* DeGeer, peach leaf-curl aphid *Brachycaudus helichrysi* (Kaltenbach) (= *xanthii* Del Guercio) and green peach aphid *Myzus* (*Neotarosiphon*) *persicae* (Sulzer).

Aphid nymphs and adults are always found crowded in clusters or colonies on tender plant parts especially ventral surface of leaves and growing shoots and occasionally on inflorescences as well. Both suck the cell sap. In case of severe infestation, the affected parts (leaves) start drying, curling and ultimately wither away. Besides, these insects also expel through their anus (not cornicles) copious quantity of honeydew which favours rapid growth of sooty mould fungus, covering the affected parts with a thin superficial black coating that in turn interferes the photosynthetic activity of the tree and consequently the growth of the tree is stunted and besides the premature fall of fruits, the quality of fruits is also adversely affected. The honeydew is also eagerly fed upon by black ants *Camponotus compressus* Linnæus. The ants carry the aphis nymphs to their nests thereby helping in dissemination of pest population from tree to tree. In addition to this direct loss caused by feeding and excreting honeydew, aphids also act as vector for transmitting various virus diseases including tristeza virus and the loss caused on this account is often tremendous and irrepairable.

Aphids are soft-bodied, pear-shaped, tiny sucking bugs, measuring less than two mm in length and are pale yellowish-green to black in colour. They have a pair of cornicles or siphunculi (erroneously called honey-tubes) arising from V abdominal segment; these are secretory ducts for excreting waxy fluid (not honeydew) and alarm pheromones for self protection against various parasites and predators.

The biology of aphids is rather complex; they have a peculiar mode of development and pronounced polymorphism. Reproduction is both asexual and sexual — mostly parthenogenetic and viviparous, the rate of multiplication being phenomenal. The females — virginpara — produce five to ten living young ones per day and 35 to 50 in her life span of three to five weeks, depending upon the prevailing climatic conditions. These young ones take one to two weeks to mature and start reproducing. During severe cold season feminism is dethroned and the race is saved by sexual reproduction. The insects tide over the unfavourable season in egg stage. Depending upon mode of reproduction, alternation of the host plants and the variation in life-cycle, due to the climatic factors, following forms of aphids have been observed :

Fundatrices* — Also known as stem mothers. Apterous or alate, viviparous parthenogenetic females that emerge from the over-wintering eggs on primary host. Antennæ, legs and sense-organs of these are not so well developed as those of the successing generations.

Fundatrigeniæ — Apetrous or alate ovoviviparous parthenogenetic females — the progeny and successive generations of fundatrices that live on primary host.

Migrantes — Alate, ovoviviparous parthenogentic females that arise from second or third generation of apterous fundatrigeniae or primary host. These individuals dis-

* In case of Adelegidoe (Aphidoidea : Homoptera), fundatrices remain unmature throughout the following Winter and in Spring they mature and lay eggs that give rise to gallicolae.

perse and migrate to other similar or secondary hosts.

Alienicolæ — Apterous, ovoviviparous, parthenogenetic females born on secondary host; these comprise of several generations and look more or less like fundatrigeniæ.

Sexuparæ (gynoparæ) — Apterous or alate ovoviviparous parthenogenetic females, offsprings of alienicolæ; these give birth to winged sexuales, either on secondary host or after migrating back to primary host.

Sexuales — Male and female progeny of sexuparæ either born on secondary host when they migrate to primary host or born of primary host. Females usually apterous; males may be alate or apterous and are generally smaller in size and lighter in colour than the females. They feed very little or not at all, develop rapidly and mate. Females lay one to several or many (upto 15) eggs, usually around the buds and twigs of current year's growth.

Toxoptera aurantii (Boyer de Fonscolombe), commonly called citrus black aphid, was collected from France on *Citrus sinensis* (Linnæus) and described by Boyer de Fonscolombe (1841). It is one of the most destructive pest, found in almost all the warmer parts of the World (CIE map No. A-131) as also in greenhouses in temperate regions. An allied species, *T. citricidus* (Kirkaldy) — citrus brown aphid — is predominantly a pest of humid region (CIE Map No. A-132) and has not been reported from Europe, North Africa and USA, *T. aurantii* is polyphagous; its main host, as the name suggests, is *Citrus* spp. while some of the secondary

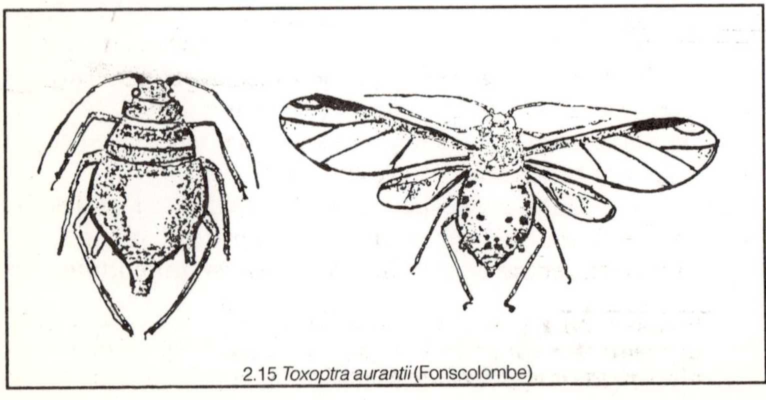

2.15 *Toxoptra aurantii* (Fonscolombe)

hosts include, custard-apple, jack-fruit, java-plum (jamun), litchi, loquat, mango, radish, sapota, tamarind and various ornamentals; while *T. citricidus* is oligophagous and feeds and breeds exclusively on Rutaceæ plants. Normally the aphid infestation occurs during the flowering time, but occasionally severe outbreaks are also observed when rainy season is followed by dry weather. Distortion of young leaves and tender shoots coupled with sooty mould coating on older leaves are the usual symptoms of attack.

Both the species are similar in appearance. Nymphs of both the species are brown in colour, those of *T. aurantii* being slightly darker, while the adults, alate or apterous, are all shiny black in colour. The only obvious diagnostic character is wing-venation — median vein of forewing has one branch in *T. aurantii* and two in *T. citricidus*. Reproduction in both species is parthenogenetic. In case of *T. aurantii*, a single life-cycle normally takes six to eight days but at 15°C it takes as long as three weeks and at 25°C, it comes down to only six days; development is retarded as temperature rises beyond 30°C. One virginapara can produce upto seven offsprings per day and over 50 young ones in its life-time.

According to Börner and Schillder (1932), coccinellid and syrphid predators as also chalcid and braconid parasites take a heavy toll of this aphid in several countries.

Aphis pomi DeGeer — green apple aphid — is obviously a major pest of apple, citrus being one of its main alternate hosts besides, peach, pear, quince and walnut. Eggs are laid in large number on growing shoots during Winter. The eggs are quite hard and can easily withstand the low temperatures. These egg hatch early in Spring and attack leaves and tender branches of citrus trees. The affected leaves get badly curled but not etiolated. As a result of this infestation, ripening of fruits is delayed and the quality adversely affected. In case of severe infestation, even the tender fruits are attacked causing premature fall of unripe and under-developed fruits.

Brachycaudus helichrysi (Kaltenbach) — peach leaf-curl aphid — is one of the most destructive pest of peach trees though almond, apricot, plum and to lesser extend citrus are also attacked. Affected leaves turn pale and curl up; blossoms wither and fruits do not develop into normal size and often drop prematurely. Eggs are cylindrical, 0.5 to 0.6 mm long and light green in colour, while nymphs and adults are green to dark green in colour. Reproduction is both, parthenogenetic as well as sexual. Young ones are apterous viviparous females and after producing three to four asexual generations, the aphids migrate to pass Summer on its alternate hosts. The migration takes place during mid May in plains and around July in cooler regions.

Myzus persicae (Sulzer) — peach green aphid — is consmopolitan in distribution (CIE Map No. A-45), northwards upto South Scandinavia, North China and Canada and southwards upto Australia, Fiji, New Zealand, Tahiti and South America. It is a polyphagous pest, *Prunus* spp., being its primary hosts. Samuel (1940) recorded this aphid on 21 host plants from Delhi alone, including almond, apple, apricot, cherry, citrus, peach, pear and plum. The affected leaves become pitted and curled; flower-buds wither and young fruits shrivel and drop prematurely. Besides doing direct damage by feeding on foliage, these aphids cause much greater loss by transmitting a large number of virus diseases. In tropics and subtropics these aphids reproduce parthenogenetically throughout the year. In temperate regions, gynoparæ appear during September-October. Each gynopara produces 5 to 15 oviparæ, which are fertilised by males and each fertilised ovipara lays 4 to 15 eggs. The eggs diapause during Winter. Hatching starts with onset of Spring. Newly hatched young ones (fundatrices) feed on swelling buds, develop rapidly and start reproducing parthenogenetically (Hill, 1975).

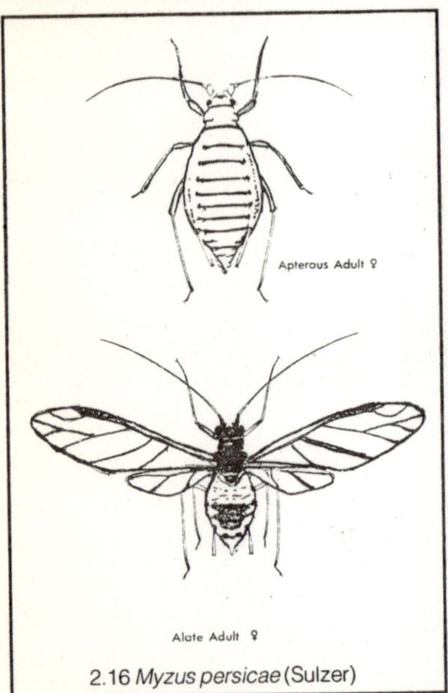

2.16 *Myzus persicae* (Sulzer)

Aphis gossypii Glover — cotton (melon) aphid — is cosmopolitan in distribution (CIE) map No. A-18) being absent only from the colder regions of Asia and Canada. Besides attacking a large number of cucurbitaceous and malvaceous plants, these aphids have also been recorded feeding on citrus, apple guava, grapevine, papaya etc. (Butani, 1979). Nymphs are yellowish- to greenish-brown in colour, while adults are variable in colour are less than two mm in length. Males are rare. Reproduction is parthenogenetic and viviparous. High humidity and cloudy weather with little rainfall are favourable factors, for rapid multiplication of this aphid (Reddy, 1968). In temperate regions overwintering takes place in egg stage and the adults normally perish during Winter.

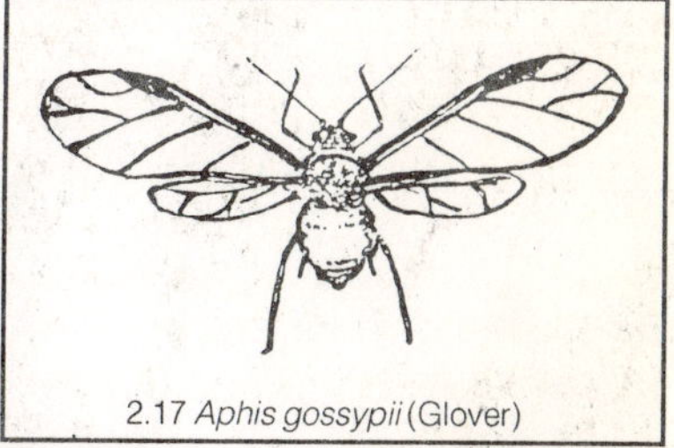

2.17 *Aphis gossypii* (Glover)

To control the pest population, Singh and Rao (1976) suggested spraying the trees with 0.02% oxydemeton-methyl or monocrotophos whereas Butani and Jotwani (1984) have recommended spraying with 0.05% acephate or dimethoate.

Aphids being migratory in habit with a wide range of host plants, the use of aphidicides can only alleviate the infestation for some time, making it imperative to give repeated applications, which in case of citrus trees may prove to be uneconomical. The use of insecticides is no doubt effective in checking the pest population but not the spread of virus diseases. Besides, frequent applications of insecticides also kill the parasites and predators available in nature, thereby reducing the chances of natural control of the pest. Hence integrated pest management is the highly desirable approach.

Coccids

Coccids come under superfamily Coccoidea (Sternorrhyncha: Homoptera) that comprises of a group of minute to small-sized, soft-bodied, sucking insects, protected by wax or through scale. Greatly specialised and degenerated with body segmentation obscure. Females discoidal, oval-spherical and gall-like, always wingless and without distinction of head, throax and abdomen; eyes poorly

C-6. Mealy bug *Drosicha mangiferae* (Green)

C-7. Mealy bug *Planococcus* Sp.

developed. Adult females mostly inactive and sedentary. Males active but without functional mouthparts and having only forewings, hindwings reduced to halteres. Parthenogenesis is frequent. The important families that includes pests of economic crops (including citrus) are — Monophlebidæ (Margarodidæ), Ortheziidæ, Pseudococcidæ, Diaspidæ, Coccidæ and Lacciferidæ. More than 50 species of coccids have been recorded feeding on various parts of citrus trees, including roots, branches, leaves, flowers and fruits; most of these occur sporadically and cause only minor loss but at least ten may be classified as pests of economic importance.

Drosicha (*Monophlebus*) *mangiferae* (Green) (Monophlebidae : Homoptera) — giant mealybug — is a most destructive pest of mango though Rahman and Latif (1944) found it feeding on 62 host plants including fruit (citrus as well) and forest trees, ornamental plants and weeds. To this long list Tandon *et al.* (1978) added eight new host plants, Though normally this is a minor pest of citrus, Saxena and Rawat (1968) have reported a serious outbreak of this pest on citrus in Madhya Pradesh.

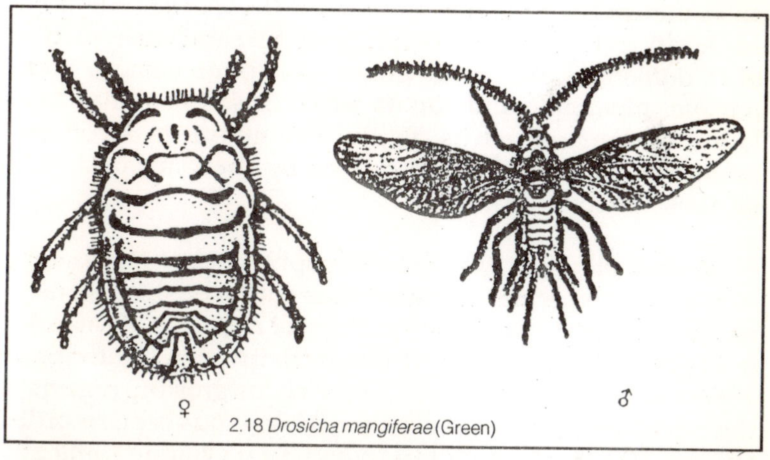

2.18 *Drosicha mangiferae* (Green)

The gravid females crawl down the host tree during Summer (April-May), enter the soil, 80 to 150 mm deep where they excrete whitish foam that forms a pouch, in which the female goes on laying the eggs for 7 to 16 days. The female dies soon after completing the oviposition. A single female may lay 400 to 500 eggs which remain in the soil in a state of diapause. Winter chilling terminates the diapause (Atwal, 1961). On hatching, the nymphs start crawling up the tree trunk and clusters of nymphs may be seen on young shoots and panicles, sucking the cell sap therefrom. There are three nymphal stages; first instar, second instar, third instar female and third instar male live for 45 to 71, 18 to 38, 15 to 26 and 5 to 10 days respectively, on mango trees (Rahman and Latif, 1944). Total life-cycle occupies 67 to 119 days in case of males and 77 to 135 days in case of females.

Eggs are oval in shape and dull white in colour. Nymphs and adult females are flat, oval, waxy-whitish in colour, often mistaken for fungal growth. Adult females are wingless while males are crimson coloured bugs with two dark brownish-black wings and cause no damage, except fertilising the females.

Other mealybugs (*Drosicha* spp.) reported on citrus trees, include *D. contrahens* (Walker), *D. dalbergiae* (Green) and *D. stebbingi* (Green), the last one has often been confused with *D. mangiferae*. All these are also polyphagous and besides mango their main host, hosts of *D. contrahens* include, apple, apricot, citrus, mulberry, pineapple and papaya; those of *D. dalbergiae* are citrus, guava, java plum (jamun), litchi, papaya, pineapple and sapota while *D. stebbingi* has been found on banana, citrus, fig, jujube, guava, jackfruit and tamarind. But the damage caused by these mealybugs to all these hosts is negligible

Icerya purchasi Maskell (Monophlebidæ : Homoptera) — citrus fluted scale, also known as cotton cushion scale, is ubiquitious in tropical and subtropical regions (CIE map No. A-51). Native to Australia, it was introduced in California in 1868 and is by now found in all the citrus growing regions of the World. Being of polyphagous nature, it has been recorded on over 100 plant species in India (Prem Chand, 1995) and though *Citrus* spp. are its main hosts, other important hosts among the fruit trees include, almond, apple, apricot, fig, grapevine, guava, mango, peach, pomegranate and walnut. Newly hatched larvae are quite active on leaf lamina and move about freely till they settle on the midrib or larger veins from which they draw their nourishment. The infested leaves become yellowing and fall prematurely. In nurseries, often the entire seedlings are killed whereas tender shoots of young plants when heavily infested turn blackish-brown and die away. The insects also excrete copious quantity of honeydew on which sooty mould grows rapidly, hindering the photosynthetic activity of the tree.

Eggs are elongated oval (0.6 × 0.2 mm) in shape, smooth and pinkish in colour. Newly hatched nymphs are also elongated oval (0.8 × 0.3 mm) in shape having prominent eyes and long, slender antennæ and legs. Soon after fixation on host plant, the entire body becomes covered with white waxy plates. Full grown nymphs are broadly oval (3.0 × 1.5 mm), reddish-brown to brick red in colour, having stout antennæ and legs, and a few conspicuous patches of wax filaments on dorsal surface. Microscopic characters of various nymphal instars have been studied and described in detail by Bodenheimer (1951). Pupae are elliptical in shape, reddish-orange in colour with dark purple eyes and yellowish-brown wing-pads and legs. Adult females are broadly oval (4.5 × 3.3 mm), brown to reddish-brown in colour, dorsal surface convex (dome-shaped) and ventral flat, with 46 lateral wax bristles and body segmentation indistinct. These are hermaphrodite having white ovisac, that is as broad as the female body; dorsally convex and distinctly fluted. It is the convex fluted mass of white wax underneath the scale that attracts the attention and characterises this insect (Talhouk, 1969). Males, though rare, are slender, about 3 mm long, reddish-purple in colour with long dark brown antennæ and two short fleshy appendages with long bristles at caudal end. Forewings are shiny metallic-blue and hal-

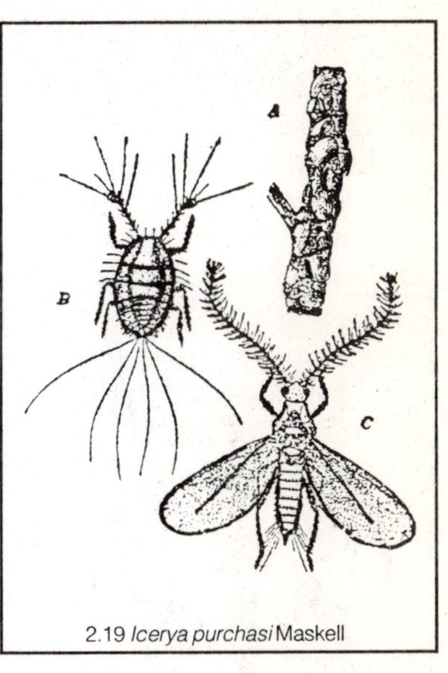

2.19 *Icerya purchasi* Maskell

teres foliated with 2 to 3 hooks at the tip; wing-span is 5 to 7 mm. Life-cycle studies conducted in Japan, reveal that incubation period is 21 to 27 days, the nymphal stages last for 59 to 92 days; while in Palestine, preoviposition period, incubation, nymphal period and total life-cycle duration last for 12 to 17, 16 to 35, 46 to 58 and 79 to 98 days respectively (Bodenheimer, 1951).

Other *Icerya* spp. reported on citrus trees include, *I. agyptiaca* (Douglas). *I. minor* Green and *I. seychellarum* (Westwood). All these are minor pests of citrus trees and of no significant importance.

Effective control of this scale has been achieved by releasing its exotic predator, ladybird beetle *Rodolia cardinalis* (Mulsant), introduced from Australia. *I. purchasi* and *R. cardinalis* are often quoted as an example of successful biological control all over the World. Some of the indigenous coccinellids predating upon this scale are, *R. fumida* (Mulsant), *R.amabilis* Kapur), *R. minima* (Kapur) and *R.breviuscula* (Weise). Subramaniam (1949) has reported a dipterous parasite *Cryptochaetum iceryae* (Williston). Besides, tachinid *Cryptochaetum melanochra* Meyrick and *Euzophara cochiphaga* (Hampson) feed on egg-masses.

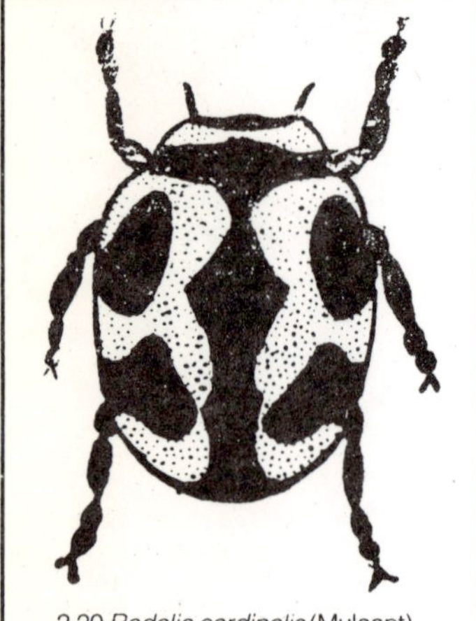
2.20 *Rodolia cardinalis* (Mulsant)

Orthezia insignis (Browne) (Ortheziidæ : Homoptera) — lantana bug, has been reported from South India, Sri Lanka, Malaysia, Africa, USA, Central and South America (CIE map No. A-73). The main host of this scale is lantana (*Lantana camara*), a highly obnoxious weed found in abundance on the hills of India. The pest has been found keeping this weed under check and attempts are being made to use this bug in developing the bio-control of the weed. Besides lantana, the insect also feeds on a number of economic crops, including citrus, but causing usually little damage. According to Hill (1975), usefulness of this insect, heavily outweight its nuisance value as pest of flowering plants.

Adult females are olive-green in colour with well developed antennæ and legs. Segmentation of body is quite distinct with white waxy plates or laminæ at the extremity; egg-sac is situated between these waxy plates. Short waxy processes are also there at the side and double row of the similar processes along the dorsum. In India, no control measures are generally adapted against this scale.

Lac insects, *Kerria* spp. and *Paratacharidina* spp. (Lecciferidæ : Coccoidea : Homoptera) are considered by and large as beneficial insects as they yield stick-lac, a commodity of great commercial value. These insects are commercially reared, mostly on jujube (*Zizyphus* spp.). Besides, being polyphagous, these insects have been reported attacking a large variety of host plants including citrus, but seldom as pests of any economic importance. The species commonly found on *Citrus* spp. include *Kerria lacca* (Kerreman) and *Paratacharidina lobata* (Green).

Females have irregular globose body with vestigial antennæ and are apterous. They live enclosed in a resinous mass secreted by the lac-resin glands that are found all over their body. A female lays 200 to 500 eggs that are extruded into the cell in a space formed by the contraction of female body. On hatching, the tiny red crawlers come out of the cell and actively move about on host plants in search of a succulent tender shoot. Once a suitable spot is located, the crawlers insert their proboscis and suck the sap therefrom, remaining stationary thereon. Once settled, the insects start secreting resinous fluid which at first appears as shining uniform coating over their bodies, but gradually it increases in thickness and extend until the secretion of adjoining larvae eventually meet and coalesce in a continuous encrustation. The effected trees lose all their vigour and vitality.

To check the infestation of lac insects, remove and destroy promptly, the affected plant parts along with the insects thereon in the initial stage of attack. This will prevent the infestation from spreading.

Ferrisiana (*Ferrisia, Pseudococcus*) *virgata* (Cockerell) (Pseudococcidae : Homoptera) — white-tailed mealybug is found throughout tropical and subtropical countries (CIE map No. A-219) causing severe damage to number of economic crops including aonla (*Emblica officinalis*), banana, citrus, custard apple, guava, grapevine and jack-fruit. It is widely distributed all over India (Ali, 1962, 1968) though only as a minor pest of citrus, Nymphs and adult females are seen clustered on ventral surface of leaves, terminal shoots and often on fruits as well, sucking the sap. The pest is found throughout the year though it prefers dry weather and prolonged period of drought may result in heavy outbreak of the pest, when the insects even more down and inhabit the roots.

Adult females are apterous, oval in shape (4 × 2 mm), covered with a number of waxy filaments all over the body, besides a pair of conspicuous longitudinal submedian dark stripes and two pronounced long waxy processes at posterior end. Reproduction is both, parthenogenetic as well as sexual, former being more common. In case of sexual reproduction, mating takes place only once and the fertilised female lays 100 to 300 eggs in 3 to 4 weeks. Egg-masses remain under the female till the young ones hatch out. Incubation period is 3 to 4 hours. Developmental period of male and female nymphs varies from 31 to 57 and 26 to 47 days respectively. Longevity of males is one to three days while that of female varies between 36 and 53 days (Nayar *et al.*, 1976).

Nipaecoccus viridis (Newstead) [= *corymbatus* (Green), *filamentosus* (Cockerell), *vastator* (Maskell)] (Pseudococcidæ Homoptera) — fluffy mealybug, is mainly pest of ornamentals. Earlier work has been done under the name *Pseudococcus filamentosus* Cockerell. Atwal (1976) has reported *Cactus* spp. *Begonia* sp., ferns, *Gardenia* sp., poinsettia and other flowering plants as its hosts. Butani, (1979) has listed it feeding on aonla (*Emblica officinalis*), citrus, custard apple, fig, grapevine, guava, jack-fruit, jujube, mango, mulberry and tamarind but only as a sporadic pest. The mealybug is found all the year round but the peak period of activity is November when clusters of these mealybugs may be seen on leaves, flowers and even fruits, sucking the call sap therefrom, besides excreting honeydew.

Nymphs are amber coloured with whitish waxy coating and filaments around the margins. Adult female is flattened and wingless having short filaments around the margin whereas males are winged with long antennæ and atrophied mouthparts (Bindra, 1970; Atwal, 1976).

Eggs are laid in clusters enclosed in a protective cottony mass. A female lays about 300 eggs in its life time. Eggs hatch in 10 to 20 days and soon they envelope themselves with white fluffy material. Female nymphs moult thrice and become full fed in 6 to 8 weeks, while male nymphs moult four times and after passing through a prepupal stage, emerge as winged adults.

To control this pest, Atwal (1976) suggested spraying the trees with 0.03% phosphamidon or 0.05% dimethoate or malathion. In nature, it is predated upon by ladybird beetle, *Rodolia cardinalis* and *Chrysoperla* species as also by a syrphid fly maggot (Atwal, 1962).

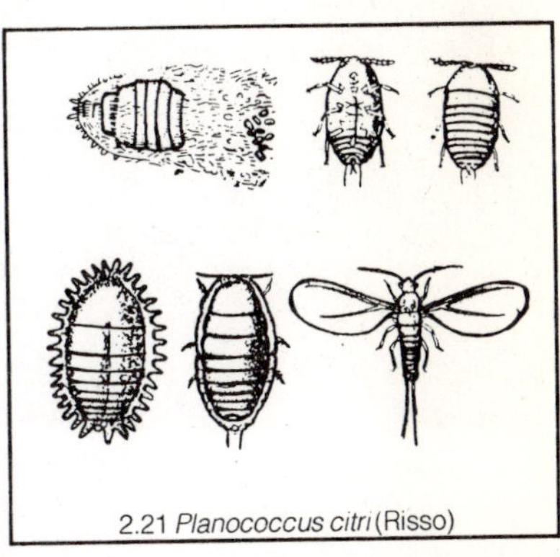
2.21 *Planococcus citri* (Risso)

Planococcus citri Risso (Pseudococcidæ : Homoptera) — citrus mealybug, is almost completely pantropical in distribution extending well into subtropical regions (CIE map No. A-43) and is also found in greenhouses in temperate countries. Its main hosts are *Citrus* spp., specially lime (*C. aurantifolia*), lemon (*C. limon*) and oranges (*C. reticulata* and *C. sinensis*); it has also been recorded feeding on fig, pine apple, sapota etc. Ayyar (1930) has listed about 50 host plants of this mealybug whereas Bodenheimer (1951) has reported more than 26 hosts from Palestine. The pest attacks leaves, tender shoots and fruits (at the base near the fruit stalks) and even the roots. As is common in scale insects, these insects also besides sucking the cell-sap, excrete honeydew on which as usual there is rapid growth of sooty mould, hindering with the photosynthetic activity of the tree. As a re-

B-2.8 *Pseudococcus adonidum* (Linnaeus) - long tailedmealy bug

B-2.9 *Aonidiella aurantii* (Maskell), scales on orange

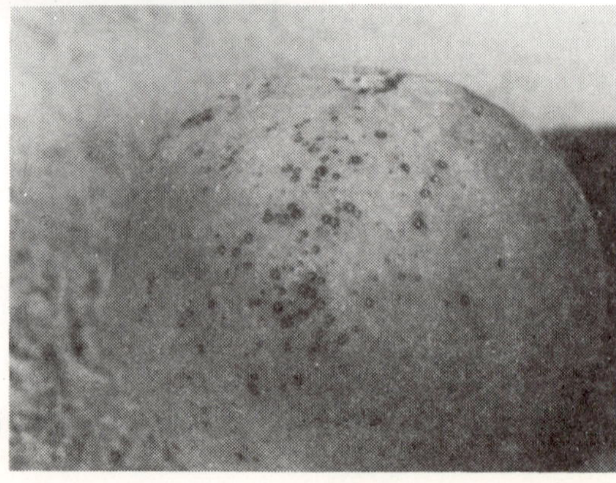

B-2.10 *Chryomphalus aonidum* (Linnaeus) on ripe orange

sult, the growth of the tree is arrested and the fruits fall prematurely.

Eggs are ellipsoidal, 0.3 to 0.4 mm long and pale creamy-yellow in colour. Freshly hatched nymphs are pale yellow in colour without any waxy coating, but slowly and gradually the waxy coat appears, covering the entire nymphs. Adult females are slightly elongate ovate, 5 to 7 mm long, covered completely with white waxy secretion and having 34 wax covered appendages round the entire periphery. A female lays 600 to 800 eggs in about 15 days, these are deposited in loose cottony masses. Eggs hatch in one to two weeks during Summer. Female nymphs undergo three moults and males four moults and there are two to three generations in a year (Pruthi and Batra, 1960).

Planococus lilacinus (Cockerell) is another polyphagous mealybug, widely distributed in the Indian subcontinent and South-east Asia (CIE map No. A-101). In India, this is a major pest of coffee specially in Karnataka; its other host plants include, citrus, custard apple, fig, guava, pomegranate, sapota and tamarind. These mealybugs feed on leaves and roots; more individuals are found in the root zone 400 to 600 mm deep in the soil, than on leaves. The infested roots develop spongy tissues and affected leaves become chlorotic and ultimately fall down.

Efficient control of these pests is still a far cry. Even fumigation with hydrocyanic acid gas (HCN) is unsatisfactory, because the killing concentration of mealybugs is higher than the tolerance of the trees (Bindra, 1967). Spraying 1% diesel-oil is effective but expensive. For checking the subterranean population of the pest, plough around the trees, 600 to 800 mm deep and mix thoroughly 5% gamma HCH dust with the soil. A parasitic wasp, *Prospaltella perniciosi* (Tower) has been successfully established in Kashmir, where it has given fairly effective control of *P. citri.*

Armoured scale insects (Diaspididæ : Coccoidea: Homoptera) is the largest group of scale insects and comprises of a large number of economically important pest species. Garg (1978) and Butani (1979) have listed 33 species occurring on citrus in India but of these, only the following three are of regular occurrence and are often classified as major pests.

Aonidiella aurantii (Maskell) — citrus red scale or California red scale — is native to India and has now spread over all the citrus growing countries of the World, except West Africa (CIE map No. A-2). It thrives best in semi-arid climate. Quayle (1938) has listed 86 host plants of this scale from different parts of the World. In India, it is found all over the country feeding on banana, citrus, fig, guava, java plum (jamun), Jujube, mulberry, peach, pear, ornamentals etc. Among *Citrus* species the pest prefers malta or sweet orange (*C. sinensis*), seville orange (*C. aurantium*), grapefruit (*C. pradisi*) and pomelo (*C. grandis*). The scales usually infest leaves and tender shoots but in case of severe infestation even the fruits and sometimes tree trunks are also attacked. Dispertion is mainly by wind — the crawlers are blown off by wind from tree to tree and even from one orchard to another. Its feeding habit is very peculiar; it injects its toxic saliva in the cell sap before ingestion, which severly affects the host tree and the young tree may even be killed due to injected toxins. Branyovitis (1953) observed that these insects spend more time in moulting than in feeding.

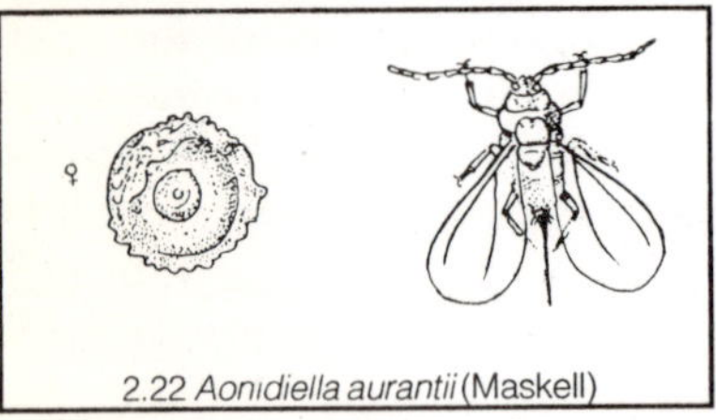

2.22 *Aonidiella aurantii* (Maskell)

All the immature stages as also the adults are found all the year round though their proportion may vary according to climatic conditions. There is no true diapause but the rate of growth slows down considerably in Winter (December to February) and the activity is maximum from August to October.

Morphology and bionomics of the species, have been studied more in other countries than in India. Berlese and Leonardi (1896) were first to describe the immature stages. Female scale is more or less circular and flattened with raised centre while the scale of male is elongated. The scales apparently appear reddish in hue, which actually is the colour of the insect, seen through the scale covering (Ebeling, 1959).

Adult females are flat and circular in shape, about 2 mm in diameter with conspicuous antennæ and rostrum is about twice the body length; these are wingless and apodous (legless). Gravid females become kidney-shaped and yellowish-orange in colour. Males are winged; emerge from beneath the scale, fly about and mate with virgin females (Talhouk, 1969). Females are ovoviviparous and produce two to three nymphs per day for more than two months. Young nymphs (crawlers) remain under protection of the mother for one to three days, then crawl about a little in search of suitable succulent spot and within a few hours settle thereon. Once settled, it seldon leaves the spot whether it is on leaf, twig or fruit and starts secreting the wax which soon covers its entire body.

As these scales feed only on cell sap in the parenchyma tissues and hardly ingest any sap from the phloem, systemic insecticides do not have any appreciable effect in checking their population. Fumigation with HCN gas has been found to be very effective in controlling this pest and is used extensively in USA, but since this method requires expert handling and sufficient technical knowledge it is not being used in India.

Chrysomphalus ficus Ashmead (= *aonidum* Linnæus) — Citrus purple scale or Florida red scale is native of Oriental region and is by now found in almost all tropical and subtropical regions of the World (CIE map No. A-4). In temperate regions, it is restricted to greenhouses. Inspite of being widely distributed, it is pest of economic importance only in hot and dry places. It is polyphagous and Bodenheimer (1951) was listed 105 host plants from Cuba (West Indies) alone. In India, besides *Citrus* spp. it has been recorded on almond, apple, avocado, banana, coconut, date-palm, grapevine, guava, jujube, mango, mulberry, pomegranate etc.

A female lays in her life-time, 50 to 150 eggs, @ 5 to 6 per day during Summer and one or two per day in Winter. Egg production is higher on fruits than on leaves. Freshly hatched nymphs are very active, move about for a day or so, in search of suitable succulent feeding spot and settle there more or less permanently; soon after settling, the nymphs get covered entirely with a scaly substance secreted by them

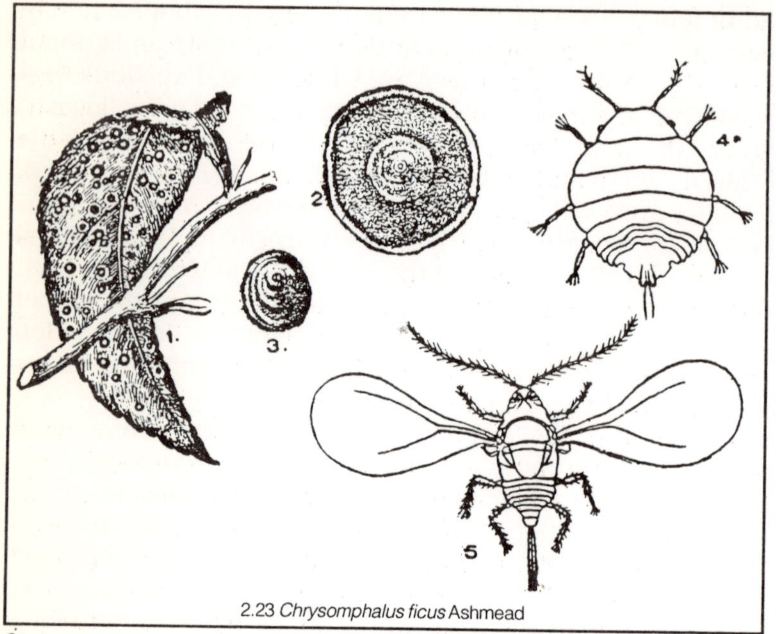

2.23 *Chrysomphalus ficus* Ashmead

for protection. The pest prefers shade during Summer and sunshine during Winter. Low temperature during Winter has been reported to cause 98% mortality (Pruthi and Mani, 1945).

Eggs are long, oval (2.0 to 2.5 × 1.0 to 1.5 mm) in shape and lemon-yellow in colour. Neoate nymphs are short and oval (1.2 × 0.2 mm) in shape and yellow in colour. Adult females are flat and pear-shaped (1.2 × 1.0 mm), broadly rounded anteriorly; pale yellow when young and getting reddish tinge with advance in age. The scaly covering is white, circular (1.5 to 2.0 mm in diameter) and nipple-shaped. Males are winged, 0.8 to 1.0 mm long, orange in colour; antennæ and legs being brown; wings whitish and wing spread 1.8 to 2.0 mm. The bionomics studied outside India, show egg period to last for one to four days and total life-cycle to occupy 6 and 10 weeks in case of males and females respectively (Quayle, 1938), with atleast five to six generations in a year.

Cornuaspis beckii (Newman), a native of tropical America is now found all over the citrus growing regions of the world (CIE map No. A-49). It attacks almost all *Citrus* spp. though it prefers sweet orange (*C. sinesis*).

Eggs are elongate (1.0 × 0.2 mm) in shape and pearly-white in colour. Neoate nymphs are elongate-ovate (0.4 × 0.2 mm) and pale yellow in colour with two small lobes and four spines protruding from posterior end. Male pupae are very conspicuous being purple in colour with antennæ, legs and wing-sheaths free from the body. Adult females are elongate (1.5 × 1.3 mm), yellowish-white in colour with terminal end reddish-fulvous. Males winged, very delicate, less than one mm long, white to pale yellowish-white in colour; antennæ and legs purple in colour; wings hyaline and wing spread 2 mm. Female life-cycle is completed in 50 days in Summer extends upto 110 days during Winter. There are four overlapping generations in a year.

Being minor pests, generally no control measures are adapted against these armoured scales. However, if and when an infestation is noticed clip off the affected parts and destroy the same promptly.

Parasiassetia (= *Saissetia*) *nigra* (Neitner) (Coccidæ : Homoptera) — black (soft) scale, is widely distributed throughout India, Africa, South-east Asia, Taiwan, Philippines, Indonesia, Australia, New Zealand and USA. Through it is commonly found on citrus trees as a minor pest, occasionally grapevines, guava and litchi trees are also attacked.

Adult scales are oval in shape, slightly dome-shaped, measuring two or three mm in length and dark brown to black in colour. Reproduction is usually parthenogenetic. A large number of eggs are laid by the mature female (without mating) and these eggs remain hidden under her body. After hatching, the crawlers creep out under the mother scale, move about a bit in search of a succulent spot and settle there more or less permanently, feeding and growing by sucking the cell-sap.

To prevent the pest infestation from spreading, remove the infested plant parts, in the initial stage of attack and burn the same immediately. Spraying with 0.8% dimethoate or monocrotophos or 0.03% fenvalerate has proved effective in controlling this pest.

In nature, it is parasitised by *Aneristus ceroplastae*

Howard, *Anicetus ceylonensis* Howard, *Anysis saissetiae* (Ashmead), *Cocophagous longifasciatus* Howard, *Encyrtus barbatus* Timberlake, *E. kotinski* (Fullaway), *Eucomys lecaniorum* (Meyrick), *Marietia leopardina* Motschulsky and *Scultellista cyanea* (Motschulsky); besides a predaceous carterpiller of *Eublemma scital* Rambur (Nair, 1975).

Saissetia coffeae (Walker) and *S. oleae* (Barnard), the soft scales, that are major pests of coffee and olive respectively, are also found occasionally on citrus trees causing some minor or insignificant damage.

Thrips

Thrips (plural-thrips/thripses)*, are minute to very small-sized insects, about one mm in length, slender and fragile having asymmetrical rasping and sucking mouthparts. Metamorphosis though incomplete, is approaching completeness. Between nymphal and adult stage there is an inactive stage called pupa, which looks very much like the nymph but has well developed wing-pads that are not there in the nymphs. Adults have two pairs of wings that are long and narrow, fringed with long marginal setae; wing venation is greatly reduced; when at rest, the wings lie horizontally along the top of abdomen. Reproduction is both sexual and parthenogenetic, both occurring simultaneously. Oviposition period of parthenogenetical females is much longer than those that of fertilised females though the latter lay more eggs. Parthenogenetically developing progeny usually consists of males only.

Thrips are found all over the World, These are mostly polyphagous attacking a vast majority of crops but prefer plant families Poaceae (Gramineae), Cyperaceae, Asteraceae (Compositae) and Leguminoseae, as these crops offer more favourable microclimatic conditions. Good many species of thrips are oligophagous mostly graminivorous whereas all cecidogenous species are strictly monophagous (Ananthakrishnan, 1973).

Both nymphs and adults feed on plant juices by puncturing, lacerating and rasping the surface of leaves, buds

* erroneously mentioned as thrip.

and blossoms; the cell sap that oozes out is then imbibed by lapping. In case of leaves, the thrips congregate usually on ventral surface specially in depressions and grooves adjacent to main veins. The affected leaves show silvery sheen and also bear small black spots of faecal matter. In case of blossom-thrips, the affected flower-buds shrink and shrivel, petal show feeding scars and ultimately the fruit setting is adversely affected.

Thrips recorded on citrus trees include, *Heliothrips haemorrhoidalis* (Bouché), *Thrips pandu* Ramakrishna and *T. nilgiriensis* (Ramakrishna) on leaves; *Frankliniella dampfi* Priesner, *Ramaswamiachiella subnudula* Karny, *Scirtothrips dorsalis* Hood, *Thrips flavus* Schrank and *T. florum* Schmutz on flowers and *Scirothrips aurantii* Faure and *S. citri* (Moulton) on fruits.

Heliothrips haemorrhoidalis (Bouché) — greenhouse thrips — is cosmopolitan in distribution (CIE map No. A-732). In temperate regions, this thrip is confined to glasshouses. In India, Ramachandran (1954) reported its infestation from the South. Though a major pest of crotons and tea, it has been reported feeding on leaves of citrus, date palm, mango, passion fruit etc. In case of citrus, the pest prefers grapefruits (*C. paradesi*) and lemons (*C. limon*). Lacerting of leaves and puncturing of tissues by several individuals results in bleach-

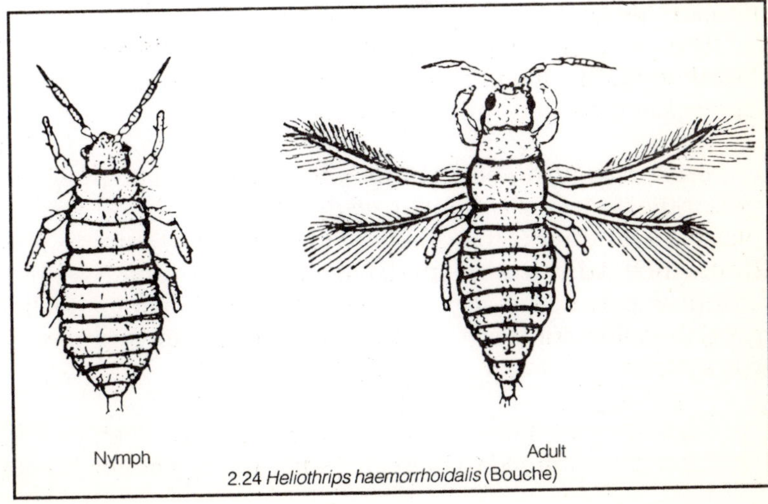

2.24 *Heliothrips haemorrhoidalis* (Bouche)

ing of green parts. Due to heavy infestation, the leaves get distorted, crinkled and mottled; the fruits become discoloured and develop cracks (Nair, 1975).

Eggs are bean-shaped and about 0.3 mm long. Freshly emerged nymphs are white, become gradually greenish-white and have red eyes. Pupae are yellow in colour. Adult females are slender, about 1.5 mm long, body brown coloured with head and thorax dark brown; legs uniformly yellow and wings yellowish with a median longitudinal pale grey band. Reproduction is by parthenogenesis. Eggs are inserted singly in leaf tissues and covered with a drop of excreta. Incubation period is 2 to 7 days and the total life-cycle is completed in 30 to 32 days at 26-28°C, 40 days at 23-25°C, two months at 20°C (Ananthakrishnan, 1971) and 12 weeks at 15°C (Hill, 1975). In South India, Sekhar and Sekhar (1964) observed the life-cycle to occupy 20 to 30 days on coffee leaves.

Scirothrips dorsalis Hood — chilli thrips — a polyphagous pest has also been found feeding on flowers and young fruits of citrus, grapevine, jujube, mango, pomegranate and tamarind. The pest is more active during dry weather. In Assam, this is a major pest of tea where its incubation, nymphal and pupal periods during Summer last for 6 to 7, 4 to 6 and 3 to 4 days respectively with life-cycle occupying 13 to 18 days (Dev, 1964). Adult longevity is 10 to 15 days and Raizada (1965) observed as many as 25 overlapping generations in a year.

Predaceous thrips, *Scolothrips indicus* Priesner and *Franklinothrips megalops* Back have been found preying upon this thrip (Ananthakrishnan, 1971).

Ramaswamieahiella subnudula Karny is a polyphagous pest reported feeding and breeding in flowers of several plants, including *Citrus medica*, mango, pomegranate and tamarind. Adults are small-sized, body orange-yellow, antennæ pale white, legs pale yellow and the wings having greyish-yellow infumation. It is a minor pest of various *Citrus* species.

Thrips flavus Schrank is cosmopolitan in distribution. In India, it is confined to high altitudes and is a major pest of ornamental plants particularly at hill stations. Among the fruit trees, it has been recorded on citrus and apple. Eggs are laid on flower buds before the buds open. On hatching the nymphs feed on vital parts of the flowers, by sucking the sap therefrom; causing distortion of flowers, as a result fruit setting is adversely affected and the fruits formed are weak which drop prematurely. Adults are informally yellow in colour, antennæ dark grey distally and wings pale yellow.

Thrips florum Schmutz — blossom thrips — is in abundance in flower buds of *Citrus* spp., specially in citron (*C. medica*) and sour oranges (*C. aurantium*). It has also been found feeding on apple and banana blossoms. Due to the feeding of these insects, the infested flower buds become smaller in size, the petals shrink and show feeding scars and ultimately the fruit setting is adversely affected.

Scirtothrips aurantii Faure — South African citrus thrips — as the name suggests is native to South Africa and is also widely distributed in Egypt (United Arab Republic) and Sudan (CIE map No. A-137). Besides *Citrus* spp., it has been found attacking various *Acacia* species. Eggs are inserted into the soft tissues of fruits. On hatching, the nymphs feed on small fruits near the peduncle; as a result of which a ring of scaly brownish tissues is formed on the fruits round the peduncle or even some irregular patches of scarred tissues may be seen on the fruits. Eggs are bean-shaped, about 0.2 mm, long, Nymphs are cigar-shaped, orange-yellow in colour and 0.6 to 0.8 mm long. Males are rare and the reproduction is by parthenogenesis. According to Hill (1975), incubation period is about one week; nymphal and pupal (including pre-pupal) periods last for one to two weeks each, while the adults may live for several weeks.

Scirtothrips citri (Moulton) — American citrus thrips — as the name suggests is a major of *Citrus* spp. in California (CIE map No. A-138). It has been reported from Eastern India, causing blemished on fruit surface (Dutta, 1966). The affected fruits often become unfit for human consumption and lose their market value.

To control thrips, spray with 0.03% dimethoate, endosulfan, phosphamidon or thiometon (Butani, 1979).

Leaf Miner

Leaf miners are tiny caterpillars of small moths — microlepidoptera — belonging to families Gracillaiidæ, Phyllocnistidæ). The caterpillars usually mine the leaves and occasionally petioles of leaves and tender shoots as well. Mingdu and Shuxin (1989) reported that damage levels due to leaf mines above 30% reduced net photosynthesis. Schaffer *et al.* (1997) demonstrated that leaf damage to citrus by leaf miner larvae can be accurately assessed by visual estimation. The percentage of leaf area damaged by leaf miner and subsequent reductions in net photosynthesis are highly correlated with mining duration and, to a lesser extent, the number of leaf miner per leaf. There is only one species reported from India on citrus, *Phyllocnistic citrella* Stainton (Phyllocnistidæ). It is commonly called citrus leaf-miner and is found all over the Orient, from Africa to Australia (CIF map No. A-274). Almost all *Citrus* species are attacked though the varieties with succulent leaves and thin cuticle are preferred (Latif and Yunus, 1951), *Citrus medica* and *C. sinensis* which have succulent leaves are heavily damaged while *C. aurantium* having thick and coarse leaves are least damaged (Pandey and Pandey, 1964). Besides *Citrus* species, it has also been reported damaging Bengal quince (bael, *Aegle mamelos*), curry leaf (*Murraya koengii*) orange-jasmine (*M. paniculata*), Arabian jasmine (*Jasaminum sambae*), cinnamon (*Cinnamomum zeylanicum*) etc. *Phyllocnistis toparcha* Meyrick, an allied species has been reported mining the leaves of grapevines in India.

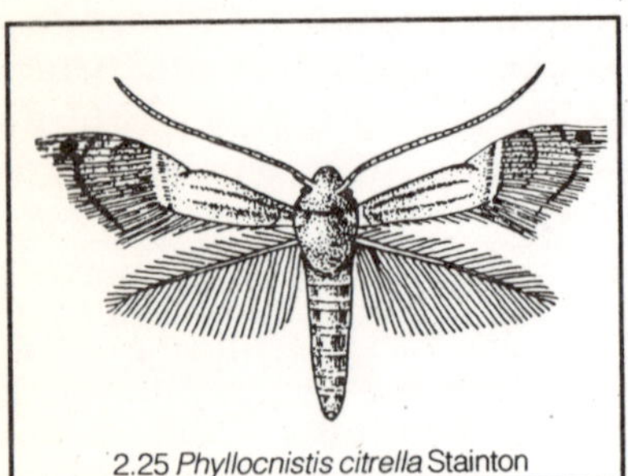

2.25 *Phyllocnistis citrella* Stainton

The caterpillars make serpentine mines in the leaves feeding on epidermal cells of the leaf leaving behind the remaining leaf tissues quite intact; they feed actually more on sap than on solid tissue (Pruthi ans Mani, 1945). The caterpillars attack only young and tender leaves, old leaves are generally avoided. The mined leaves fade, get distorted and ultimately dry up. Occasionally, when the pest population increases rapidly, the succulent tender shoots specially of oranges and grapefruits are also mined (Hutson and Pinto, 1934). As a result the growth specially of young trees is retarded considerably and in nurseries the saplings may even die away. According to Hutson and Pinto (1934), the pest attack also encourages the development of citrus canker disease. The pest is active throughout the year except during severe cold (December to February); through much of the injury is caused early in Spring and the pest population always decreases during Summer.

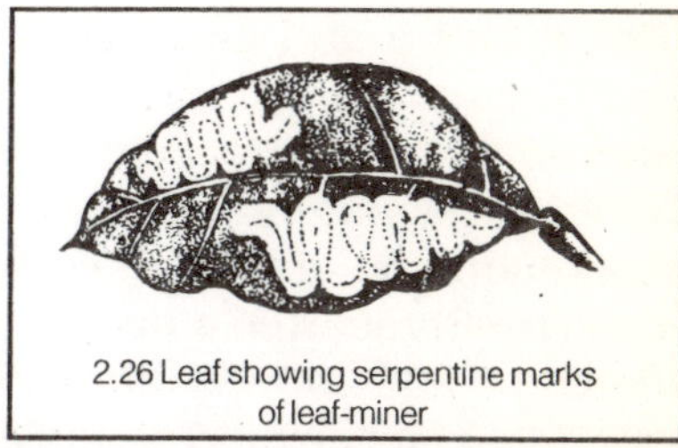
2.26 Leaf showing serpentine marks of leaf-miner

Eggs are minute, about 0.3 mm long, flattened, broadly-oval in shape and greenish-yellow in colour. Full grown caterpillars are cylindrical in shape, about 5 mm long, dull greenish-yellow in colour and apodous (legless). Adults are tiny, silvery-white moths having heavily fringed wings. Forewings have brown stripes and a prominent black spot near the apical margin while hind wings are pure white; wing-spread is four to five mm. Mating takes place mostly at night or at dawn; 14 to 24 hours after the emergence of female moths. Eggs are laid singly, attached to leaves and twigs; a female lays 36 to 76 eggs in 2 to 6 days. Incubation period is two to ten days. Caterpillar stage lasts for 5 to 10 days during Autumn (August to October) and 13 to 20 days during Winter (November to January). Pupae are found in white cocoons lying near the margin of mined leaves, the edge of which is turned over and the inside lined with silken webbing. Pupal period is 5 days (Summer and Autumn) to 25 days (Winter) and the entire life-cycle occupies 20 days during Spring extending upto 60 days in Winter. There are 9 to 13 overlapping generations in a year and adults hibernate during Winter.

To prevent the carry over of the pest, prune heavily the affected parts during Winter and burn the same. Fumigation with hydrocyanic acid gas (HCN) is quite effective but requires great care and technical aid. Nayar *et al.* (1976) recommended 0.25% neem cake extract. Borle and Kharat

(1977), Lakra and Gupta (1977), Singh and Rao (1977) as also Thammi Raju *et al.* (1977) found most of the insecticides (organophosphatic) to be effective, 0.05% chlorpyriphos, dimethoate, methyl-demeton, monocrotophos, phosalone, phosphamidon, quinalphos and thiometon to be better. Rae *et al.* (1996) reported a narrow range petroleum oil applied at a rate of 500 ml/100 liters of water provided control of leafminer equivalent to cartap and methomyl. Abamectin plus petroleum spray oil at a rate of 1.5 gm (a.l.) plus 50 ml/100 liters of water, respectively, provided complete control of leafminer. To keep the pest under check, spraying should be repeated at weekly interval during flushing period. It will be advantageous, if atleast two different insecticides are sprayed alternatively.

In nature the caterpillars are parasitised by number of chalcidoids specially during September-November. *Eurytema* species has been reported from Sri Lanka (Hutson and Pinto, 1934) and *Ageniaspis* species from Java (Voute, 1932). According to Prem Chand (1995) caterpillars are also parasitised by *Cirrospiloideus phylloenistis* (Narayan) and *Scotolinx quadristiriata*, Rao and Rema.

FLOWER FEEDERS

Blossom Midge

Dasineura (*Pectinociplosis*) *citri* (Rao and Grover) (= *D. citri* Grover and Prasad) (Cecidomyiinæ: Cecidomyiidæ: Diptera) is the most common gall-midge often found on citrus buds and flowers. The infested buds become distarted as they grow and show a slit-like opening on one side (Prasad, 1973). The adults are orange coloured minute flies that lay minute whitish eggs between the sepals and petals of the bud. Incubation period is 32 to 40 hours while larval and pupal periods last for 9 to 11 and 4 to 6 days respectively.

No separate control measures are adopted against this pest. Chemical control measures to check aphids and whiteflies, controls the midges as well.

Flower Moth

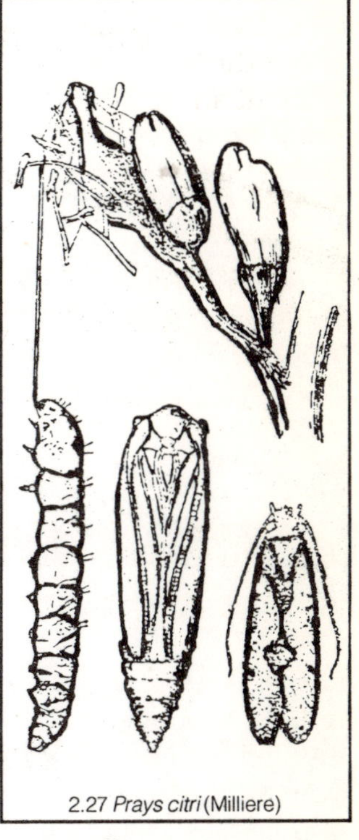
2.27 *Prays citri* (Milliere)

Prays citri (Milliere) — citrus flower moth — is widely distributed all over southern Europe, Syria, Israel, Indian subcontinent, Malaysia, New South Wales and Philippines (Ebeling, 1959). The only economic host recorded so far is *Citrus* spp., flowers of citron (*C. medica)* and sour lime (*C.jambhiri*) being preferred. According to Pruthi and Mani (1945), lemons (*C.limon*), limes (*C.aurantifolie*), oranges (*C.reticula*) and pomelo (*C. decumana*) are the potential fruits that are destroyed in a large number.

Eggs are laid singly during night on buds and flowers. On hatching, the tiny caterpillars bore inside the fruits and feed blow the rind, next to pulp, causing gall-like growth or swelling. As many as 20 galls may be found on a single fruit. Full fed caterpillars come out to pupate. The entry and exit holes in the rind of fruits permit entry to bacteria and fungi, causing rotting and premature fall of fruits. Such type of damage is found in India, Malaysia and Philippines (Garcia, 1939) while in western countries the young caterpillars spin and web together the flowers, feed on calyces and other floral parts and pupate with these webbed flowers. Dry heat is unfavourable for the pest activity whereas colder climate accelerates the activity, hence maximum activity is during January-February.

Eggs are subelliptical, measuring on an average 0.2 × 0.1 mm and yellowish in colour. Full grown caterpillars are 40 to 45 mm long, subcylindrical in shape, semitransparent and covered with fine hair. Pupae are 5 to 6 mm long and chocolate-brown in colour; their anal end is fastened to the host plant by silken thread. Moths are greyish-brown with broadly fringed long narrow wings; forewings bear numerous irregular markings and a conspicuous marginal fringe of hair while the hind wings are uniformly grey, membranous and without any spots. Wing spread is 9 to 11 mm (Manalae, 1924). Total life-cycle in Italy has been observed to be 92 days at 20.5°C with three generations in a year whereas in Philippines a single life-cycle is completed in 69 days at 27°C with five generations in a year.

Hand-picking of caterpillars and infested blossoms and their mechanical destruction may be carried out to check the spread of this pest. In Philippines, Garcia (1939) observed red ants, *Oecophylla smaragdina* (Fabricius) preventing the fruit damage on those trees where these ants had made their nests.

FRUIT PESTS

Fruit Sucking Bugs

Five phytophagous pentatomids, *Cappaea taprobanensis* (Dallas), *Chrysocoris grandis* (Thunberg), *Nezara viridula* (Linnæus), *Rhynchocoris humeralis* (Thunberg) and *Vitellus orientalis* Distant; two coreids *Dasynus antennatus* (Kirby) and *Leptoglossus australis* (Fabricius) (= *membranceus* Fabricius) and one lygaeid but, *Spilostethus pandurus* (Scopoli) have been recorded from India, sucking the juice from citrus fruits; but none of these is of any significant importance as the damage caused is usually negligible.

Cappaea taprobanensis (*Dallas*) (Pentatomidæ Heteroptera) has been reported from Indian subcontinent and Indonesia. It is a minor pest of oranges (*C. reticulata*), occurring regularly in the hills of northern India. Nymphs and adults may often be seen crowded on tree trunks.

Chrysocoris grandis (Thunberg) (Pentatomidæ : Heteroptera) has been reported from eastern India, Myanmar, South-east China and Japan. In India, it is recorded as a minor pest of citrus from Assam and adjoining states. Eggs are laid in July and the nymphs of this generation such the sap from leaves and flower-buds. Fruits are attacked by the nymphs of second generation and the affected fruits fall prematurely — thus the second generation is comparatively more destructive than the first generation. There are only two generations in a year, the bugs overwinter in adult stage.

Nezara viridula (Linnæus) (Pentatomidæ : Heteroptera) — the green stink bug — is cosmopolitan in distribution (CIE map No. A-27). The polyphagous pest is found all over India, causing damage to number of fruits and vegetables. Eggs are laid in batches of 40 to 60 eggs and a single female lays upto about 300 eggs, glued in rafts on ventral surface of leaves. The freshly hatched nymphs continue to stay in their egg-shells, without feeding, till the first moulting; thereafter they come out of egg-shells and attack the unripe fruits. The feeding punctures cause local necrosis coupled with brown spotting and ultimately resulting in deformation of the affected fruits; such fruits taste bitter. Subsequently, bacterial and fungal invasions also follow, causing rotting of fruits and their premature shedding.

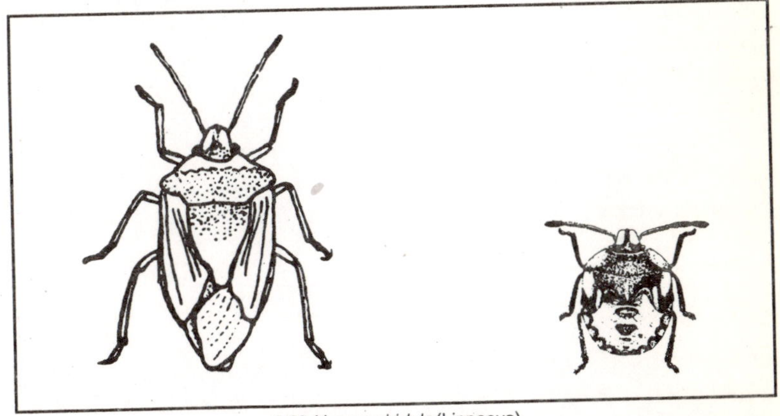

2.28 *Nezara viridula* (Linnaeus)

Eggs are barrel-shaped. 1.0 to 1.5 mm long, white when freshly laid, becoming pinkish before hatching, Adults are green shielded bugs, 13 to 18 mm long, colour varying from apple-green to reddish-brown. Egg period is 4 to 7 days, nymphal development takes about a month and there are four to five overlapping generation in a year.

Rhynchocoris humeralis Thunberg (Pentatomidæ : Heteroptera) — citrus stink bug — has been reported from

India, China, Taiwan and Myanmar. In India, it occurs regularly in Assam, Kumaon and eastern Himalayas as a major pest of oranges (*C. reticulata*) (Chowdhury and Majid, 1954), Freshly hatched nymphs feed gregariously on tender fruits but later they disperse and attack more and more ripening fruits. Adults bugs also feed on developing fruits sucking the juice of these fruits. The affected fruits. invariably catch secondary infection of bacteria and fungi, start rotting and ultimately fall down prematurely.

Eggs are laid during May to September in 2 to 3 batches of 14 to 15 eggs each, on dorsal surface of leaves (Nair, 1975). Incubation and nymphal periods are about a week and a month respectively; overwintering is in adult stage. According to Luh Nien Tsin (1936), these bugs have only one generation in a year in China.

Vitellus orientalis Distant (Pentatomidæ : Heteroptera) has been reported as a minor pest from South India where it punctures fruits of oranges (*C. reticulata*). The puncturing appears to hasten rotting of fruits and also attraction for fungi (Pruthi and Mani, 1945).

Leptoglossus australis (Fabricius) (= *membranceus* Fabricius) (Coreidæ : Heteroptera), causes considerable damage to citrus fruits specially in South India, South Africa and Malaysia, while in Queensland, it is pest of passion fruit. Nymphs and adults congregate on fruits, pierce the rind and suck the juice. Adults are more destructive than nymphs. The infested fruits do not develop, remain stuncted and ultimately dry and fall prematurely. Besides citrus, these bugs also attack pomegranate (Jadhav *et al.*, 1976) and cucurbit fruits including melons.

2.29 *Leptoglossus australis* (Fabricius)

Eggs are cylindrical in shape and light brown in

colour. Nymphs are reddish in colour when young but with subsequent moults become brown and then black. There are two spine-like projections on head in-between the eyes. Thoracic plate and the abdominal segments are projected laterally on either sides to form black spines. Full grown nymphs are 12 to 16 mm long, bear wing-pads and their hind tibia are flattened. Adults are 20 to 25 mm long, dull black in colour; antennæ are 4-segmented that are alternatively black and yellow in colour; mesothoracic shield has a yellow line anteriorly and just like nymphs, its lateral sides remain projected into a spine on either side. There is a minute yellow spot in the centre of each forewing and hind flat tibia. Incubation and nymphal duration last for, on an average, 7 and 52 days; adults live for 10 and 14 days in case of females and males respectively (Visalakshi *et al.*, 1980).

Spilostethus pancdurus ((Scopoli) Lygæidæ : Heteroptera) is a brightly coloured bug that often occurs in large number but inspite of that, its status as pest is doubtful. These bugs are polyphagous and have been reported feeding on apple, citrus, grapevine, java-plum (jamun), litchi, mango, peach, pear and plum fruits. Due to their feeding, the fruits start fermenting and fall down prematurely.

No separate control measures are generally required for these bugs. However, when serious, dusting with 4% endosulfan dust or spraying with 0.05% endosulfan or 0.2% carbaryl can effectively control the pest population.

Fruit Sucking Moths

A large number of noctuid moths have been observed attacking citrus fruits all over the World. The caterpillars of most of the species are no doubt leaf defoliators but these are generally found feeding on other host plants, usually the wild ones (menispermaceous and anacardiaceous); of course sometimes even the economic crops, such as castor, jujube, pomegranate etc. are also attacked and defoliated. It is only the adult moths that are classified as pests, as they suck the juice of ripening fruits. The moths are big-sized, robust, having a well developed proboscis with dentate tips, with which they pierce the ripening fruits. These moths are nocturnal in habit and are often seen hovering around the trees in the orchard after dusk, especially during rainy season. The punctured fruits are soon attacked by bacteria and fungi, as a result, the fruits rot and drop down prematurely. Of the various species of fruit sucking moths. *Othreis* species, specially *O. materna* (Linnæus) and *O. fullonia* (Clerck) are comparatively more destructive and cause on an average 3 to 5 per cent damage every year.

Others materna (Linnæus) – the most common citrus fruit sucking moth, has been reported from the Indian subcontinents and Indonesia. In India besides citrus, it has been found on grapes, mango and pear fruits. Eggs are oval in shape and shining pale green in colour. Freshly hatched caterpillars are slender, thread-like, 3 to 4 mm long and pale greenish in colour. Full grown caterpillars are stout, robust, 50 to 60 mm long, velvety-blue with yellow patterns on dorsal and lateral sides and having a hump at anal end. Pupation takes place in transparent pale whitish cover enclosed in leaf-fold. Pupae are stout, smooth, 18 to 22 mm long, reddish-brown in colour with anal cremster conical having two small hooks. Moths have head and thorax greenish-grey and abdomen orange. Forewings greyish-green with numerous faint striated reddish-lines and three rufous spots in the middle; hind wings having apical area blotched with rufous and a round black spot in the centre and a marginal black band. Wing spread is 90 to 100 mm. Egg, caterpillar and pupal stages last for 8 to 10, 18 to 35 and 14 to 18 days respectively.

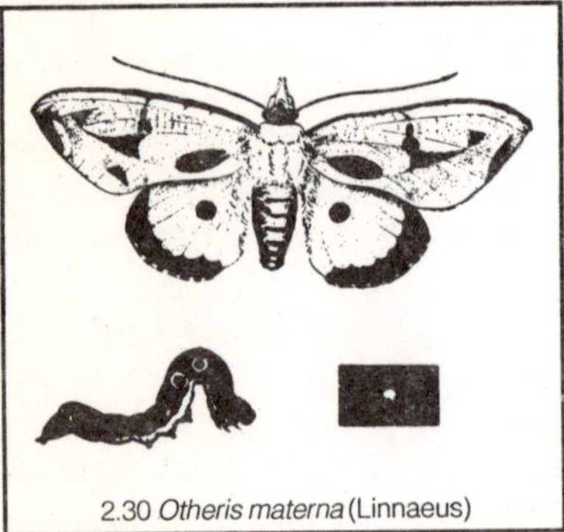
2.30 *Otheris materna* (Linnaeus)

Othreis fullonia (Clerck), another common citrus fruit sucking moth that is widely distributed throughout the Orient, extending from Africa to New Guinea and Australasia. Besides citrus, these moths are also found feeding on fruits

2.31 *Otheris fullonica* (Linnaeus)

of banana, grapes, guava, mango, peach, pear, plum, pomegranate etc. Moths have reddish-brown head and thorax and orange coloured abdomen with greenish tinge. Forewings are variegated and striated with reddish-brown colour, a triangular white mark is usually present; hind wings orange in colour having a kidney-shaped black blotch in the middle. Wing-splan is 80 to 90 and 90 to 100 mm in case of males and females respectively. According to Chowdhury and Majid (1954), egg, caterpillar and pupal periods occupy 3 to 4, 13 to 17 and 12 to 18 days respectively with two to three generations in a year.

Other species of *Othreis* are comparatively less common and confined mostly to citrus fruits. Description of these moths as also other noctuid moths, recorded sucking the citrus fruits in India have been given in detail by Hampson (1896).

Othreis aurantia Moore has been reported from Indian sub-continent and Indonesia. Moths are comparatively bigger than other moths of *Othreis* species and have head and thorax ferruginous, suffused with plum colour and slight purple tinge or having green patches; hind wings are orange coloured having large black lunule and a submarginal patch. Wing expanse is 110 to 120 mm.

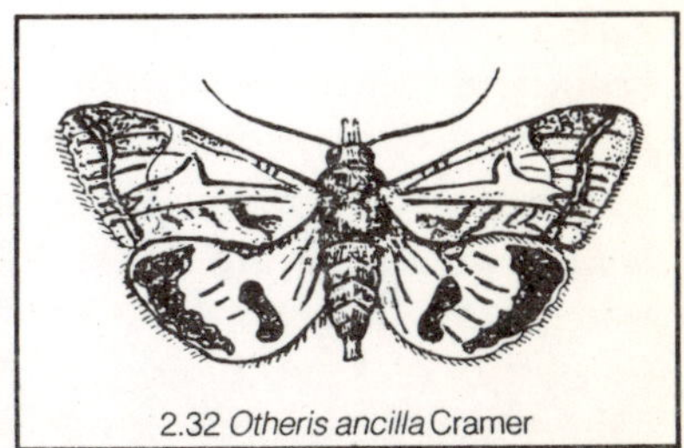
2.32 *Otheris ancilla* Cramer

Othreis ancilla Cramer is widely distributed in Indian sub-continent. These moths are the smallest among the *Othreis* species recorded on citrus in India. Head and thorax are fulvous-brown and abdomen orange. Forewings are orange-green in colour, suffused with purple-reddish brown tingle and striated with rufus; hind wings have large black lunule in the centre and a narrow submarginal band with wavy edges. Wing-spread is 68 to 75 mm.

Othreis cocalus Cramer have been reported from Indian sub-continent and Indonesia. Moths are slightly bigger than *O.ancilla* but smaller than those of *O.hypermaestra* and differ from these moths in having no black spots on hind wings and the marginal band extends upto anal angle. Wing-span is 82 to 88 mm.

Othreis discrepans (Walker) is found in East India, Singapore. Thailand and Indonesia, Moths have head and thorax greyish-brown with purplish bloom while their abdomen is orange coloured. Forewings are purplish-greyish-brown irrorated with fuscous and green patches at the base; hind wings are orange coloured, having a large black lunule and a broad black marginal band. Wing-span is 90 to 100 mm (Hampson, 1896).

Othreis hypermnestra Cramer is common throughout Indian sub-continent. Moths have head and thorax yellowish-green in colour and abdomen orange; forewings are yellowish-green having dark striae and some grey patches; hind wings are orange coloured having a black spot and marginal band from apex to vein 2. Wing expanse is 86 to 92 mm.

Othreis slaminia Fabricius is widely distributed in the entire Oriental region from Malagasy to Australia. Moths have plum-coloured head, green thorax and orange abdomen. Forewings are yellow coloured and hind wings orange with a large black lunule. Wing expanse is 80 to 92 mm and 90 to 104 mm in case of males and females respectively.

Othreis tyrannus Guenee is found along the entire Himalayan range, including Nepal, Sikkim (India), South China and Japan. These moths are more or less, as big as those of *O. aurantia*; have head and thorax dark reddish-brown in colour and abdomen orange. Forewings are chestnut-red or green suffused and striated with rufus; hind wings with a very large black lunule. Wing expanse is 110 to 120 mm.

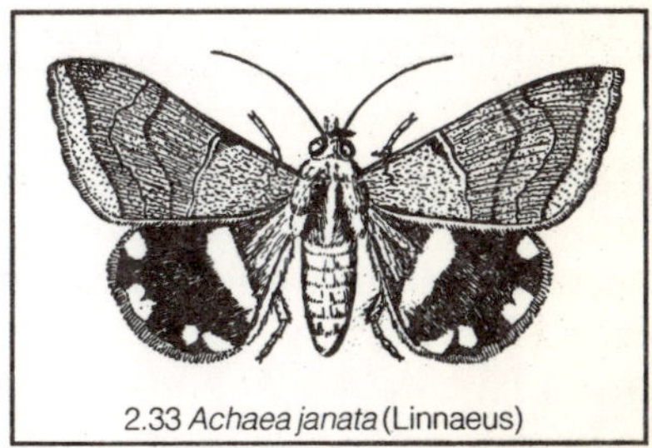
2.33 *Achaea janata* (Linnaeus)

Achaea janata (Linnæus) (Noctuidæ : Lepidoptera) — castor semilooper — as the name suggests, is major pest of castor, widely distributed from Africa to Australia. The caterpillars are leaf defoliators and besides castor have been also found feeding on leaves of grapevines, guava, jujube and pomegranate while the moths have been reported by Pruthi and Batra (1960) as serious pest of oranges, sucking juice of the fruits. Besides citrus fruits, these moths have also been reported on grapes, guava and mango fruits (Butani, 1979).

Eggs are laid singly on ventral leaf surface, one to six eggs per leaf. A female lays, on an average, 400 eggs in her life time. Eggs are hemispherical in shape and bluish-green in colour, ridged with 40 to 45 striae. Caterpillars are semiloopers 55 to 65 mm when full grown and show conspicuous colour variation, some are grey with red or brown lateral stripes while others are bluish-grey specked with blue-black and having yellow lateral stripes. Freshly formed pupae are glistering dark green, later becoming brown. Moths are stout, pale yellowish-brown with wavy lines on forewings and black hind wings having a medial white band and three large white spots on outer margin. Wing expanse is 50 to 65 mm. Incubation, caterpillar and pupal periods on castor last for 3 to 5, 9 to 23 and 7 to 26 days respectively (Pandey *et al.*, 1966). A single life-cycle is completed in 3 to 5 weeks and previposition period extends from 6 to 21 days. There are five to six overlapping generations in a year and hibernation is in pupal stage.

Anomis fulvida (Guenée) is found throughout Indian subcontinent, China, Indonesia, Pacific Islands and Australasia. Moths have head and thorax ferruginous-reddish-brown and abdomen reddish-fuscous in colour; forewings dark ferruginous with 2 to 3 white spots in the middle, hind wings vinous-red to dark brown in colour Wing expanse is 50 to 58 mm.

Anua (Ophiusa) coronata (Fabricius) is present all over the Indian sub-continent as also in Indonesia and Australasia. Moths have head and thorax pale reddish-brown and abdomen orange in colour; forewings irrorated with dark specks and having a black patch near outer angle, hind wings orange coloured with broad medial and submarginal fuscous black band not reaching inner margins. Wing expanse is 82 to 96 mm.

Anua mejanesi (Guenée) has been reported from India and West Africa. Moths have head and thorax reddish-chocolate and abdomen orange in colour. Forewings are reddish-chocolate irrorated with dark specks while hind wings have basal area whitish, suffused with fuscous. Wing expanse is 50 to 58 mm.

Anua tirthaca (Cramer) is widely distributed in South Africa, Malagasy and Indian sub-continent. The moths have head and thorax greenish-yellow and abdomen orange in colour. Forewings are also greenish-yellow with slightly dark striate, hind wings are orange with broad submarginal black band not reaching costa. Wing expanse is 64 to 80 mm.

Calpe bicolor Moore is common in Kangra valley in India. Moths have head and thorax reddish-brown in colour irrorated with grey while the abdomen is yellowish-orange; forewings have silvery sheen, underside being yellowish-orange. Wing expanse is 52 to 60 mm.

Calpe emerginata (Fabricius) is found all over Indian sub-continent including Sri Lanka, Myanmar and China. Moths have head and collar region fiery orange; thorax reddish-brown and abdomen fuscous. Forewings are reddish-brown suffused with purplish tinge, hind wings are ochreous-white suffused with fuscous towards outer margin. Wing expanse is 38 to 46 mm.

Calpe fasciata Moore is commonly found in Sikkim and Kangra valley of India. Moths differ from those of *C. bicolor* and *C.ophideroides* in having no reddish tinge and underside of wings being brown. These are also smaller in size than the other two species, wing spread being 46 to 54 mm.

Calpe ophideroides Guenée is confined to foot hills of Himalayas, Assam to Kashmir including Nepal as also Singapore and Malaysia. Moths are similar to those of *C.bicolor*, except that forewings are reddish-brown in colour, suffused with grey and having numrous fine pale striae; hind wings are yellowish-orange. Moths are bigger in size than the above three species, having wing spread of 66 to 74 mm.

Ercheia cyllaria Cramer has been recorded from Indian sub-continent including Sri Lanka and Myanmar as also Indonesia. Moths have head and thorax pale reddish-brown and abdomen fuscous black; forewings suffused with fuscous and streaked with black and hind wings fuscous-black with three medial white spots. Wing expanse is 50 to 60 is mm.

Erebus hieroglyphica (Drury) has been recorded from Indian subcontinent, Malagasy, Malaysia, Philippines and Indonesia. Moths are big and blackish-brown in colour; forewings black with a conspicuous whorl-shaped jet black mark. Wing expanse is 85 to 95 mm. The moths are usually found swarming on fallen fruits; these seldom attack the fruits on trees. They suck the oozing juice from fallen trees and do not make any fresh punctures on these fruits. They are also attracted to syrupy solution split on ground or the trunk of the trees, which they easily lick, with their well developed proboscis.

Lagoptera (Ophiusa) dotata (Fabricius) is found throughout Indian subcontinent including Myanmar and Sri Lanka as a minor pest of grapes and citrus. Moths are bronze-brown in colour having forewings irrorated with white specks and prominent marginal grey band with a wavy line on it; hind wings dark fuscous with margin and cilia whitish. Wing expanse is 72 to 82 mm.

Lagoptera honesia (Hubner) in widely distributed in Indian subcontinent and is also found in Philippines. Moths have head and thorax reddish-chestnut and abdomen crimson coloured. Forewings are reddish-chestnut, slightly irrorated with dark specks while the hind wings are crimson red with black submarginal medial patch. Wing expanse is 84 to 94 mm.

Lagoptera submira (Walker) has been recorded from Pakistan, India, Bangladesh and Myanmar. The moths differ from those of *L. dotata* is being much darker reddish-brown; forewings with costal and medial areas suffused with bluish-white and hind wings dark reddish-fuscous without any white medial band. Wing expanse is 70 to 80 mm.

Pericyma glaucinans (Westwood) has been reported from Congo, South Africa, Indian subcontinent and Indonesia. The moths are fuscous. Forewings with sinuous subbasal black line and hind wings with numerous fine oblique lines. Wing expanse is 36 to 42 mm.

Parallelia algira (Linnæus) occurs throughout the Indian sub-continent, China, Japan, Africa and Mauritius. The caterpillars are major pest of castor, while adult moths such the juice of citrus fruits causing sporadically severe damage in various parts of India.

Pelamia frugalis (Fabricius) is very common throughout the Oriental region as also in Africa and Australia. The caterpillars feed on leaves of several grasses, paddy, sorghum, sugarcane etc. whereas the moths pierce and such the juice of citrus fruits specially in South India (Pruthi and Mani, 1945).

Pelochyta astrea Drury is found through Indian subcontinent including Myanmar as also in Taiwan. The moths are very conspicuous having white-fuscous coloured body with a number of small round black spots, one pair on head, two pairs on collar, one pair each on pro-, meso- and metathorax and one pair at the base of each forewing. Both pairs of wings are hyaline, forewings having margins and apical areas pale fuscous whereas the hind wings have marginal fuscous band. Wing-span is 50 to 54 mm and 64 to 72 mm case of males and females respectively.

Polydesma quenavadi Guenée is widely distributed in the Indian subcontinent, besides being found in Borneo and Australia (Fletcher, 1920). The moths are brownish-grey in colour having forewings with wavy lines and marginal series of black specks and hind wings with basal area whitish, outer brownish and having wavy lines. Wing spread being 40 to 56 mm.

Remigia frugalis (Fabricius) is found in the entire Oriental region from Africa to Australia. The moths are greyish-brown; forewings having diffused dark markings and submarginal series of black specks, hind wings ochreous-fuscous with diffused submarginal lines. Wing expanse is 36 to 50 mm.

Serrodes inara Cramer has been reported from Africa, Indian subcontinent including Sri Lanka and Myanmar as also Indonesia and Australia. The moths are stout and greyish-brown; forewings pale olivaceous-grey and hind wings fuscous. Wing spread is 52 to 74 mm.

Sphingomorpha chlorea Cramer has been reported from Africa, UAR (Egypt), Indian subcontinent etc. The moths have ochreous-white head brown thorax and abdomen with ochreous-white spots. Forewings are reddish-brown with dark striae and hind wings fuscous-brown having an ochreous patch with black striae at centre of outer margin. Wing expanse is 60 to 84 mm.

Control of these moths is rather difficult, as the immature stages, specially caterpillars, are found feeding on other host plants and not citrus and the moths appear in citrus orchards only after dusk and suck the fruit juice only during the night. Systematic destruction of alternate host plants in the vicinity of orchards helps to check the pest population. Creating smoke in the orchard's after sunset may keep the pest (moths) at bay; bagging the fruits has also been suggested. Prasad (1992) suggested baiting with malathion 0.05%

+ 1% crude sugar (jaggery) + juice from fresh fruits. These methods may be effective in reducing the pest damage but are rather cumbersome and not practicable on large scale, besides being expensive. Three triweekly spryings with 0.1% malathion, commencing from the time fruits are half ripe upto about one month prior to harvest also help.

Fruit-Flies

Fruit-flies are not serious pests of citrus fruits except in limited areas specially in cooler subtropical regions. Hayes (1966) reported heavy infestation in Sikkim where the ground under the trees was covered with fallen infested fruits.

The species of fruit-flies (Thypetidæ : Diptera) recorded on citrus fruits in India include, ber fruit-fly *Dacus correctus* (Bezzi), melon fruit-fly *D. cucubitae* (Coquillett), guava fruit-fly *D. diversus* (Coquillett), mango fruit-fly *D. dorsalis* (Hendel), *D. scutellaria* (Bezzi), cucumber fruit-fly *D. tau* (Walker) (= *hageni* de Meijere) and peach fruit-fly *D. zonatus* (Saunders) (Butani, 1979a), Gard (1978) has also added Ethiopian fruit-fly *D. ciliatus* (Loew) (= *brevistylus* Bezzi). Besides *Callantra icariiformis* Enderlein has been reported from Sikkim (Enderlein, 1920) and *C. minax* (Enderlein) from West Bengal (Nath, 1972). All these are invariably named after their major host. Citrus fruits are not the major hosts of these flies though *Dacus dorsalis* and *D. ciliatus* often reported causing severe loss in certain areas.

The immature stages of the various fruit-flies, look alike and are hard to differentiate whereas adults can be distinguished from one another. Apart from colour variation. *D. cucubitae* and *D. tau* slightly bigger in size with costal band and anal stripes well developed; *D. correctus* and *D. zonatus* are smaller in size with costal band, incomplete and no anal stripes whereas *D. diversus* and *D. dorsalis* are medium sized with narrow costal bands and anal stripes (Kapur, 1972).

Eggs are laid in the soft skin of ripening fruits. On hatching, the maggots bore further into the fruits and feed on soft pulp. Unripe fruits are seldom attacked as the flies are not able to puncture the hard rind of unripe fruits for oviposition. The infested fruits show depressions with dark greenish punctures, get deformed and in conjunction with bacterial and fungal activity, these fruits rot and fall down prematurely. Full fledged maggots come out of these fallen fruits, to pupate in the soil.

Daccus dorsalis (Hendel) — Oriental fruit fly — as the name suggests, is found all over the oriental region from Australia and Hawaii to Pakistan (CIE map No. A-109). Kapoor (1970) has listed a long list of its host plants including apple, bael (*Aegle marmelos*), banana, chillies, *Citrus* spp., coffee, eggplant, fig, guava, jack-fruit, jujube, loquat, mango, peach, persimmon, plum, pomegranate, quince, sandal wood and sapota. According to Narayanan and Batra (1960), these flies breed profusedly on guava March, migrate to apricot, loquat and plum during April-May, then to peach and fig in June, mango during June-August and thereafter to *Citrus* spp. During early Winter, the flies shift to apple, pear and even unripe banana.

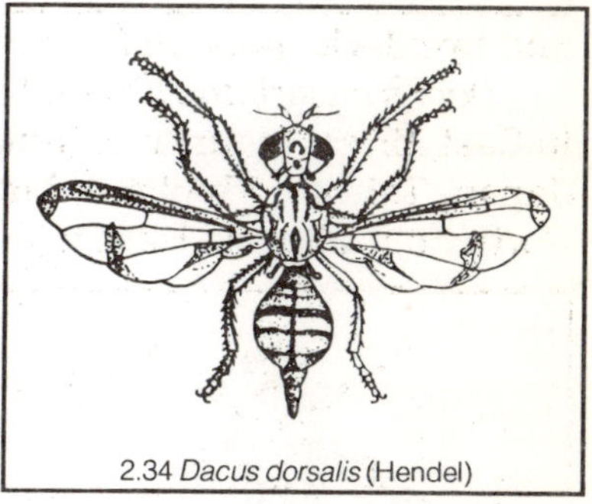

2.34 *Dacus dorsalis* (Hendel)

Bionomics have been studied by Juneja (1948) and Shah *et al.* (1948). Preoviposition period is two to five days, oviposition lasts for about a month during which a female may lay 150 to 200 eggs in clusters of two to 15 eggs. Incubation period is one to three days during Summer (March-April); maggot development takes on an average six days during Summar extending upto 19 days in Winter. Pupation usually takes place 80 to 100 mm below the soil surface and pupal period ranges from six days (Summer) to 44 days (Winter). Bess and Haramoto (1977) have reported that the lowest average temperature at which the immature stages can develop is about 14°C and temperature above 21°C is necessary for then flies to attain sexual maturity without undue prolongation in preoviposition period.

Dacus ciliatus (Loew), commonly called Ethiopian melonfly is of African origin, now widely distributed besides Africa reported from European countries, Middle East, Pakistan, India etc. Its main hosts are various cucurbits while the al-

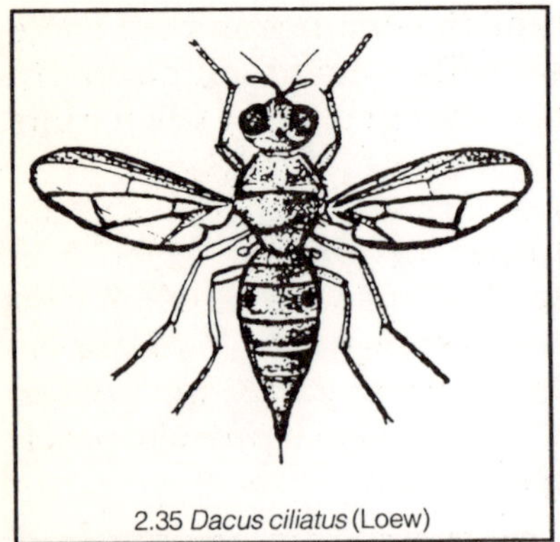

2.35 *Dacus ciliatus* (Loew)

ternate hosts include, citrus, apple etc. Eggs are shiny white, slightly curved and on an average 2.5 mm long. Maggots are whitish in colour and measure on an average eight mm in length. Pupae are cylnindrical in shape, brownish to ochraceous in colour and about 5.5. mm long. Adult flies are ferruginous-brown in colour, having hyaline wings and two dark spots on 3rd abdominal segment.

Dacus cucurbitae (Coquillett) — melon fruit-fly — is found in East Africa, Mauritius, Indian subcontinent, China, South Japan, Taiwan, Thailand, Malaysia, Indonesia, Philippines, North Australia and Hawaii (CIE map No. A-640). Batra (1953) and Gupta (1960) have listed more than 70 host plants from India, including citrus, date palm, guava, mango, papaya and peach; while from outside India, it has been reported attacking apple, avocado, Chinese-melon, custard apple, fig, mango, melons, pear and strawberry (Narayanan and Batra, 1960).

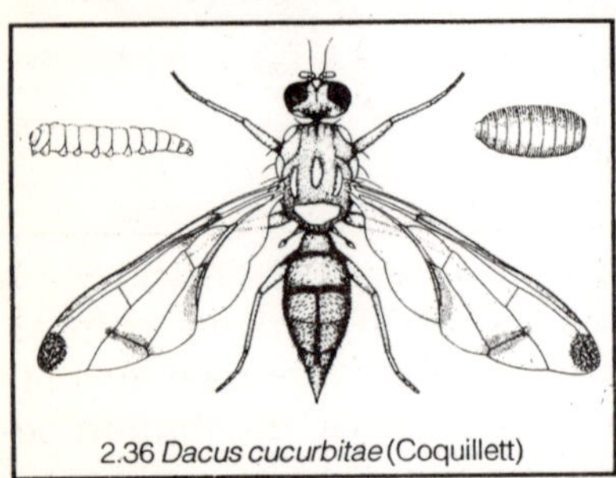

2.36 *Dacus cucurbitae* (Coquillett)

Eggs are cylindrical with slight curve, about one mm long and white in colour. Full grown maggots are pale white in colour and 8 to 10 mm long. Adults are reddish-brown flies having lemon-yellow, curved, vertical markings on the thorax. Wings are transparent with brown bands and grey spots at the apex. Abdomen of male is spherical and that of female conical (Ranjhan, 1949).

The life-span during monsoon days is about 10 days and extends upto three months during Winter. Lall and Sinha (1959) observed life-cycle to occupy 12.5 to 13.2 days during June-July (Monsoon) and 30.3 to 34.1 during Winter (December-January); the flies do not breed during warmer months (May and September) and over-winter in adult stage.

Dacus diversus — guava fruit-fly — is widely distributed in Indian subcontinent. Though guava and loqat are its preferred hosts, the flies also attack banana, citrus, java plum (jamun), mango, nutmeg, papaya and some gourds. In Summer, the flies breed exclusively in lower buds of cucurbitaceous plants and migrate to orchards during Winter. The pest is active throughout the year except during severe cold (January-February) when it over winters in adult stage in the folds of guava and loquat leaves. During July-August, egg, maggot and pupal durations last for one to four, four to five and six to seven days respectively and longer during October-November (Batra, 1953).

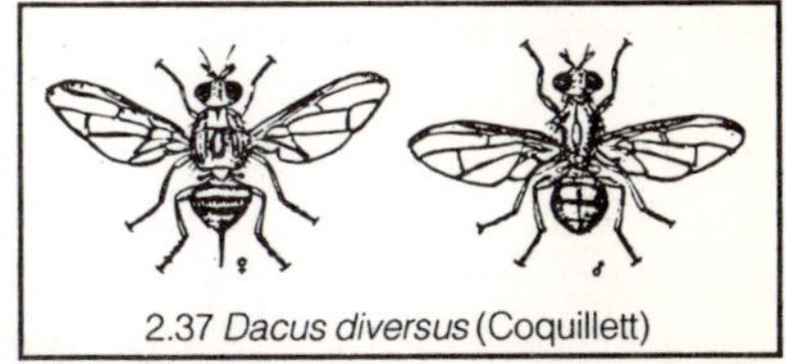

2.37 *Dacus diversus* (Coquillett)

Eggs ae smooth, shiny white, 1.0 to 1.5 mm long and slightly curved. Full grown maggots are pale creamy in colour, cylindrical in shape and 5 to 8 mm long. Pupae are barerel-shaped, ochraceous in colour and 5 to 6 mm long. Adult flies are smoky brown having greenish-black thorax with yellow markings. Male flies have wing expanse of 9-11 mm and females 12 to 14 mm.

Dacus tau (*caudatus* Fabricius, *hageni* de Meigre) — cucumber fruit-fly has been reported from Indian subcontinent, Malaysia, Taiwan, Philippines and Indonesia (Kapoor, 1970) damaging fruits of citrus, mango, melons, mulberry, sapota and a large number of cucubitaceious vegetables.

Eggs are about 3 mm long and pointed at micropyle end. Full grown maggots measure 8 to 11 mm in length and are pale white in colour. Pupae are cylindrical in shape, reddish-brown in colour and 5 to 6 mm in length. Adult flies are slightly bigger than those of *D. cucurbitae* and dark ferruginous in colour. Incubation period is 24 to 40 hours, maggots become full grown in 3 to 7 days and pupal period is 6 to 12 days; a complete life-cycle occupies 12 to 18 days and there are several generations in a year (Batra, 1968). The pest breeds throughout the year (including Winter) and normally there is no diapause; but if the Winter is severe, the pest may hibernate in adult stage.

Dacus zonatus — peach fruit-fly — a major pest of peach has been reported damaging fruit of apple, bael (*Aegle marmelos*), cherry, citrus, cucurbits, eggplant, fig, guava, jujube, mango, pear, pomegranate, sapota, tomato etc. Besides Indian subcontinent, the pest is widely distributed in South-east Asia. It is active all the year round, except during severe Winter (January-February), when it overwinters in pupal stage.

Adult flies are yellowish-red having a pale yellow band on 3rd tergite. Wings have incomplete costal band and anal band, wing-span is 10 to 12 mm. Females have red ovipositor with black tip. Eggs hatch in 3 to 4 days, maggot and pupal development takes about a week each in Summer and upto two weeks each in Winter; adult longevity is one to four months.

As fruit-flies are mostly minor pests of citrus trees, normally no control measures are adopted against these pests. Anyhow, to avoid infestation of fruit-flies, harvest the fruits before ripening. To check the carry over of the pest, collect and destroy all fallen and infested fruits; plough around the trees during Winter to expose and skill the pupae. Satisfactory control of adult flies is still a far cry.

Fruit Borers

Prays endocarpa Meyrick (Yponomentidæ : Lepidoptera) — citrus rind borer — a minor pest of citrus and bael (*Aegle marmelos*) fruits, is closely related to citrus flower moth, *P. citri* (Milliere) but the caterpillars of *P. endocarpa* feed strictly on fruits and do not attack the blossoms. Eggs are laId singly on fruits. On hatching, the caterpillars bore into the rind of those fruits. Gall-like swellings are caused by the feedings and lignified tissue may extend down the fruit pulp and thereby make the fruits unfit for human consumption. The caterpillars however remain in the rind. As many as 84 moths have been observed to emerge from a single small

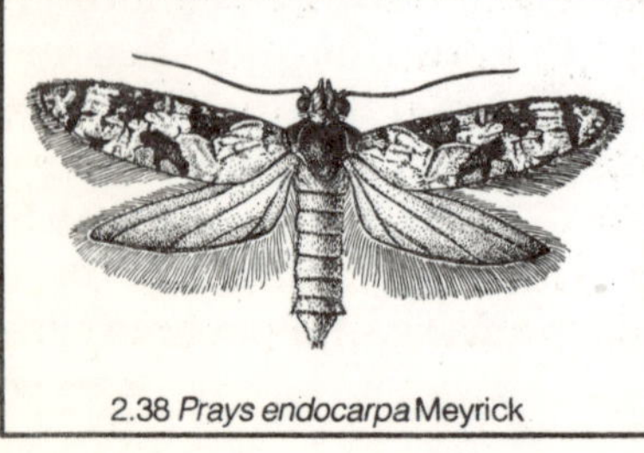

2.38 *Prays endocarpa* Meyrick

lemon (Pagden, 1931).

Generally no control measures are adopted against these borers. However, the infested fruits must be removed and destroyed promptly. If need be spray with 0.05% endosulfan or monocrotophos.

Helicoverpa (*Heliothis*) *armigera* (Hübner) (Noctuidæ : Lepidoptera) — gram pod boer — is a major pest of various grams and tomato in India that also attacks castor, cotton, cowpea, groundnut, hemp, indigo linseed, maize, millets, okra, safflower, sorghum, tobacco etc. (Jotwani and Butani, 1981). It is widely distributed in the tropics, subtropics and warmer temperate regions of the Old World (CIE map No. A-15), extending as far as Japan and Germany (Hill, 1975). The pest is more active during Winter than in Summer or monsoon months. The caterpillars thrust only a part of their body inside the host fruits or pods and feed on the inner contents as far as they can reach, thereafter move to another pod or fruit. Caterpillars are also cannibalistic — freshly hatched ones may eat the eggs lying around while the older ones prey on young caterpillars (Butani, 1977).

Eggs are dome-shaped and shinning yellow in colour. Full grown caterpillars are apple-green in colour and measure 40 to 50 mm in length. Pupae are dark brown and 11 to 14 mm long. Adults are yellowish-brown, stout moths of medium-size, wing-span being 34 to 44 mm. Incubation, caterpillar and pupal periods have been recorded to occupy 6 to 8, 15 to 28 and 10 to 25 days respectively. A single life-cycle is completed in 28 to 60 days and there are three to five generations in a year.

Hand-picking of caterpillars and their mechanical destruction in the early stage of infestation can keep the pest population under check. In case of severe attack, dusting with 5% carbaryl dust or spraying with 0.2% carbaryl or 0.04% lindane or 0.05% endosulfan has been reported to be effective.

Dichocrocis punctiferalis (Guenée) (Pyraustisdæ : Lepidoptera) — shoot and capsule borer — is a polyphagous pest reported from Indian subcontinent, China, Japan, Malaysia, Indonesia and Australia. Among the fruit trees, citrus, guava, mango, mulberry, peach, plum and pomegranate have been listed as its hosts (Butani and Jotwani, 1975). Eggs are laid on fruits, but in absence of fruits even on buds and shoots. On hatching the caterpillars bore into the fruits, buds or the shoots and feed within, throwing out the frassy matter which keep on hanging to the affected parts.

Eggs are pinkish in colour, flat and ovoid in shape and about 0.5 mm in diameter. Full grown caterpillars are 25 to 35 mm long, dark pinkish-brown in colour with spiny wrats all over. Moths are medium-sized (wing-span 30 to 35 mm), brownish-yellow with numerous dots on the wings. Incubation, caterpillar and pupal periods last for 5 to 6, 14 to 20 and 7 to 10 days respectively, adult longevity is 3 to 5 days and total life-cycle occupies 28 to 35 days, longer during Winter than in Summer. Hibernation is in caterpillar stage.

Destroy all infested fruits, buds and shoots and in case os severe infestation, spray 0.05% endosulfan or fenitrothion.

Cryptophlebia (*Argyroploce*) *illepida* (Butler) (Torticidæ : Lepidoptera), a polyphagous pest that has been recorded attacking number of fruits including bael (*Aegle marmelos*), citrus, jujube, litchi, tamarind and wood-apple (*Limonia acidissima*), but mostly as a minor pest. It is active from August to November. The caterpillars bore into the developing fruits and tunnelling through the pulp they attack the seeds, feed on inner contents making the seeds hollow; the tunnels get filled up, with excreta of the caterpillars and the fruits start rotting due to secondary infestation of bacteria and fungi; ultimately the fruits drop down prematurely. The entry holes of the infested fruits are seen plugged with the excreta of the caterpillars.

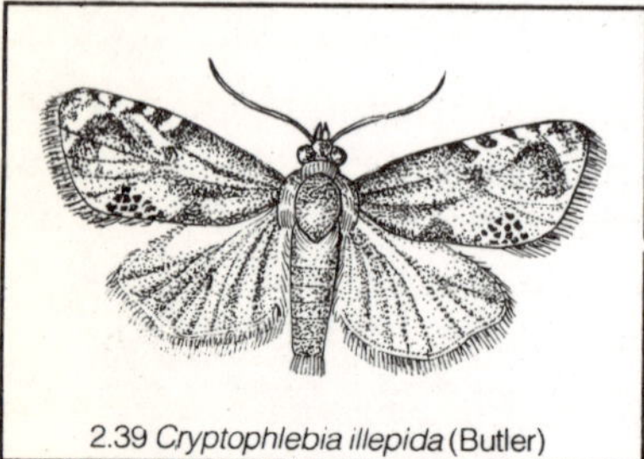
2.39 *Cryptophlebia illepida* (Butler)

Adults are small brownish-black moths having brownish wings and a dark brown spot at lower margin of forewings. Wing spread is 15 to 18 mm.

Virachola isocrates Fabricius (Lycænidæ : Lepidoptera) — pomegranate (anar) butterfly — a regular and most injurious polyphagous pest specially of pomegranate fruits. It has a very wide range of host plants including aonla (*Emblica officinalis*) apple, citrus, guava, jujube, litchi, loquat, mulberry, peach, pear, plum, sapota and tamarind. It is a minor pest of citrus fruits. Eggs are laid singly on calyx of flowers. On hatching, the caterpillars bore inside the developing fruits and are found feeding on the pulp (and seeds) just below the rind. The infested fruits are subsequently attacked by bacteria and fungi, resulting in offensive smell coming out of the entry holes along with excreta of the caterpillars. The affected fruits have no market value.

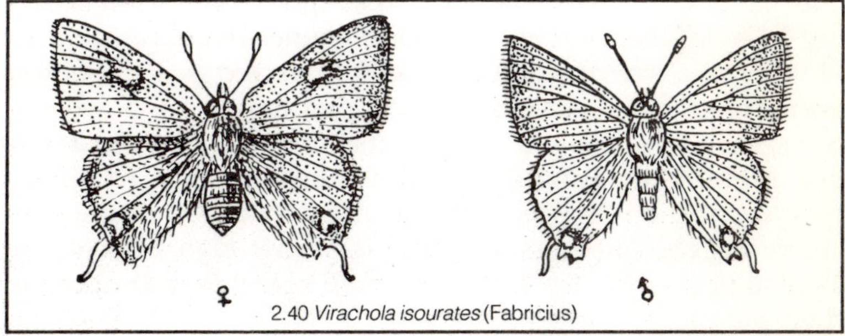

2.40 *Virachola isourates* (Fabricius)

The pest breeds throughout the year, normally without any hibernation. Egg, caterpillar and pupal stages, on pomegranate trees, last for 7 to 10, 18 to 47 and 7 to 34 days respectively with four overlapping generations in a year (Lal, 1952).

ANTS AND WASPS

Red Ant, *Oecophylla smaragdina* (Fabricius) (Formicidæ : Hymenopetra), has been reported from the entire Orinetal region extending from Australia to Africa (Atwal, 1963). The ants web and stitch together a few leaves usually at the top of the branches and built their nests on various trees, including, citrus, jackfruit, java plum (jamun), litchi, mango and sapota. These nests, though not airtight, are certainly water-proof and the leaves also remain green as the same are not detached from the tree. The behaviour and habitat of

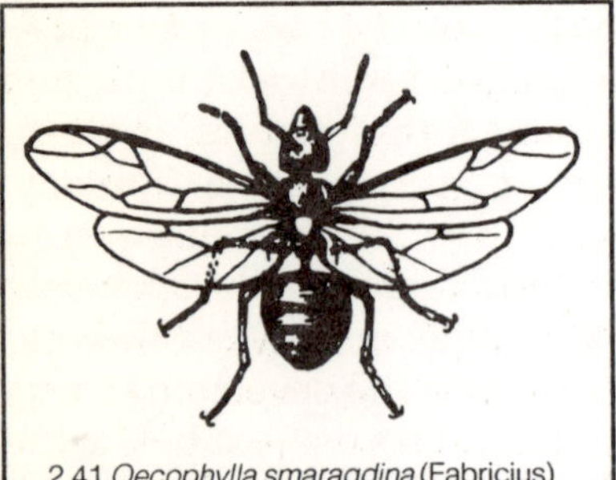
2.41 *Oecophylla smaragdina* (Fabricius)

these notorious and vivacious insects has been fully described by Aitken (1889) and Green (1900). These are carnivorous insects and very ferocious. Though they do not cause any direct damage, but are a great nuisance to the workers specially those who have to climb up the trees, and are often badly bitten by these ants. On the other hand, in several countries these are considered as beneficial as they prey upon hoppers, caterpillars, moths, beetles etc., that may be harmful pests. Garcia (1935) reported these ants preying on nymphs of citrus green bug *Phynchocoris serratus* in Philippines; whereas Voute (1935) observed that in China and Indonesia, these ants feed on mango pulp weevil, *Stenochetus frigidus* (Fabricius). The ants also feed on honeydew excreted by aphids, mealybugs and coccids; for which they carry the nymphs of these insects to their nests thus spreading the infestation of aphids, mealybugs and coccids as also protecting these insects from being preyed upon by their parasites and predators.

Red ants are active throughout the year, though the activity declines during the monsoon months but again gets a flip on sunny days. Eggs are oval in shape and whitish in colour. Full grown larvae are also whitish in colour and measure 9 to 11 mm in length. Pupae are pure white in colour and 7 to 9 mm long. Adult males are 6 to 7 mm long while the queens may measure upto 18 mm. Workers are wingless, sterile females, yellowish-red to rusty-red in colour 9 to 11 mm long; head roundly quadrangular and antennæ elbowed; thorax elongated, pronotum convex, anteriorly narrowed into a collar; measonotum constricted and narrow; metanotum rounded above; legs slender and comparatively very long; pedicel elongated and incrassate in the middle; abdomen short and oval (Bingham, 1903), Sexually functional males and females are winged and usually mate on the wings, during their nuptial flights. The fertilised females (queens) soon shed their wings and start constructing a new nest. David (1961) has studied the bionomics of this species and reported incubation, larval and pupal periods to occupy 4 to 8, 10 to 17 and 5 to 7 days respectively.

To control these ants, it is recommended that their nests be removed and destroyed mechanically (David, 1961); alternatively, the nests shouls be dusted by any chlorinated hydrocarbon (Tirumala Rao *et al.*, 1954) or sprayed to run-off point with 0.1% lindane (Butani and Tahiliani, 1974).

The hornet, *Vespa orientalis* Linnæus (Vespidæ : Hymenoptera) and house-wasp *Polistes olivaceus* (Fabricius) (Vespidæ) are often seen hovering and damaging ripe fruits in the orchards. These insects are also a great nuisance to the pickers and other workers in the orchards. *V. orientalis* wasps are mainly predaceous on various other insects including honeybees (pollinators), lepidopterous larvae etc. In addition these wasps are avid feeders on any sweet material especially honeydew secreted by aphids and scale insects. Besides citrus these wasps are also found in grapevines, orchards of mulberry, peach, pear, plum, pomegranate etc. *P. olivaceus* wasps are also found in houses and offices damaging books and files.

Vespa orientalis adults are smooth and light chestnut coloured with 3rd and 4th abdominal segments pale-sulphur-yellow; wings are flavo-hyaline and wing spread is 47-51 mm workers) and 54 to 58 mm (females).

Polistes olivaceus adults are smooth and bright yellow or fulvous-brown in colour; wings are ferruginous or flavo-hyaline and the wing expanse in 36 to 40 mm (workers) and 46 to 48 mm (females).

The wasps being minor pests, no control measures are usually adopted against these. Ram Dass and Sarabhai (1979) observed rice meal-moth *Corcyra cephalonica* Stainton and red flour bettle, *Tribolium castaneum* Herbest feeding on the combs of *Polistes olivaceus* as well as on dead grubs of *P. olivaceus* in the comb cells.

MITES

Mites (Acarina: Arachnida) are arthropods; their body comprises of two distinct regions — cephalothorax (head and thorax fused together) and abdomen; also known as proterosoma and systerosoma respectively. The most characteristic and diagnostic character is the presence of four pairs of legs in adult stage. Mites may be predacious or phytophagous (plant feeding). The phytophagous mites belong mainly to two familes — Eriophyidæ and Tetranychidæ. The mites cause damage to plants — wild and cultivated — by sucking plant sap. Eriophyids cause gall formation or a valvety mat-like covering on the leaves and panicles; some species also act as vectors of plant diseases.

Tetranychid mites are very small, generally red, yellow or greenish in colour; possess a pair of jointed chelicerae and pedipalpi forming the mouthparts. Eggs hatch into larvae, called protonymphs, these have only three pairs of legs. Protonymphs moult into the second stage, known as deutonymph, which moult into tritonymphs and finally into the adult stage. Eriophyids differ from tetranychids in having only two nymphal instars. The body of eriophyid mite is vermiform with long tapering abdomen; ther are only two pairs of legs, located near the anterior and of the body of nymphs as well as adult.

The phytophagous mites reported so far on citrus from India include, *Eutetranychus orientalis* (Klein), *Panonychus citri* (Mc-Gregor), *Schizoetranychus hindustanicus* (Hirst), *Tetranychus fijiensis* Hirst, *Tenuipalponychus citri* Channa Basavanna and Lakkundi *Brevipalpus deleoni* Pritchard and Baker, *B. californicus* (Banks), *B. phoenicis* (Geijakes), *Phyllocoptruta cleivorus* (Ashmead), *Aceria sheldoni* (Ewig), *Floracarus fleshneri* Keifer (Channa Basavanna and Nageshachandra, 1977). Of these *E. orientalis* is the only species recorded as a major pest of citrus, though *B.*

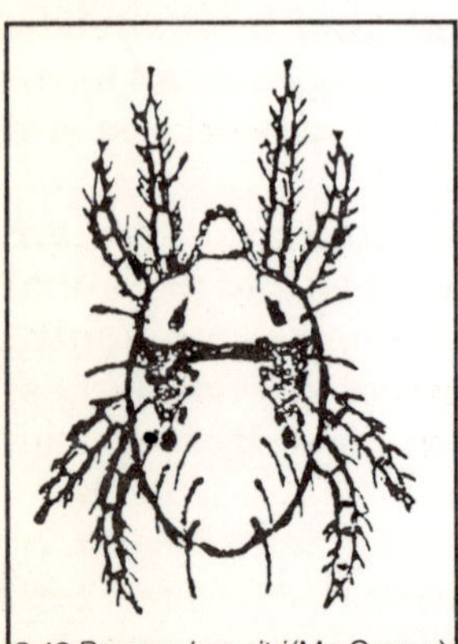

2.42 *Panonychus citri* (Mc-Gregor)

californicus and *Phyllocoptruta oleiverus* have also been reported regulatory.

Eutetranychus orientalis (Klein) (Tetranychidae : Acarina) citrus mite was originally described by Zachar (1928) under manuscript name *Anychus orientalis*. This was validated by Klein (1936) supported by Sayed (1946). Rahman and Sapra (1940) described the same species as *Anychus ricini* which Pritchard and Baker (1955) considered as synonym of *Eutetranychus banksi* (McGregor). Later, Baker and Pritcher (1960) mentioned *Orientalis* as valid species under *Eutetranychus* and placed. *A. ricini* as its synonym. Prasad (1974) suggested *E. banksi* also as synonym of *E.orientalis*.

E.orientalis is a polyphagous mite, having a very wide range of host plants. Among fruit trees, it has been reported infesting almond, citrus, jujube, papaya, peach, pear and sapota. The active larvae, protonymphs, deutonymphs and adults suck the sap of leaves predominantly from upper surface; the affected leaves become chlorotic and finally drop down. The attack is more severe in young plants (nurseries and freshly transplanted seedings). The pest is active during May to September. High temperature (27° to 41° C) and low relative humidity (30 to 40 per cent) is very conducive for rapid multiplication of this mite during May-June while the heavy rains cause significant reduction in mite population durig July-August.

Biology of *E. orientalis* has been studied by various workers on different host plants. On an average larval, protonymphal and deutonymphal development takes two to three days each; egg to adult stage lasts for about seven days and incubation period is four to five days (Basu and Channa Basavanna, 1972; Rasmy, 1977).

Eggs are laid singly in linear fashion on dorsal surface of leaves. Each egg is circular, flattened and disc-shaped. Larva is somewhat globular with three pairs of legs and measures, on an average, 0.2 mm in length. Protonymph is about 0.25 mm long pale brown in colour, dorsoventrally flattened and has four pairs of legs, which are shorter than the body, Deutonymph is more or less similar to protonymph, except that it is slightly bigger in size, being 0.3 mm long. Adult male is smaller than female, being flattened, with narrow and pointed opisthosoma; legs greatly enlarged, first pair much longer than the body. Female is oval in shape, flattened and having broad opisthosoma; legs being shorter than those of male. Average size of male and female is 0.28 × 0.18 mm and 0.41 × 0.28 mm respectively (Dhooria and Butani, 1984).

Interesting observations were reported by Yang *et al.* (1995a) about frequency distribution of citrus rust mite [*Phyllocoptera oleivora* (Ashmead)], fruits on the north quadrant of the tree were found to have the highest mean surface damage followed by the east, south, and west quadrants. Yang *et al.* (1995b) also studied the relationship between population, density of citrus rust mite and damage to 'Hamlin' sweet orange (*C. sinensis*) fruit. An mathematical-model was developed describe the relationship between cumulative damage and cumulative mite days. The rust mite disperse maximum up to 135 m from infested grove (Bergh and McCoy, 1997).

Increase in flower and young fruit abscission caused by citrus bud mite, *Aceria shddoni* (Ewing) feeding in the axillary buds of lemon, *C. limon* were studied by Phillips and Walker, (1997) and observed abscission rates of both flowers and fruits increased significantly with increasing bud mite-caused distortion.

Citrus mite has been successfully controlled by through soil application of a good systemic acaricide. Bindra *et al.* (1970) reported effective control of this mite by applying phorate @ 20 g a.i. per tree (six years old). Foliar spray wth dicofol and tetradifon were found to check this pest (Elkady *et al.*, 1977). Singh *et al.* (1977) recorded 0.05% dicofol to be the best. Omoto *et al.* (1995) observed a positive relationship between the number of dicofol application made per year and the frequency of resistance to dicofol. Dhooria and Sandhu (1975) observed 0.25% tetradifen or monocrotophos to be more effective. Under field conditions, this mite is attacked by some pathogens and predators. Garsen *et al.* (1979) from Israel reported fungus *Hirsutella thompsoni* parasitising this mite. Besides, several predatory mites, some insects, specially *Stethorus pauperculus* Weise and *Scolothrips indicus* Priesner have been reported by Channa Basavanna and

Nagesha Chandra (1977) preying upon *E. orientalis* and *P. citri.* Citrus mites are also controlled in nature by various predacious mites (Phytoseiidæ) including, *Euseius stipulatus* (Athias-Henriot), *E. finlandicus* (Oudemans), *Amblyseius swirskii* Athias-Henriot, *A. potentillae* Germar and *Iphiseius degenerans* (Berlese).

BIBLIOGRAPHY

ABBAS, H.M., M.S. KHAN AND H. HAQUE, 1955 : Black fly of citrus in Sind and its control. *Agriculture Pakistan*, **6** (1): 5-23, Karachi.

ABRAHAM, E.V., 1957 : A note on the control of the citrus butterfly, *Papilio demoleus* L. *South Indian Horticulture*, **5** : 25-29, Coimbatore.

ADACHI, T., 1994, Development and life cycle of *Anoplophora malasiaca* (Thomson) (Coleoptera: Cerambycidae) on citrus tree under fluctuating constant temperature regimes. *Applied Entomology and Zoology*, **29** (4) : 385-497, Tokyo.

AHLAWAT, Y.S., A. VARMA, N.K. CHAKRABORTY AND K.P. SRIVASTAVA, 1993 : Biotypes of vectoral capacity of *Diaphornia citri* in transmitting citrus greeting BLO. p. 222-232. *In* : *Proceedings of Second Agricultural Science Congress, 1993.* Indian Academy of Agricultural Sciences, IARI., New Delhi.

AHMAD, T., 1945 : *Spilostothus pandurus* Scop. as a pest of cultivated fruits in India, Indian *Journal of Entomology*, **7** (1 & 2) : 240-241, New Delhi.

AITKEN, E.H.A., 1889 : The red ant. *Journal of Bombay Natural Society*, **4** : 151-52, Mumbai (Bombay).

ALAM, M. ZAHURUL, 1962 : Insects and Mites Pests of Fruits and Fruit Trees in East Pakistan and Their Control, 115 pp., East Pakistan Government Press, Dacca.

ALAM, M. ZAHURAL, AHMAD ALAUDDIN, ALAM SHAMSUL AND ISLAM AMEERUL, MD, 1964 : A Review of Research, Division of Entomology (1947-64) : 272 pp. Department of Agriculture, East Pakistan, Dacca.

ALBRIGO, L.G., G.E. BROWN AND P.J. FELLERS, 1970 : Pest and Internal quality of oranges as influenced by grove applications of pinolene and benlate. *Proceedings of Florida Station Horticultural Society*, **83** : 263-267, Gainesville, Florida.

ALCOCK, A., 1903 : Notes on insect pests from Entomological section, Indian Museum-II. Insect pests of fruit trees. *Indian Museum Notes*, **5** (3) : 117-127, Calcutta.

ALI, S., 1925 : Citrus psylla and how to control it. *Seasonal Notes, Punjab Agricultural Department*, **2** (3) : 30-31, Lahore.

ALI, S. MOHAMMAD, 1957 : Some biological studies on *Pseudococcus vastator* Maskell. *Indian Journal of Entomology*, **19** (1) : 54-58, New Delhi.

ALI, S. MOHAMMAD, 1962: Coccids affecting sugarcane in Bihar, *Indian Journal Sugar Cane Research*, **6** (2) : 72-75, New Delhi.

ALI, S. MOHAMMAD, 1968: Coccids (Coccidae : Hemiptera : Insecta) affecting fruit plants in Bihar (India). *Journal of Bombay Natural History Society*, **65** (1) : 120-137, Mumbai (Bombay).

ANANTHAKRISHNAN, T.N., 1969 : Thysanoptera. Zoological Monograph No. 1, 171 pp. Publication and Information Directorate, Council of Scientific and Industrial Research, New Delhi.

ANANTHAKRISHNAN, T.N., 1971 : Thrips (Thysanoptera) in Agriculture, Horticulture and Forestry — Diagnosis, bionomics and control. *Journal of Science and Industrial Research*, **30** (3) : 113-146, New Delhi.

ANANTHAKRISHAN, T.N., 1973 : Thrips : Biology and Control, 120 pp. The MacMillan Company of India, New Delhi.

ANANTHANARAYANAN, K.P. AND E.V. ABRAHAM, 1959 : The slug caterpillar, *Parasia lepida* Cram. and its control. *Journal of Bombay Natural History Society*, **53** (2) : 205-209, Mumbai (Bombay).

ANONYMOUS, 1950 : Bark eating caterpillar on amla tree. *Agriculture and Animal Husbandry, Uttar Pradesh*, **5** : 30, Allahabad.

ANONYMOUS, 1989 : Batting the citrus black fly. *Citrograph;* **75** (2) : 29, Los Angeles, California.

ANTRAM, CHAS B., 1986 : Butterflies of India, 226 pp. Reprinted by Periodical Expert Book Agency, New Delhi.

ARGOV, YAEL, 1986 : Biological control of citrus whitefly, *Dialeurodes citri*, *Phytoparasitica*, **14** : 164, Rehovot, Israel.

ARGOV, YAEL, 1988 : Biological control of citrus whitefly, *Dialeurodes citri*. (Ashmead) (Homoptera : Aleyrodidae). p. 1169-1175. *In* : *Proceedings of the Sixth International Citrus Congress, Tel Aviv, Israel, March 6-11, 1988*; Balaban Publishers, Rehovot (Israel).

ARGOV YAEL AND Y. ROSSLER, 1986 : The introduction of *Encarsia lahorensis* (Howard) (Hymenoptera : Aphelinidae) into Israel for the control of the citrus whitefly, *Dialeurodes citri* (Ashmead) (Homoptera: Aleyrodidae). *Israel Journal of Entomology*, **20** : 1-5, Rehovot (Israel).

ARORA, G.L. AND S.K. GILOTRA, 1960 : The biology of *Odontotermes obesus* (Rambur) (Isoptera). *Research Bulletin, Punjab University (New Series)*, **10** (3-4) : 247-255, Chandigarh.

ASHBY, S.F., 1915 : Notes on diseases of cultivated crops observed in 1913-1914. Bulletin, Department of Agriculture, Jamaica, **2** : 229-327, Kingston.

ASHMEAD, W.H., 1879 : The injurious and beneficial insects found on oragne tree in Florida. *Canadian Entomologist*, **1**: 159-160, Ontario.

ASHMEAD, W.H., 1885 : The orange Aleurodes (*Aleurodes citri*, n. sp.), *Florida Despatch*, **2** (42) : 704, Gainesville, Florida.

ARROW, G.J., 1917 : The Fauna of British India including Ceylon and Burma, Coleoptera : Lamellicornia, part 2, Taylor and Francis, London.

ASI, A. AND N. ALI, 1970 : Effect of growth regulators on physico-chemical characters and fruit quality of Kinnow mandarin. *Pakistan Journal of Science*, **22** : 233-238, Karachi.

ASTHANA, E.B. AND J.P. NAYAK, 1947 : Survey of the incidence of *Indarbela quadrinotata* WLK. on oranges in the Central Provinces and Berar, *Nagpur Agricultural College Magazine*, **22** (1) : 14-21, Nagpur.

ATWAL, A.S., 1962 a : Insect pests of citrus in Punjab. I-Biology and control of citrus psylla, *Diaphorina citri* Kuwayama (Hemi : Psyllidae). *Punjab Horticultural Journal,* **2** (2) : 104-108, Patiala.

ATWAL, A.S., 1962 b : Insect pests of citrus in Punjab. II-Biology and control of citrus whitefly, *Dialeurodes citri* Ashmead (Hemi: Aleyrodidae). *Punjab Horticultural Journal,* **2** (3) : 149-152, Patiala.

ATWAL, A.S., 1962 c : Insect pests of citrus in Punjab. III-Biology and control of citrus mealy bug, *Pseudococcus filamentosus* Ckl. (Hemi : Coccidae). *Punjab Horticultural Journal,* **2** (4) : 230-232, Patiala.

ATWAL, A.S., 1963 a : Insect pests of citrus in Punjab. IV-Biology and control of fruit sucking moth. *Ophideres fullonica* L. (Lepidoptera). *Punjab Horticultural Journal,* **3** (1) : 43-45, Patiala.

ATWAL, A.S., 1963 b : Insect pests of mango and their control. *Punjab Horticultural Journal,* **3** (2-4) : 235-258, Patiala.

ATWAL, A.S., 1964 a : Insect pests of citrus in Punjab. V-Biology and control of citrus caterpillar, *Papilio demoleus* L. (Lep. Papillionidae). *Punjab Horticultural Journal,* **4** (1) : 40-44, Patiala.

ATWAL, A.S., 1964 b : Insect pests of citrus in Punjab. VI-Biology and control of citrus and leaf miner, *Phyllocnistis citrella* Station (Lep. Phyllocnistidae). *Punjab Horticultural Journal,* **4** (2) : 100-103, Patiala.

ATWAL, A.S., 1964 c : Insect pests of citrus in Punjab. VII-Biology and control of citrus mite. *Punjab Horticultural Journal,* **4** (3 & 4) : 142-144, Patiala.

ATWAL, A.S., 1965 : Review of agricultural research in the Punjab from 1947 to 1962. Entomology (crop pests and their control). *Punjab Agricultural University Journal,* **4** (6-7): 1-46, Ludhiana.

ATWAL, A.S., 1976 : Insect pests of ctirus. p. 195-213. *In* : Agricultural Pests of India and South-east Asia : Kalyani Publishers, Ludhiana.

ATWAL, A.S., J.D. CHOUDHARY AND M. RAMZAN, 1970 : Studies on development and field population of citrus psylla, *Diaphorina citri* Kuwayana (Psyllidae : Homoptera), *Journal of Research, PAU,* **7** (3) : 333-338, Ludhiana.

ATWAL, A.S. AND G.S. JOSAN, 1962 : Citrus pests round the year. *Punjab Horticultural Journal,* **2** (1) : 8-16, Patiala.

ATWAL, A.S. AND G.C. VERMA, 1967 : Role of insect pests in citrus decline. *Himachal Horticulture,* **8** (3-4) : 13-16, Shimla.

ATWAL, A.S. AND G.C. VERMA, 1968 : Studies on the control of citrus psylla, *Diaphorina citri* Kuwayama (Hemiptera : Pysllidae) by foliage spray and soil application. *Journal of Research,* PAU. **5** (2) : 240-243, Ludhiana.

ATWAL, A.S. AND G.C. VERMA, 1970 : Studies on the control of citrus leaf miner *Phyllocnistis citrella* Stainton (Gracillariidae: Lepidoptera). *Journal of Research, PAU,* **7** (1) : 55-57, Ludhiana.

AUBERT, B., 1990 : Integrated activities for the control of huanglungbin greening and its vector, *Diaphornia citri* Kuwayana in Asia. p. 133-144. *In* : Proceedings of the Fourth International Asia-Pacific Conference on Citrus rehabilitation.

AYYAR, RAMAKRISHNA, T.V., 1992 : The weevil fauna of South India with special reference to species of economic importance. *Agricultural Research Institute, Pusa, Bulletin No. 125,* 21 pp., Calcutta.

AYYAR, RAMAKRISHNA, T.V., 1924 : Short notes on some South Indian insects

Report of Proceedings of 5th Entomological Meeting; Pusa (Bihar), February 1923 : 263-269, Calcutta.

Ayyar, Ramakrishna, T.V., 1932 : An annotated list of the insects affecting the important cultivated plants in South India. *Madras Department of Agriculture, Bulletin No. 27,* 95 pp., Chennai (Madras).

Ayyar, Ramakrishna, T.V., 1938 : An annotated conspectus of the insects affecting fruit crops in South India. *Madras Agricultural Journal,* **26** : 341-351, Coimbatore.

Ayyar, Ramakrishna, T.V., 1940 (reprinted 1963) : Handbook of Economic Entomology for South India, 516pp. Madras Government Press, Chennai (Madras).

Ayyar, Ramakrishna, T.V., 1941 : Notes on some South Indian mealy bugs. *Indian Journal of Entomology,* **3** (1) : 107-113, New Delhi.

Ayyar, Ramakrishna, T.V., 1943 : Notes on some fruit sucking moths of Deccan. *Indian Journal of Entomology,* **5** (1-2) : 29-33, New Delhi.

Back, E.A., 1912 : Notes on Cuban whiteflies with description of two new species. *Canadian Entomologist,* **44** : 145-153, Ontario.

Badawi, A., 1981 : Studies on some aspects of the biology and ecology of the citrus butterfly,. *Papilio demoleus* L. in Saudi Arabia (Papilionidae : Lepidoptera), *Zeitschrift fur Angewandte Entomologie,* **91** : 286-292, Berlin.

Bajpai, R.N., 1955 : A simple method of controlling fruit piercing moths (*Othreis fullonica* Linn. and *O.materna* Linn.), *Science and Culture,* **20** (8) : 387, Calcutta.

Baker, A.C., 1923 : An undescribed orange pest from Honduras. *Journal of Agricultural Research,* **25** : 253-254.

Banerjee, S.N. and A.L. Mookerjee, 1962 : Some observations on the control of citrus psylla, *Diaphornia citri* Kuw. (Psyllidae : Hemiptera). *Science and Culture,* **28** (1) : 32-33, Calcutta.

Baptist, B.A., 1944 : The fruit piercing moth (*Othreis fullonica*) with special reference to its economic importance. *Indian Journal of Entomology,* **6** (1-2) : 1-13, New Delhi.

Barlow, E., 1900 : Notes on insect pests from the entomological section, Indian Museum. *Indian Museum Notes,* **4** (3) : 119, Calcutta.

Bar-zakay, I., B.A. Peleg and Ch. Chen, 1988 : The spherical mealybug infesting citrus in Israel. p. 1075-1082.*In* : *Proceedings of the Sixth International Congress, Tel Aviv, Israel, March 6-11, 1988,* Balaban Publishers, Rehovot (Israel).

Basile, M. and G. Russo, 1988 : Accumulated of Phenamiphos residues in Lemon fruits from treated groves. p.1103-1106. *In: Proceedings of the Sixth International Congress, Tel Aviv, Israel, March 6-11, 1988* Balaban Publishers, Rehovot (Israel).

Basu, A.N. and S.N. Banerjee, 1958 : Aphids of economic plants of West Bengal, *Indian Agriculturist,* **2** : 89-112, Calcutta.

Basu, A.N., D.K. Nath and P.B. Chatterjee, 1969 : Insects occurring on orange plant (*Citrus reticulata* Blanco) in Darjelling districts, West Bengal. *Proceedings Zoological Society, Calcutta,* **22** : 169-178, Calcutta.

Batra, H.N., 1976 : Assessment and evaluation of plant protection progress. *Pesticides,* **10** (4) : 20-29, Mumbai (Bombay).

BATRA, H.N. AND K. KUMAR, 1958 : Technique for rearing of *Indarbela tetraonis* (Moore) (Metarbelidae : Lepidoptera) in the laboratory, *Indian Journal of Entomology,* **20** (3) : 240-241, New Delhi.

BATRA, H.N., G.R. SHARMA AND K. KUMAR, 1968 : Protect roses and citrus from Red scale. *Indian Horticulture,* **13** (1) : 31-33, New Delhi.

BATRA, R.C., 1989 : Insect pest damage on citrus trees. *Proceedings National Seminar-cum-Workshop on Citrus Cult, Ludhiana, 2-14, January, 1989* : 158-169, Ludhiana.

BATRA, R.C., 1990 : Insect pest problems in citrus nursery with special reference to Kinnow. *Proceeding of Citrus Show-cum-Seminar on Prospects and Problems of Kinow cultivation at Abohar. Punjab Agricultural University, 6-7 January, 1989* : 189-196, Ludhiana.

BATRA, R.C. AND G.S. SANDHU, 1986 : Chemical control of citrus leaf-miner in nursery. *Punjab Horticultural Journal,* **26** (1-4) : 31-33, Patiala.

BATRA, R.C. AND G.S. SANDHU, 1988 : Management of citrus leaf-miner in the nursery. *Punjab Horticultural Journal,* **28** (3-4) : 133-138, Patiala.

BATRA, R.C., G.S. SANDHU AND D.R. SHARMA, 1993 : Host plant resistance to insect pests. p. 1649-1660. *In* : Advances in Horticulture — Fruit crops (Eds. K.L. Chada and O.P. Pareek), **3** : Malhotra Publishing House, New Delhi.

BATRA, R.C., G.S. SANDHU, S.C. SHARMA AND RAGHBIR SINGH, 1988 : Biology of citrus leaf-miner on some citrus rootstock and its relation with abiotic factors. *Punjab Horticultural Journal,* **28** (1-2) : 25-30, Patiala.

BATRA, R.C., G.S. SANDHU AND A.S. SOHI, 1987 : Outbreak of California red scale on citrus and suppression through coccinellid predators in Punjab. *Bulletin of Entomology,* **28** (2) : 161-162, New Delhi.

BATRA, R.C. AND D.R. SHARMA, 1989 : Biology, ecology and population dynamics of citrus leaf folder. p. 152-157. *In: Proceedings of National Seminar-cum-Workshop on Citurs Cult., Ludhiana 2-14 January,* Ludhiana.

BATRA, R.C. AND D.R. SHARMA, 1990 a : Bionomics of citrus leaf-roller — A review. *Pestology,* **14** (2) : 21-23, Mumbai (Bombay).

BATRA, R.C., D.R. SHARMA, AND Y.R. CHANANA, 1989 : Relative efficacy of different insecticides for the control of citrus whitefly, *Dialeurodes citri* Ashmead. *Indian Journal of Entomology,* **40** (1) : 122-125, New Delhi.

BATRA, R.C., D.R. SHARMA, RAGHBIR SINGH AND S.N. SINGH, 1990: Status of citrus insects and mite pests and their natural enemies in different agroclimatic zones of Punjab. *Indian Journal of Horticulture,* **47** (3) : 331-336, New Delhi.

BATRA, R.C. AND D.S. SIDHU, 1993 : Studies on chemical control of citrus whitefly. *Punjab Horticultural Journal,* **33** (1-2) : 25-29, Patiala.

BATRA, R.C., B.S. SOHI AND O.S. BINDRA, 1972 : Spraying schedule for controlling insect pests of citrus in the Punjab. *Punjab Horticultural Journal,* **12** (4) : 237-239, Patiala.

BATRA, R.C., D.K. UPPAL AND B.S. SOHI, 1970 : Indexing the genetic stock of different species of citrus against citrus leaf-miner and citrus psylla. *Indian Journal of Horticulture,* **27** : 76-79, Bangalore.

BEATTIE, G.A.C., Z.M. LIV, D.M. WATSON, A.D. CLIFT AND L. JIANG, 1995 : Evolution of petroleum spray oils and polysaccharides for control of *Phyllocnistis citrella* Stainton *Journal of the Australian Entomological*

Society, **34** (4): 349-353, Brisbane, Queensland.

BEATTIE, G.A.C., V. SOMSOCK, D.M. WATSON, A.D. CLIFT AND L. JIANG, 1995 : Field evaluation of *Steinernema carpolapsa* (Weiser) (Rhabditila; Steinernematidae) and selected pesticides and enhancers for control of *Phyllochistis citrella* Stainton. *Journal of the Australian Entomological* Society, **34** (4): 335-342, Brisbane, Queenoland.

BEAUMONT, C.H., 1930 : Scale insects of citrus and their control by fumigation. *Journal of Department of Agriculture, South Australia*, **33** (7) : 618-624.

BEDFORD, E.C.G., 1978 : Citrus pests in Republic of South Africa. *Scientific Bulletin, Department of Agriculture, Technical Service, Republic of South Africa*, **391** : 252 pp., Pretoria.

BEESON, C.F.C., 1941 : The Ecology and Control of Forest Insects of India and Neighbouring Countries, 1007pp. The Vasant Press, Dehradun.

BEHURA, B.K., 1978 : Biology of aphids. Presidential address. *Proceedings of 65th Indian Science Congress, Part II* : 21-44, Calcutta.

BEN-DEV, Y., 1988 : The scale insects (Homoptera : Coccidae) of citrus in Israel-diversity and pest status. p. 1075-1082. *In*: *Proceedings of the Sixth International Citrus Congress, Tel Aviv, Israel, March 6-11, 1988* : Balaban publishers, Rehovot (Israel).

BERGER, E.W., 1909 : Whitefly studies in 1908. *Bulletin, Agriculture Experimental Sation, Florida, No. 97* : 68-74, Gainesville, Florida.

BERGER, E.W., 1910 : Whitefly studies in 1909, *Bulletin, Agriculture Experimental Station, Florida, No. 113* : 1-28, Gainesville, Florida.

BERGER, E.W., 1917 : Whiteflies of citrus. *Monthly Bulletin, California State Commission Horticulture, Sacramento*, **6** : 298-307, California.

BERGH, J.C. AND C.W. WCCOR, 1997 : Aerial dispersal of citrus rust mite (Acari; Eriophyidae) from Florida citrus groves. *Environmental Entomology*, **26** (2): 256-264, Lanham, Maryland.

BERLESE, A. AND G. LEONARDI, 1896 Le cocciniglie Italiane viventi sugliagrame. *Rev. Patol. Veg.*, **3** : 49-171; 44-179, 195-292, Paris.

BESS, H.A. AND F.H. HARAMOTO, 1977 : *Dacus dorsalis* Hend. p. 526-528. *In* : Diseases, Pests and Weeds in Tropical Crops. Verlag Paul Parey.

BHAMBURKAR, B.L., 1972 : Citrus pests and their control. All India Agricultural Seminar, Nagpur : 160-165, Nagpur.

BHAT, S.S., 1929 : Jamburi (*Citrus medica* var *limenom*) the stock variety of citrus trees. *Poona Agricultural College Magazine*, **21** : 114-117, Pune.

BHATNAGAR, V.S. AND R.C. SAXENA, 1975 : Efficacy of *Bacillus thuringiensis* Berliner for the control of bark boring caterpillar, *Indarbela quadrinatata* Walker, *Pesticides*, **9** (5) : 38-39, Mumbai (Bombay).

BHUMANNAVAR, B.S. AND S.P. SINGH, 1981 : Population dynamics of citrus leaf-miner, *Phyllocnistis citrella* Stainton. *National Symposium on Strategies of Pest Management, IARI, New Delhi (December 21-23, 1981)* : 37, New Delhi.

BHUMANNAVAR, B.S. AND S.P. SINGH, 1983 a : Some observations on the biology and habits of orange shoot borer, *Oberea lateapicalis* Pic, (Coleoptera : Lamiidae). *Entomon*, **8** (4): 331-336, Thiruvananthapuram (Trivandrum).

BHUMANNAVAR, B.S. AND S.P. SINGH, 1983 b : Studies on population dynamics of citrus leaf-miner, *Phyllocnistis citrella* Station (Lepidoptera : Phyllocnistidae). *Entomon.,* **8** (4) : 397-400. Thiruvananthapuram (Trivandrum).

BHUMANNAVAR, B.S. AND S.P. SINGH, 1985 : Studies on population dynamics of citrus psylla, *Diaphorina citri* Kuwayama (Psylidae : Hemiptera). *Entomon,* **10** (1) : 63-66, Thiruvananthapuram (Trivandrum).

BHUMANNAVAR, B.S. AND S.P. SINGH, 1986 a : Seasonal incidence and population of black citrus aphid, *Toxoptera aurantii* (B. de F.) on Coorg mandarin. *Entomon,* **11** (4) : 223-226, Thiruvananthapuram (Trivandrum).

BHUMANNAVAR, B.S. AND S.P. SINGH, 1986 b : Studies on the population dynamics of oriented red mite of citrus, *Eutetranychus orientalis* (klein). *Entomon,* **11** (4) : 223-226, Thiruvananthapuram (Trivandrum).

BHUMANNAVAR, B.S., S.P. SINGH AND V.V. SULLADMATH, 1988 : Evaluation of citrus germplasm for resistance to the oriental red mite, *Eutetranychus orientalis* (Klein) under tropical humid south Indian conditions. *Tropical Pest Management,* **34** (2) : 193-198, London.

BHUMMANNAVAR, B.S., S.P. SINGH AND V.V. SULLADMATH, 1989 : Field evaluation of citrus germplasm for resistance to the black aphid, *Toxoptera aurantii* (Boy.) under tropical humid conditions. *Insect Science Application,* **10** (1) : 81-88, Nairobi (Kenya).

BINDRA, O.S., 1957 a : Important insect enemies of citrus fruit plants in Madhya Bharat region and their control. *Gwalior College of Agriculture Journal,* **1** : 11-17, Gwalior.

BINDRA, O.S., 1957 b : Insect pests of citrus and their control. *Indian Journal of Horticulture,* **14** (2) : 89-96, Bangalore.

BINDRA, O.S., 1957 c : Insect pests of citrus and their control. *Assam Department of Agriculture (Fruit series),* **14** (5) : 80-98, Shillong.

BINDRA, O.S., 1966 : Role of insects and other animals in relation to citrus decline. *Punjab Horticultural Journal,* **6** (1-2) : 108-116, Patiala.

BINDRA, O.S., 1967 : Fighting pests of commercial fruits. *Indian Horticulture,* **11** (4) : 71-74, New Delhi.

BINDRA, O.S., 1969 : Controlling citrus pests. *Indian Horticulture,* **13** (2) : 28-31, New Delhi.

BINDRA, O.S., 1970 : Insects. p. 64-78. *In;* Citrus Decline in India — Causes and Control : PAU/Ohio State University/United States Agency for International Development, Ludhiana.

BINDRA, O.S. AND H.K. CHHABRA, 1968 : Control of citrus nematodes may halt decline of your orchard. *Progressive Farming, PAU,* **5** (2) : 17-18, Ludhiana.

BINDRA, O.S. AND B.S. SOHI, 1969 : Save your orchard from citrus psylla, *Progressive Farming, PAU,* **6** (1) : 11-12, Ludhiana.

BINDRA, O.S. B.S. SOHI AND R.C. BATRA, 1973 : Note on the comparative efficacy of some contact and systemic insecticies for the control of citrus psylla in Punjab. *Indian Journal of Agricultural Sciences,* **43** : 1087-1088, New Delhi.

BINDRA, O.S., G.C. VARMA, G.S. SANDHU AND H.K. CHHABRA, 1970 : Studies on the control of citrus pests by soil application of systemic insecticides. *Journal of Research, PAU,* **7** (2) : 197-202, Ludhiana.

BINGHAM, C.T., 1903 : The Fauna of British India including Ceylon and Burma. Hymenoptera-2, Ants and Cuckoo-wasps, 506 pp. Taylor & Francis, London.

BINGHAM, C.T., 1907 : The Fauna of British India including Ceylon and Burma. Butterflies, **2** : 417-419, Taylor and Frances, London.

BIRAT, R.B.S., 1969 : Fight mites the new way, *Indian Horticulture,* **13** (4) : 29-30, New Delhi.

BLANCHARD, EVERARD, E., 1944 : Descriptions by anotaciones de afidoideos Argentinos. *Acta Zoologie,* **2** : 15-62, Lilloana.

BLONDEAU, R. AND J.C. CRANE, 1890 : The cultivated oranges and lemons etc. of India and Ceylon. W.H. Allen & Co., London.

BLUMBERG, D. AND E. SWIRSKI, 1988 : Colonisation of *Metaphycus* spp. (Hymenoptera : Encyrtidae) for control of the Mediterranean black scale, *Saisetia oleae* (Oliver) (Homoptera : Coccidae) in Israel. p. 1209-1213. *In: Proceedings of the Sixth International Citrus Congress, Tel Aviv, Israel, March 6-11, 1988* : Balaban Publishers, Rehovto (Israel).

BODENHEIMER, F.S., 1951 : Citrus Entomology in the Middle East, 663 pp. Dr. W. Junk publishers, The Hague (The Netherlands).

BORLE, M.N. AND S.B. KHARAT,. 1967 : Evaluation of some important insecticides for the control of whitefly (*Aleurocanthus woglumi* Ashby) in pupal stage of citrus. *Indian Journal of Entomology,* **33** (3) : 370-371, New Delhi.

BORLE, M.N. AND S.B. KHARAT, 1977 : Record of three species of whiteflies from central Maharashtra on citrus plantation. *International Symposium on Citriculture, Bangalore; Abstracts* : 39; Horticultural Society of India, Bangalore.

BORLE, M.N. AND P.S. KHODASKAR, 1977 : Pesticidal trials against citrus leaf-miner, lemon butterfly and citrus mites in Vidarbha region. *International Symposium on Citriculture, Bangalore; Abstracts* : 35; Horticultural Society of India, Bangalore.

BÕRNER CARL AND F.A. SCHILLDER, 1932 : Aphidoidea : blattlause. *In* : Sorauer's Handbeich der Pflanzenkrankheiten, **5** (2) : 551-673.

BOSE, S.K., 1953 : An entomogenous fungus on citrus whitefly, *Dialeurodes citri* Ashm. *Proceedings of 40th Indian Science Congress, Lucknow, III*: 79, Calcutta.

BOYCE, A.M., 1946 : Insects and mites and their control. *In*: The Citrus Industry, 2, chapter 14, University of California Press, Berkeley.

BOYER, DE FONSCOLOMBE, E.L., 1841 : Descripto des pucerons qui se trouvent aux environs d'Aix. Annals of Society Entomologie, France, **10** : 157-198, Paris.

BRANYOVITS, F., 1953 : Some aspects of the Biology of armoured scale insects. *Endeavour,* **12** : 202-209, London.

BRINK, T. AND P.H. HEWITT, 1993 : Parasitoids of white powdery scale *Criprolecaniium andersoni* (Newsted) (Hemiptera : Coccidae), a pest of citrus. *International Journal Pest Management*, **39** (1) : 99-102, London.

BUDHATHOKI, K. AND P.M. PRADHANANG, 1992 : Production constraints of Mandarin in Western Hills of Nepal. *Acta Horticulture*, **292** : 51-59, The Hague.

BUITENDAG, C.H. AND W. NAUDE, 1994 : Fruitfly control; development of new fruitfly attractant and correct bait administration. *Citrus Journal*, **4** (1) : 22-25.

BUTANI, DHAMO K., 1970 : Bibliography of Aleyrodidae-II. *Beiträge zur Entomologie,* **20** (3-4) : 317-335, Berlin.

BUTANI, DHAMO K., 1973 a : Les ravageurs et les maladies des Citrus en Inde. *Fruits,* **28** (12) : 851-856, Paris.

BUTANI, DHAMO K., 1973 b : Insect pests of fruit crops—4 : Citrus. *Pesticides,* **7** (12) : 23-26, Mumbai (Bombay).

BUTANI, DHAMO K., 1975 : Parasites et maladies du manguier en Inde. *Fruits,* **30** (2) : 91-101, Paris.

BUTANK, DHAMO K., 1976 : Insect pests of fruit crops and their control—21 : Pomegranate. *Pesticides,* **10** (6) : 23-26, Mumbai (Bombay).

BUTANI, DHAMO K., 1977 a : Insect pests of vegetables—7 : Tomato. *Pesticides,* **11** (1) : 11-16, Mumbai (Bombay).

BUTANI, DHAMO K., 1979 a : Insect pests of citrus and their control, *Pesticides,* **13** (4) : 15-21, Mumbai (Bombay).

BUTANI, DHAMO K., 1979 b : Insects and Fruits, 415 pp. Periodical Expert Book Agency, New Delhi.

BUTANI, DHAMO K., 1993 : Mango : Pest Problems, 290 pp. Periodical Expert Book Agency, New Delhi.

BUTANI, DHAMO K. AND R.A. AGARWAL, 1982 : Insect pests of citrus and their control. *Citrus Seminar, Indian National Science Academy, Abstracts,* New Delhi.

BUTANI, DHAMO K. AND M.G. JOTWANI, 1975 : Trends in control of insect pests of fruit crops in India. *Pesticides Annual, 1975* : 139-149, Mumbai (Bombay).

BUTANI, DHAMO K. AND M.G. JOTWANI, 1984 : Insects in Vegetables, 356 pp. Periodical Expert Book Agency, New Delhi.

BUTANI, DHAMO K. AND R.P. SRIVASTAVA, 1983 : Insect pests of tropical and subtropical fruits. p. 189-211. *In* : Agricultural Entomology, **2** : All India Scientific Writers' Society, New Delhi.

BUTANI, DHAMO K. AND B.D. TAHILIANI, 1974 : Red ants on mango in South Gujarat. *Entomologists' Newsletter,* **4** (7): 37-38, New Delhi.

CAMPBELL, D., 1917 : Blackfly of citrus. *Journal of Jamaica Agricultural Society,* **16** : 50, Kingston.

CAPOOR, S.P., D.G. RAO AND S.M. WISWANATH, 1967 : *Diaphorina citri* Kuwayama, a vector of the greening disease of citrus in India. *Indian Journal of Agricultural Sciences,* **37**: 572-576, New Delhi.

CARMEAN, L.W., 1984 : Sen Josquin valley citrus thrips and the pesticide treadmill. *Association of Applied Insect Ecologists,* **4** (4) : 1.

CATLING, H.D., 1970 : Distribution of the psyllid vectors of citrus greening disease, with notes on the biology and bionomics of *Diphorina citri. FAO Plant Protection Bulletin,* **18** : 8-15, Rome.

CHADDA, K.L.N.S. RANDHAWA, O.S. BINDRA, J.S. CHOHAN AND L.C. KNORR, 1970 : *Citrus Decline in India, Causes and Control,* 97 pp. PAU/OSU/USAD Publication, Ludhiana.

CHAKRABORTY, N.K., P.K. PANDEY, S.N. CHATERJEE AND A.B. SINGH, 1976 : Host preference in *Diaphorina citri* Kuwayama, vector of greeining disease of citrus in India. *Indian Journal of Entomology,* **38** (2) : 196-197, New Delhi.

CHANNA BASAVANNA, G.P. AND B.K. NAGESHA CHANDRA, 1977 : Present status of our knowledge of mites on citrus in India. *International Symposium on Citriculture, Bangalore, Abstract No. 37;* Horticultural Society of India, New Delhi.

CHATURVEDI, P.L., 1950 : Biology and control of fruit sucking moths of citrus in the Uttar Pradesh. *Agriculture and Animal Husbandry, Uttar Pradesh,* **1** (7 & 8) : 21-23, Lucknow.

CHEEMA, G.S., S.S. BHAT AND NAIK, K.C., 1954 : Commercial Fruits of India with Special Reference to Western India, 422pp. Macmillan & Co. Ltd., Calcutta.

CHEEMA, S.S. AND S.P. KAPUR, 1975 : *Murraya paniculate* Linn., a new host of *Diaphorina citri* Kuwayama. *Current Science,* **44** (7) : 249, Bangalore.

CHEN, FONG-GE, 1936 : Notes on scale-insects of Citrus in several districts of East Chekiang with description of new species. *Entomology and Phytopathology,* **4** (11) : 208-228, Hangchow.

CHERIAN, M.C., 1942 : Our present position with regard to control of fruit pests. *Madras Agricultural Journal,* **30** (1) : 14-17, Coimbatore.

CHERRY, R. AND R.V. DOWELL, 1979 : Predators of citrus blackfly (Hom : Aleyrodidae). *Entomophaga,* **24** : 385-391, Paris.

CHILDRENS, C.C., 1994 : Feeding injury to 'Robinson' tangerine by *Brevipalpus* mites (Acari : Tenuipalpidae) in Florida and evaluation of chemical controlon citrus : *Florida Entomology,* **47** (2): 265-271, Gaineville, Florida.

CHITRIV, A.J., 1970 : Lemon butterfly, *Papilio demoleus* (Lepidoptera : Papilionidae). *Pesticides,* **4** (11) : 13-16. Mumbai (Bombay).

CHOPRA, R.L., 1932 : Citrus whiteflies and their control. *Punjab Agriculture Department,* **10** (2) : 22 pp., Lahore.

CHOWDHURY, S., 1954 : Control of citurs trunk borer *Monohamnus versteegi* Rits. Science and Culture, **19** (9) : 457-458, Calcutta.

CHOWDHURY, S. AND S. MAJID, 1954 : Handbook of Plant Protection, 117 pp. Department of Agriculture, Assam Publication, Shillong.

CILLERS, C.J. AND E.C.G. BEDFORD, 1984 : Citrus mealybugs. Farming in South Africa, Citrus series, H-23/1984.

CLAUSEN, C.P., 1915 : Mealybugs on citrus trees. *California Agriculture Experimental Station, Bulletin No. 258* : 22 pp California.

CLAUSEN, C.P., 1917 : Citrus Insects of Japan. *United States Department of Agriculture, Technical Bulletin No. 15,* Washington, D.C.

CLAUSEN, C.P., 1931 : Two citrus-miners of the Far East. *United States Department of Agriculture, Technical Bulletin No. 252,* 13 pp., Washington, D.C.

CLAUSEN, C.P., 1933 : The citrus insects of tropical Asia. *United State Department of Agriculture, Circular No. 266* : 25-30, Washington, D.C.

CLAUSEN, C.P. AND P.A. BERRY, 1932 : The citrus blackfly in Asia and the importation of its natural enemies into tropical America. *United States Department of Agriculture, Technical Bulletin No. 320* : 58 pp., Washington, D.C.

COCKERELL, T.D.A., 1903 : Whitefly (*Aleurodes citri*) and its allies. *Florida Agriculture Experimental Station, Bulletin No. 67* : 662-666, Gainesville, Florida.

COLLINES, P.J. T.M. LAMBKIN AND K.P. DNARUK, 1994 : Suspected resistance to methidathion in *Aonidiella aurantii* from Queensland *Journal of the Australian Entomological Society* **33** (4) : 325-326, Brisbane, Queensland.

Corbett, G.H., 1935 a : Three new Aleyrodes (Hem.) (*Aleurocanthus punjabensis, Aleurolobus citrifolii* on citrus in Punjab and *Traileurodes mossopi* on beans in S. Rhodesia). *Stylops*, **4** : 8-10, London.

Corbett, G.H., 1935 b : Malayan Aleyrodidae, *Journal of the federated Malay State Museum*, **17** : 722-852, Kaula Lampur.

Corbett, G.H., 1936 : New Aleurodidae (Hem.), *Proceedings of Royal Entomological Society*, **5** (8) : 18-22, London.

Corbett, G.H., 1939 : A new species of Aleurodidae from India (*Aleurocanthus husaini* sp. n., on citrus in Punjab). *Indian Journal of Entomology*, **1** : 69-70, New Delhi.

Costa, Lima, A., 1928 : Contribuicao as estudio des aleyrodideos da subfamila Aleurodicinae (A contribution to the study of the aleurodids of the subfamily Aleurodicinae). *Supplement Memoir, Institute Oswaldo Cruz*, **4** : 128-140, Rio de Janeiro.

Cramer, P., 1779-83 : Papillons Extiques de l'Asie de l'Afrique et de l'Amerique, 4 volumes. Amesterdam and Utrecht.

Crawford, D.L., 1914 : A Monograph of the Jumping Plant-lice or Psyllidae, of the New World. United States National Museum Bulletin No. 85, Washington, D.C.

Crawford, D.L., 1919 : The jumping plant-lice of palaeotropics and South Pacific Islands. Family Psyllidae or Chermidae, Homoptera. *Phillippines Journal of Science*, **15** : 139-208, Manila.

Curtis, C.E., P.J. Landlot and J.D. Clarck, 1989 : Disruption of Naval orange worm (Lepidoptera : Pyralidae) mating in large-scale plots with synthetic sex pheromones. *Journal of Economic Entomology*, **78** : 1425-1430, Lanham, Maryland.

Dahiya, K.K., R.K. Lakra, A.S. Dahiya and S.P. Singh, 1994: Bioefficiency of some insecticides against citrus psylla, *Diaphorina citri* Kuw. (Psyllidae : Homopetera). *Crop Research*, **8** (i) : 137-140), Hisar.

David, B.V. and T. kumarswami, 1975 : Elements of Economic Entomology, 507 pp. Popular Book Depot, Chennai (Madras).

David, B.V. and T.R. Subramaniam, 1976 : Studies on some Indian Aleurodidae, *Records of Zoological Survey of India*, **70** : 133-233, Calcutta.

David, B.V. and S. Vijayaraghavan and P.S. Narayanaswamy, 1963: *Indarbela tetraonis* Moore — a new bark borer pest of curry leaf tree and its control. *Madras Agricultural Journal*, **50** (5) : 195-199, Coimbatore.

David, Leela, A., 1961 : Notes on the biology and habits of the red tree ant *Oecophylla smaragdina* (Fabricius). *Madras Agricultural Journal*, **48** (2) : 51-57, Coimbatore.

David, S.K., 1960 : Three ornamental plants as new hosts of the black citrus aphid. *South Indian Horticulture*, **8** (1-2) : 32, Coimbatore.

De Geer, Charles, 1773 : Second memoire. Des pucerons. *Memoires pour servir a l'histoire des insects*, **3** : 19-129, Hesselberg, Stockholm.

Deshmukh, S.N. and D.O. Garg, 1984 : Past and Present Status of the Citrus whitefly *Dialeurodes citri* Ashmead (Hemiptera: Aleyrodidae), 42 pp. Hindustan Insecticides Limited, New Delhi.

Dev, H.N., 1964 : Preliminary studies on the biology of the Assam thrips, *Scirtothrips dorsalis* Hood on tea. *Indian Journal of Entomology*,

26 : 184, New Delhi.

DHARA, JOTHI, B.A. VERGHESE AND P.L. TANDON, 1989 : Ecological studies on citrus psylla *Diaphorina citri* Kuwayama with reference to its spatial distribution and sampling plan, *Entomon,* **14** (3-4) : 319-324, Thiruvananthapuram (Trivandrum).

DHARA, JOTHI, B., A. VERGHESE AND P.L. TANDON, 1990 : Evaluation of different plant oils and extracts against citrus aphid, *Toxoptera citricida* (Kirkalidy), *Indian Journal of Plant Protection,* **18** : 251-254, Hyderabad.

DHARMARAJU, E. AND C.S. REDDY, 1975 : Insect pest associated with citrus decline. p.46-54. *In* : Citrus Decline in Andhra Pradesh: Andhra Pradesh Agricultural University, Hyderabad.

DIETZ, H.F. AND J. ZETEK, 1920 : The blackfly of citrus and other subtropical plants. *United States Department of Agriculture, Bulletin No. 885,* 55 pp., Washington, D.C.

DHOORIA, M.S. AND DHAMO K. BUTANI, 1984 : Citrus mites, *Eutetranychus orientalis* (Klein) and its control. *Pesticides,* **18** (19) : 35-38, Mumbai (Bombay).

DORGE, S.K., M.N. BORLE, B.R. SONAWANE AND L.M. NAIK, 1968: Pests of fruit crops and their control in Maharashtra. *Technical Bulletin No. 15,* Government of Maharashtra, Pune.

DORGE, S.K. AND V.P. DALAYA, 1964 : Psylla (*Diaphorina citri* Kuw.) a serious pest of citrus in Maharashtra. *Parbhani Agricultural College Magazine,* **5** (1) : 5-8, Parbhani.

DOZIER, H.L., 1924 : Insect Pests and Diseases of the Satsuma Orange. *Gulf Coast Citrus Exchange, Educational Bulletin No. 1,* 103 pp.

DOZIER, H.L., 1927 : An undescribed whitefly attaching citrus in Puerto Rico. *Journal of Agricultural Research,* **34**: 853-855, Washington, D.C.

DOZIER, H.L., 1928 : The new Aleyrodid (citrus) pests from India and the South Pacific. *Journal of Agricultural Research,* **36** : 1001-1003, Washington, D.C.

DOZIER, H.L., 1932 a : The identity of certain whitefly parasites of the genus *Eretmocer* Held. with description of new species (Hymenoptera : Aphelininæ). *Proceedings of Entomological Society,* **34** : 112-118, Washington, D.C.

DOZIER, H.L., 1932 b : Two undescribed chalcid parasites of the wooly whitefly *Aleurothrixus floccosus* (Maskell) from Haiti. *Proceedings of Entomological Society,* **34** : 118-122, Washington, D.C.

DOZIER, H.L., 1934 : Description of new genera and species of African Aleyrodidae. *Annals and Magazine of Natural History,* **14** (1) : 184-192, London.

DUMBLETON, I.J., 1956 : New Aleyoridae (Hemiptera : Homoptera), New Caledonia, *Proceedings of the Royal Entomological Society, London (B),* **25** : 129-141, London.

DUTTA, S., 1966 : Horticulture in the Eastern Region of India, 264 pp. Directorate Extension, Ministry of Food and Agriculture, New Delhi.

EBELING, WALTER, 1959 : Subtropical Fruit Pests, 436 pp. University of California Press, Los Angeles.

EL-KADY, M.H., M.E.S. NASAR AND I.H. HEYEEL, 1977 : Sensitivity of citrus brown mite, *Eutetranychus orientalis* (Klein) to certain pesticides on citrus tree. *Agricultural Research Review,* **55** (1) : 99-101; Cairo.

Elmer, H.S., 1968 : Citrus flat mite injury patterns on some citrus varieties. *Journal of Economic Entomology,* **61** : 1689-1692, Lanham, Maryland.

Elmer, H.S. and O.L. Brawner, 1988 : The effect of citrus thrips and citrus red mite on navel orange fruit production in California's San Jaoquin valley. p.1089-1905. *In: Proceedings of the Sixth International Citrus Congress, Tel Aviv, Israel, march 6-11, 1988* : Balaban Publishers, Rehovot (Israel).

Elmer, H.S., O.L. Brawner and W.H. Ewart, 1973 : Scarring and silvering on central valley navels. *Citograph,* **58** : 335-338, Los Angeles, California.

Elmier, H.S., O.L. Brawner and W.H. Ewart, 1980 : Effect on citrus red mite on navels in central valley. *Citograph,* **65**: 155-157, Los Angeles, California.

Elmer, H.S., W.H. Ewart and O.L. Brawner, 1975 : Effect of citrus thrips populations on navel orange fruit yield and on tree growth in central California. *Citrograph,* **61**: 9-10 & 20-22, Los Angeles, California.

Enderlein, G., 1909 : *Udamoselis, eine neue* Aleurodiden — Gottung. *Zoologische Anzeiger,* **34** : 230-233, Leipzig.

Enderlein, G., 1920 : Zur Kenntnis tropischer Frucht-Bohfliegen, *Zoologische Jahrbacher, Jena,* **43** : 336-360, Leipzig.

Essig, E.O., 1932 : The original description of *Dialeurodes citri* (Ashmead) *Journal of Economic Entomology,* **25** : 1207-1208, Madison.

Essig, E.O., 1949 : Aphids in relation to quick decline and Trisreza of citrus. *Pan-Pacific Entomologist,* **65** (1) : 13-22, San Fransisco.

Fabricious, Johann Christian, 1792-98 : Entomologia Systematica, **I** (1792), **2** (1793), **3** and **4** (1794) and **Supplement** (1798), Hotniae.

Fahmy, I., 1956 : The economic importance of mites affecting fruit trees with special reference to citrus in Egypt (Acarina). *Bulletin of Entomological Society of Egypt,* **40** : 449-450, Cairo.

Fawcett, H.S., 1948 : Biological control of citrus insects by parasitic fungi and bacteria. *In* : The Citrus Industry, **2**, Chapter 13, University of California Press, Berkeley.

Febris, G.F., 1923 : Observations on the Chermidae. *Canadian Entomologist,* **55** : 88-92 & 250-256, Ottawa, Ontario.

Ferris, G.F., 1925 : Observations on the Chermidae. *Canadian Entomologist,* **57** : 46-50, Ottawa, Ontario.

Ferris, G.F., 1926 : Observations on the Chermidae. *Canadian Entomologist,* **58** : 13-20, Ottawa, Ontario.

Fitzpatrick, E.G., H.R. Cherry and R.V. Dowell, 1979 : Effect of Florida citrus pest control practices on the citrus blackfly (Homoptera : Aleyrodidae) and its associated natural enemies. *Canadian Entomologist,* **111** : 731-734, Ottawa, Ontario.

Fletcher, T. Bainbrigge, 1914 : Some South Indian Insects and Other Animals of Importance, 565 pp. Superintendent Government Press, Madras.

Fletcher, T. Bainbrigge, 1916 : One Hundred notes on Indian insects. *Imperial Agricultural Research Institute, Pusa, Bulletin No. 59,* 39 pp., Calcutta.

Fletcher, T. Bainbrigge, 1917 : Fruit trees. *Report of Proceedings of 2nd*

Entomological Meeting, Pusa (Bihar), February 1917 : 209-257, Calcutta.

FLETCHER, T. BAINBRIGGE, 1920 : Annotated list of Indian crop pests. *Report of Proceedings of 3rd Entomological Meeting, Pusa (Bihar), February 1919*, **I** : 33-314, Calcutta.

FLETCHER, T. BAINBRIGGE, 1921 : Annotated list of Indian crop pests. *Agricultural Research Institute, Pusa (Bihar) Bulletin No. 100*, 246 pp., Calcutta.

FREEBORN, S.B., 1931 : Citrus scale distribution in the Mediterranean Basin. *Journal of Economic Entomology*, **24** (5) : 1025-1031, Madison.

GAHAN, C.J., 1906 : The Fauna of British India including Ceylon and Burma. Coleoptera-I (Cerambycidae), 329 pp. Taylor and Francis, London.

GANDHI, S.R., 1959 : Oranges, Lemons and Limes in India. *Ministry of Food and Agriculture, Bulletin No. 15* : 53-66, New Delhi.

GANGULI, R.N. AND M.R. GHOSH, Biology of *Papilio demoleus* pest of citrus in Tripura State. *Indian Agriculturist*, **11** : 13-18, Calcutta.

GARCIA, G.E., 1935 : A field study on the citrus green bug, *Phynchocoris serratus* Donovan. *Philippines Journal of Agriculture*, **6** (3) : 311-325, Manila.

GARCIA, C.E., 1939 : The citrus rind borer and its control. *Philippines Journal of Agriculture*, **10** (1) : 89-92, Manila.

GARG, D.O., 1978 a : Insect Pests of Citrus Fruits, 176 + liv, pp. Agri-Horticultural Publishing House, Nagpur.

GARG, D.O., 1978 b : Pests that cause citrus decline. *Intensive Agriculture*, **16** (9) : 26-27, New Delhi.

GARG, D.O., 1981 : Dark season for orange growers of Vidarbha. *Intensive Agriculture*, **19** (6) : 18, New Delhi.

GARSEN, V., R. KENNETH AND I.A. MUTTATH, 1979 : *Hirsutella thompsonii*, a fungal pathogen of mites. *Annals of applied Biology*, **91** (1) : 29-40, Essex, England.

GEORGALA, M.B., 1975 : Possible resistance of red scale *Aonidiella aurantii* (Mask.) to corrective spray treatments. *Citrus and Subtropical Fruit Journal*, **504** : 5-17, Nelspruit 1200, South Africa.

GEORGALA, M.B., 1979 : Investigations on the control of the red scale *Aonidiella aurantii* (Mask.) and citrus thrips *Scirtothrips aurantii*. Faure on citrus in South Africa. *University Stellenbosch Annale Series A_3* (Landbou), **1** : 1-179.

GEORGALA, M.B., 1984 : Preventive programme options for the control of red scale. *Citrus and Subtropical Fruit Journal*, **606** : 9-12, Nelspruit 1200, South Afirca.

GEORGALA, M.B., 1988 : Control of red scale *Aonidiella aurantii* (Mask.) in South Africa influenced by its resistance to organo phosphates. p.1097-110. *In: Proceedings of the Sixth International Citrus Congress, Tel Aviv, Israel, March 6-11, 1988* : Balaban Publishers, Rehovot (Israel).

GHAI, SWARAJ, 1964 : Mites. In : Entomology in India : p. 385-396. Silver Jubilee Number, Entomological Society of India, New Delhi.

GHAI, SWARAJ, 1983 : Mites pests of crops: p. 319-339. *In* : Agricultural Entomology, All India Scientific Workers' Society, New Delhi.

GHOSE, S.K., 1961 : Studies on some coccids (Coccoidea : Hemiptera) of economic importance of West Bengal, India. *Indian Agriculturist*, **5**

(1) : 57-58, Calcutta.

GHOSE, S.K., 1965 : Control of orange coccid, *Labioproctus polei* Green (Margarodidae : Hemiptera). *Indian Journal of Agricultural Sciences,* **35** : 32-38, New Delhi.

GHOSH, C.C., 1914 : The lemon caterpillars, *Papilio demoleus* Linn. and *Papilio pammon* Linn. Life-histories of Indian insects-V. Lepidoptera (Butterflies). *Memoirs of Department of Agriculture, India (Entomological series), Pusa,* **5** (1) : 33-52, Calcutta.

GHUGUSKAR, H.T., S.G. RADKE, M.N. BORLE AND G.B. OHEKAR, 1983 : Efficacy of some insecticides against nymphs of citrus blackfly. *Punjabrao Krishi Vidyapeeth, Research Journal,* **7** (2) : 41-44, Akola.

GIESELMANN, M.J., C.A., HENRICK, R.J. ANDERSON, D.S. MORENO AND W.L. ROELOPS, 1980 : Response of male California red scale of sex pheromone isomers. *Journal of Insect Physiology,* **26** : 179-182, Oxford.

GLOVER, P.M., 1935 : An account of the occurrence of *Chrysomphalus aurantti* Mask. and *Laccifer lacca* Kerr, on grape-fruit in Ranchi district, Chota Nagpur, with a note on the chalcidoid parasites of *Aspidiotus orientalis* Newst. *Journal of Bombay Natural History Society,* **38** (1) : 151-153, Mumbai (Bombay).

GOLDI, E.A., 1886 : Beitrage zur Kenntnis der kleinen und kleinsten Gliederthierwelt Brasiliens. II. Neue brasilianische Aleurodes — Arten. *Mitteilungen der Schweizerischen entomologischen Gasellschaft Schaffhausen,* **7** : 241-280, Bern.

GOLDING, F.D., 1945 : Fruit piercing Lepidoptera in Nigeria. *Bulletin Entomological Research,* **36** (2) : 181-184, London.

(VAN DER) GOOT, P., 1912 : Ulber einige wahrscheinhlich neue blattlausarten aus der Sammlung des Naturhistorischen Museums in Hamberg. *Mitteilungen Naturhist Meseums,* **29** : 273-286, Hamburg.

GOUX, L., 1940 : Contribution a l'etude des Aleurode (Hem. Aleyrodidae) de la France. II-Description dedeux especes nouvelles de Marseille. *Bulletin de la Societe Entomologique de France,* **45** : 45-48, France.

GOWDEY, C.C., 1921 : The citrus blackfly (*Aleurocanthus woglumi*), *Jamaica Department of Agriculture, Entomological circular No. 3,* 11 pp., Kingston.

GRAFTEN, C.B., A. ELLER AND N.O. CONNEL, 1995 : Integrated citrus control reduces secondary pests. *California Agriculture.* **49** (2) : 23-28, Berkeley.

GREANY, P.D., R.E.MC DONALD, P.E. SHAW, W.J. SCHROEDER, D.F. HOWARD, T.T. HATTON, P.L. DAVIS AND G.K. RASMUSSEN, 1987: Use of gibberellic acid to reduce grapefruit susceptibility to attack by the carribean fruitfly *Anastrepha suspensa* (Diptera : Thephritidae). *Tropical Science,* **27** : 261-270, London.

GREANY, P.D., S.C. STYER, P.L. DAVIS, P.E. SHAW AND D.L. CHAMBERS, 1983 : Biochemical resistance of citrus to fruit flies. Demonstration and elucidation of resistance to the carribian fruit fly *Anastrepha suspensa. Entomologia Experimentalis et Applicata,* **34** : 40-50, Amsterdam.

GREEN, ERNEST E., 1900 : Note on web spinning habit of the red and *Oecophylla smaragdina. Journal of Bombay Natural History Society,* **13**

: 181, Mumbai (Bombay).

GROVER, P. AND MADHU BAKSHI, 1976 : On the study of cocoons of some Gallmidges (Cecidomyiidae : Diptera). *Cecidologia indica*, **11** (2 & 3) : 83-115, Allahabad.

GROVER, P. AND S.N. PRASAD, 1966 : Studies on Indian gallmidges XVII-*Sasyneura citri*, the blossom midge. *Cecidologia indica*, **1** (1) : 25-31, Allahabad.

GUNTHER, F., 1969 : Insecticide residue in California citrus fruits and products. *Residue Review*, **8** : 1-119, New York.

GUPTA, B.P. AND N.K. JOSHI, 1973 : Bionomics of citrus whitefly *Dialeurodes citri* Ashmead at Chaubattia in U.P. hills. *Progressive Horticulture*, **5** : 77-83, Ranikhet, Uttar Pradesh.

GUPTA, B.P. AND K.C. JOSHI, 1977 : Evaluation of some insecticides for the control of the citrus whitefly (*Dialeurodes citri* Ashmead). *Indian Journal of Horticulture*, **34** (1) : 34-39, Bangalore.

GUPTA, K.M. AND A. HAQ, 1959 : Citrus whitefly (*Dialeurodes citri* Ashmead) and its control in the hills of Uttar Pradesh. *Proceedings 46th Indian Science Congress, Delhi, III* : 502-503, Calcutta.

GUPTA, O.P. AND B.S. CHUNDAWAT, 1975 : Grape fruit, a promising citrus fruit in Northern India. *Financing Agriculture*, **7** (1) : 25-28, New Delhi.

GUPTA, O.P. AND B.S. CHUNDAWAT, 1975 : Lemon — a prospective fruit. *Financing Agriculture*, **8** (2) : 23-27, New Delhi.

GUPTA, R.L., 1952 : A brief note on some important pests of citrus. *Madhya Pradesh Government Bulletin, Five years of Citrus Research Fruit Scheme in M.P.*, 18-19, Bhopal.

GUPTA, R.L., 1954 : Life-History of *Tonica zizyphi* Stn. the citrus leaf roller, *Indian Journal of Entomology*, **16** (1) : 20-23, New Delhi.

GUPTA, R.L. AND S.A. JOSHI, 1955 a : Plant Protection in Madhya Pradesh, 50 pp, Department of Agriculture, Bhopal.

GUPTA, R.L. AND S.A. JOSHI, 1955 b : Progress of citrus pest control. *Department of Agriculture, M.P., Bulletin No. 50* : 20-22, Bhopal.

GUPTA, R.L. AND S.A. JOSHI, 1956 : Review of plant protection work in Madhya Pradesh. *FAO, Plant Protection Bulletin*, **8**: 17-25, Rome.

GUPTA, S.K., 1985 : Handbook of Plant Mites of India, 520 pp. Zoological Society of India, Calcutta.

GUPTA, S.K., M.S. DHOORIA AND A.S. SIDHU, 1971 : A note on predators of citrus mite in Punjab. *Current Science*, **37** (10) : 484, Bangalore.

GUPTA, S.K., A.S. SIDHU AND GURDIP SINGH, 1971 : Occurrence of a tenuipalpid mite on citrus in the Punjab and its control. *Indian Journal of Entomology*, **33** (1) : 30-33, New Delhi.

GUPTA, V.K. AND N.K. SHARMA, 1988 : Tree Protectioin, 477 pp., Society of Tree Scientists, Solan.

HALDEMAN, S.S., 1850 : On four new species of Hemiptera of the genera *Ploiaria, Chermes* and *Aleurodes* and two new Hymenoptera, parasitic on the last named genus. *American Journal of Science*, **9** (2) : 108-111, New Haven (USA).

HALL, W.J., 1930 : Notes on the control of some of the more important insect pests of citrus in Southern Rhodesia. *Rhodesia Agricultural Journal*, **27** (7) : 737-747, Salisbury.

HAMEED, S.F., N.P., KASHYAP AND D.N. VAIDYA, 1975 : Host records of insect pests of crops. *FAO, Plant Protection Bulletin,* **23** (6) : 191-192, Rome.

HAMPSON, (SIR) G.F., 1894 : The Fauna of British India including Ceylon and Burma. Moths, **2** : 376-562. Taylor and Francis, London.

HANEY, P.B., 1988 : Identification, ecology and control of the ants in citrus. A World Survey. p.1227-1251, *In: Proceedings of the Sixth International Citrus Congress, Tel Aviv, March 6-11, 1988 :* Balaban Publishers, Rehovot (Israel).

HANIFA, A. MOHAMED AND S. PALANISAMY, 1977 a : Evaluation of certain insecticides against citrus leaf-miner, *Phyllocnistris citrella* on acid lime. *International Symposium on Citriculture, Bangalore, Abstracts* : 34, Horticultural Society of India, Bangalore.

HANIFA, A. MOHAMED AND S. PALANISAMY, 1977 b : Chemical control of citrus caterpillar, *Papilio demoleus* Linn. *International Symposium on Citriculture, Bangalore; Abstracts*: 36, Horticultural Society of India, Bangalore.

HAYES, W.B., 1945 : The citrus industry in Sikkim. *Indian Journal of Horticulture,* **3** : 49-55, Sabour, New Delhi.

HAYES, W.B., 1957/1966 : Fruit Growing in India, 502 + 12 pp. Kitabistan, Allahabad.

HAFETZ, A., S. KRONENBERG, B.A. PELEG AND I. BAR-ZAKAY, 1988: Mating disruption of the California red scale *Aonidiella aurantii* (Homoptera : Diaspididae). p. 1121-1127. *In* : *Proceedings of the Sixth International Citrus Congress, Tel Aviv, Israel, March 6-11, 1988* : Balaban Publishers, Rehovot (Israel).

HATTING, V. 1994 : IPM on citrus in Southern Africa. *Citrus Journal,* **4** *: 21-24.*

HELY, P.C., 1933 : Citrus red-scale. Experiments with liquified hydrocyanic acid gas fumigation. *Agricultural Gazette of New South Wales,* **44** (11) : 823-828, Sydney.

HEPPNER, J.B. 1995 : Citrus leaf minor (Lepidoptera : Gracillaridae) on fruits in Florida. *Florida Entomologist,* **78** (1) : 183-184, Gainesville, Florida.

HILL, DENNIS S., 1975/1983 : Agricultural Insect Pests of the Tropics and Their control, 516pp. 746 pp. Cambridge University Press, Cambridge.

Hill, Dennis S. and Jeremy D. Hill, 1994 : Agricultural Entomology, 635 pp. Timber Press Inc., Porland, Oregon.

HUFFAKER, CARL B., 1988 : Ecology of insect pest control. p.1047-1066. *In: Proceedings of the Sixth International Congress, Tel Aviv, Israel, March 6-11, 1988* : Balaban Publishers, Rehovot (Israel).

HOFFMANN, W.E., 1928 : A stink bug injurious to citrus in South China. *Proceedings of Third Pan-Pacific Congress, Tokyo, 1926,* **2** : 2030-2037, Tokyo.

HOFFMANN, W.E., 1935 : *Diaphoriania citri* Kuw. (Homoptera: Chermidae) a citrus pest in Kwantung. *Longman Science Journal,* **15** (1) : 127-132, Canton.

HUSAIN, M.A., 1927 : A warning to citrus growers. *Seasonal Notes, Punjab Agricultural Department,* **4** (1) : 27, Lahore.

HUSAIN, M.A. AND ABDUL WAHID KHAN, 1930 : The Aleyrodidae on citrus and their control in the Punjab. *Proceedings of 17th Indian Science Congress*

Allahabad, III : 51-52, Calcutta.

HUSAIN, M.A. AND ABDUL WAHID KHAN, 1932 : Food plants of *Dialeurodes citri* Ashmead (Aleyrodidae). Is Jasmine one of them? *Indian Journal of Agricultural Sciences,* **2** (3) : 242-253, New Delhi.

HUSAIN, M. AFZAL AND ABDUL WAHID KHAN, 1945 : The citrus Aleyrodidae (Homoptera) in the Punjab and their control. *Memoirs of Entomological Society of India, No. 1,* 41 pp., New Delhi.

HUSAIN, MOHAMMAD AFZAL AND DINA NATH, 1927 : The citrus psylla (*Diaphorina citri* Kuw.) (Psyllidae : Homoptera). *Memoirs of the Department of Agriculture in India, Entomological series,* **10** (2) : 27 pp., Calcutta.

HUSAIN, M.A. AND U.B. MATHUR, 1923 : A preliminary list of parasites of economic importance bred in the Punjab. *Report of Proceedings of 5th Entomological Meeting, Pusa (Bihar), February 1922* : 119-121, Calcutta.

HUTSON, J.C. AND M.P.D. PINTO, 1934 : Two caterpillar pests of citrus. *Tropical Agriculture,* **83** (3) : 188-193, Paradeniya.

JADHAV, L.D., D.S. AJRI, M.V. KADAM AND S.K.DORGE, 1976 : Leaf, footed plant bug on pomegranate in Maharashtra. *Entomologists Newsletter,* **6** (10) : 57, New Delhi.

JALALI, S.K. AND S.P. SINGH, 1990 : A new record of *Ooencyrtus papilionis* Ashmead (Hymenoptera : Encyrtidae) on eggs of *Papilio demoleus* (Linn.) from India. *Journal of Biological Control,* **4** (1) : 59-60, Bangalore.

JAMES, D.G. 1994 : The development of suppression tactices for *Biprorulus bibax* (Heteroptera : Pentatomidae) as part of an integrated pests management programme in citrus in Inland South Eastern Australia. *Bulletin Entomoligical Research,* **84** (1) : 31-38, London.

JANDU, ARJAN SINGH, 1942 : Biological notes on the butterflies of Delhi. I-Papilionidae and Pilridae. *Indian Journal of Entomology,* **4** (2) : 201-204, New Delhi.

JOSHI, N.A. AND B.D. TAHILIANI, 1957 : Citron cultivation in Palanpur. *The Farmer, July 1957* : 27-20, Mumbai (Bombay).

JOTWANI, M.G. AND DHAMO K. BUTANI, 1980 : Insects pests of crops. p. 417-551. *In* : Handbook of Agriculture : Indian Council of Agricultural Research, New Delhi.

JOTWANI, M.G. AND DHAMO K. BUTANI, 1981 : Pest problems of national importances. p.301-310. *In* : Indian Pesticide Industry : Vishvas Publications, Mumbai (Bombay).

KALIDAS, POTINENI, 1990 : Incidence of citrus blackfly and its control. *Proceedings of the International Citrus Symposium, 5 to 8 November, 1990,* Guangzhou (China).

KALIDAS, POTINENI, 1991 : Insect pests of citrus nursery, orchard and their management. *Farmer and Parliament, October, 1991* : 2, New Delhi.

KALIDAS, POTINENI, 1993 a : Control of citrus blackfly using monocrotophos — A demonstration. *Pestology,* **17** (7) : 26-32, Mumbai (Bombay).

KALIDAS, POTINENI, 1993 b : Comparative efficacy of convential and new insecticides against citrus blackfly, *Aleurocanthus woglumi* Ashby *(Homoptera : Aleyrodidae). Indian Journal of Agricultural Sciences,* **63** (8) : 526-528, New Delhi.

KALIDAS, POTINENI, 1995 : Role of ecological factors on the survival and mortality of citrus blackfly. *Indian Journal of Horticulture,* **52** (2) : 117-120, New Delhi.

KALTENBACH, J.H., 1843 : Monographie der Femilien der Pflanzenlause Aachen, 1823 : xliii + 223 pp.

KAPOOR, V.C., 1972 : Identification of common fruit flies (Tephritidae : Diptera) of India, *Entomologists Record,* **84**: 165-169, London.

KAPUR, S.P., R.C. BATRA AND HARMINDAR KAUR, 1995 : Biochemical basis of resistance against Greening and leaf-miner in citrus. *Journal of Mycology and Plant Pathology,* **25** (1 & 2): 63-64, Udaipur.

KATOLE, S.R., R.K. MAHAJAN AND U.P. BARKHADE, 1993 a : Field efficacy of fenvalerate and some insecticides against blackfly nymphs on citrus. *Pestology,* **17** (11) : 5-6, Mumbai (Bombay).

KATOLE, S.R., S.K. THAKARE AND R.K. MAHAJAN, 1993 b : Management of citrus blackfly nymphs with some insecticides and plant oils. *Journal of Maharashtra Agricultural Universities,* **18** (1) : 68-71, Pune.

KATOLE, S.R., L.H. MESHRAN, R.K. MAHAJAN AND W. MANKAR : 1994. Relative toxicity of some pesticides to blackfly pupae on citrus. *Crop Research Journal,* **76** : 105-108, Hisar.

KATYAL, S.L., 1969 : Recent trends in citrus research. *Indian Horticulture,* **13** (2) : 3-4, New Delhi.

KATYAL, S.L. AND C.P. DUTTA, 1976.: Horticultural research and development in hill areas. *Indian Horticulture,* **21** (2): 11-12, New Delhi.

KAJUR, HARMINDER, S.P. KAPUR, R.C. BATRA AND S.N. SINGH, 1994: Correlation of biochemical composition in different strains of rough lemon to Greening disease and leaf-miner. *Indian Journal of Horticulture,* **51** (1) : 56-58, Bangalore.

KHADER, J.B.M., M.D. ABDUL AND S. BALASUBRAMANIAN, 1977 : In mango and citrus — watch for root borers. *Indian Horticulture,* **22** (3) : 4-6, New Delhi.

KHAN, A.W., 1942 : The citrus Aleyrodidae (Whiteflies of citrus) in the Punjab and their control. *Punjab Fruit Journal,* **6** (24) : 1218-1219, Lahore.

KHAN, M. AND I. HAQ, 1954 : Nomenclature of *Citrus* species in the Punjab. *Punjab Fruit Journal,* **18** : 41-45, Lahore.

KHANNA, K.L., L.N. NIGAM, S.B.D. AGARWALA AND DHAMO K. BUTANI, 1956 : Termites infesting sugarcane and their control. I: p.908-936, *In: Proceedings International Society of Sugar Cane Technologists, 9th Congress, New Delhi,* New Delhi.

KHANNA, S.S. AND Y.D. PANDE, 1966 : Bionomics and control of *Phyllocnistis citrella, Allahabad Farmer,* **40** (5) : 203-209, Allahabad.

KHARE, J.L., 1921 : Some citrus pests in Nagpur district. *Department of Agricultural, Central Provinces, Nagpur, Bulletin No. 10,* 14 pp., Nagpur.

KHURANA, A.D. AND O.P. GUPTA, 1972 : Bark eating caterpillars pose a serious threat to fruit trees. *Indian Farmers' Digest,* **5** (4) : 51-52, Pantnagar.

KIRKALDY, G.W., 1907 : On some peregine Aphidae in Oahu (Hem). *Proceedings of Hawaiin Entomological Society,* **1** : 99-102, Hawaii.

KLEIN, N.Z., 1936 : Contributions to the knowledge of the red spiders in Palestine. *Agricultural Research Station Bulletin* No. **21** : 3-36, Rehovot (Israel).

KNAPP, J.L. AND H. ANSON MOYE, 1988 : Aldicarb and metabolite residues in nectar following commercial application to Florida, USA, Citrus.

Citriculture — p.1107-1111. *In: Proceedings of the Sixth International Citrus Congress, Middle East, Tel Aviv, 6-11 March, 1988,* Balaban Publishers, Rehovot (Israel).

KODAMA, G., 1931 : Studies on *Aleurocanthus spiniferus* Quaint. *Kagoshima-Ken,* 38 pp. Kyushu, Japan (in Japanese), Summary in *Review of Applied Entomology,* **20 A** : 113, London.

KOTINSKY, J., 1907 : Aleyrodidae of Hawaii and Fiji, with description of new species. Board Commission Agriculture, Hawaii, Division of Entomology, Bulletin No. 2 : 93-103, Hawaii.

KRISHNAMOORTHY, A., 1988 a : Host range, development and sex ratio of *Leptomastix dactylopii* on different stages of citrus mealybug, *Planococcus citri. Journal of Biological Control,* **2** (1) : 8-11, Bangalore.

KRISHNAMOORTHY, A., 1988 b : Development and voracity of *Chrysopa lacciperda* Kimmis on *Planococcus citri* (Risso) *Journal of Biological Control,* **2** (2) : 97-98, Bangalore.

KRISHNAMOORTHY, A., 1989 : Effect of cold storage on the emergence and survival of the adult exotic parasitoid *Leptomastix dactylopii* How. (Hym. Encyrtidae). *Entomon,* **14** (3 & 4) : 313-318, Thiruvananthapuram (Trivandrum).

KRISHNAMOORTHY, A., 1990 : Evaluation of permanent establishments of *Leptomastrix dactylopii* against citrus mealybug in India. *Fruits,* **45** (1) : 29-32, Paris.

KRISHNAMOORTHY, A. AND M. MANI, 1988 a : Records of green lacewings, preying on mealybugs in India. *Current Science,* **58** : 466, Bangalore.

KRISHNAMOORTHY, A. AND M. MANI, 1988 b : *Coccidoxenoides peregrina* a new parasitoid of *Planococcus citri* in India. *Current Science,* **58** : 466, Bangalore.

KRISHNAMOORTHY, A. AND M. MANI, 1989 : Recovery of an exotic parasitoid, *Leptomastrix dactylopii* How. from *Planococcus citri* (Risso) infesting some horticultural crops. *Journal of Biological Control,* **3** : 125, Bangalore.

KRISHNAMOORTHY, A. AND M. MAINI, 1994 : Manage mealybug in citrus orchards. *Indian Horticulture,* **29** (1) : 12-13, New Delhi.

KRISHNAMOORTHY, A. AND M. MAINI, 1996 : Record of new hyperparasitoids on exotic parasitoid *Leptomastix dactylopii* How., parasitizing citrus mealybug *Planococcus citri* (Risso) in India. *Entomon,* **21** : 111-112, Thiruvantanthapuram (Trivandrum).

KRISHNAMOORTHY, A. AND D. RAJAGOPAL, 1995 : Comparable toxicity of pesticides to natural enemies of California red scale, *Aonidiella aurantii* (Maskell). *Pest Management in Horticultural Ecosystem,* **1** (2) : 71-79, Bangalore.

KRISHNAMOORTHY, A. AND S.P. SINGH, 1986 : Record of egg parasite, *Trichogramma chilonia* on *Pyrilla* spp. in citrus. *Current Science,* **55** (9) : 461, Bangalore.

KRISHNAMOORTHY, A. AND S.P. SINGH, 1988 : Observational studies on the occurrence of parasitoids of *Papilio* spp. in citrus. *Indian Journal of Plant Protection,* **16** (1) : 79-81, Hyderabad.

KRISHNAMOORTHY, P.N., S.P. SINGH, N.S. RAO, S.V. SARODE, LALITHA ANAND AND M.D. AWASTHI, 1991 : Residues of monocrotophos in mandrin fruits. *Pestology,* **15** (1) : 33, Mumbai (Bombay).

KUCHANWAR, D.B., M.G. HARDAS AND B.K. SHARANAGAT, 1983 : Collateral hosts of *Aleurocanthus woglumi* Ashby. *Nagpur Agricultural College Magazine,* **55** : 69-70, Nagpur.

KUNHI KANNAN, K., 1923 : The lime (citrus) tree borer *Chelidonium cinctum* (Guer.). *Mysore Agricultural and Experimental Union,* **5** : 129-131, Bangalore.

KUNHI KANNAN, K., 1928 : The large citrus borer of South India, *Chelidonium cinctum* (Guer.). *Bulletin, Department of Agriculture, Mysore, Entomological Series, No. 8,* 24 pp., Mysore.

KUSHWAHA, K.S., 1955 : External morphology of the soldier of *Odontotermes obesus* (Rambur). *Current Science,* **24** (6): 203-204, Bangalore.

KUSHAWAHA, K.S., 1956 a : External morphology of the termite, *Odontotermes obesus* (Rambur) (Isoptera : Termitidae). Part I — Soldier. *Records of Indian Musuem,* **54** : 209-227, Calcutta.

KUSHWAHA, K.S., 1956 b : External morphology of the termite, *Odontotermes obesus* (Rambur) (Isoptera : Termitidae). Part II — Alate and Workers. *Records of Indian Museum,* **54** : 229-250, Calcutta.

KUSHWAHA, K.S., 1959 a : A preliminary account of external morphology of workers and alate of *Odontotermes obesus* (Rambur). *Current Science,* **28** (7) : 298-299, Bangalore.

KUSHWAHA, K.S., 1959 b : Chaetotaxy of the soldier, worker and alate of the termite, *Odontotermes obesus* (Rambur). *Current Science,* **28** (10) : 415-417, Bangalore.

KUSHWAHA, K.S., 1972 : Termite pests of fruit trees and grasses in India. p.58-61. *In:* Termite Problems in India. Council of Scientific and Industrial Research, New Delhi.

KUWANA, S.I., 1911 : The whiteflies of Japan. *Pomona College Journal of Entomology,* **3** : 620-627, Claremont, Cal.

KUWANA, S.I., 1927 : On the genus *Bemisia* (Family Aleyrodidae) found in Japan, with description of new species. *Annotationes Zoologiche Japaneses,* **9** : 245-253, Tokyo.

KUWANA, S.I., 1928 : Aleyrodidae or whitefly attacking citrus fruits in Japan. *Scientific Bulletin, Ministery of Agriculture and Forestry, Japan,* **I** : 43-78, Tokyo.

KUWAYAMA, S., 1907 : *Transactions of Sapparo Natural History Society,* **2** : 100, Sapparo.

LAKRA, R.K. AND D.S. GUPTA, 1977 : Incidence of citrus leaf-miner, *Phyllocnistis citrella* Stainton on different cultivars of citrus and its chemical control. *International Symposium on Citriculture, Bangalore. Abstracts* : 33. Horticultural Society of India, Bangalore.

LAKRA, R.K. AND D.S. GUPTA, 1983 : Incidence of citrus leaf-miner, *Phyllocnistis citrella* Stainton (Gracillariidae: Lepidoptera) on different cultivars of citrus and its control. p.345-351, *In: Proceedings of International Citrus Symposium Bangalore, December, 1977*, Bangalore.

LAKRA, R.K. D.S. GUPTA AND W.S. KHARUB, 1977 : Studies on chemical control of citrus psylla, *Diaphorina citri* Kuwayama (Pysllidae : Homoptera) in Haryana. *International Symposium on Citriculture, Abstracts* : 35, Horticultural Society of India, Bangalore.

LAKRA, R.K., D.S. GUPTA AND W.S. KHARUB, 1983 : Studies on chemical control of citrus psylla, *Diaphorina citri* Kuwayama (Psyllidae :

Homoptera) in Haryana.p. 364-369. *In* : *Proceeding of International Citrus Symposium, Bangalore, December, 1977* : Bangalore.

LAKRA, R.K., ZILE SINGH AND W.S. KHARUB, 1983 : Population dynamics of citrus psylla, *Diaphorina citri* Kuwayama in Haryana. *Indian Journal of Entomology,* **45** (3) : 301-310, New Delhi.

LAKRA, R.K., ZILE SINGH AND W.S. KHARUB, 1984 : Population dynamics of citrus leaf-miner, *Phyllocnistris citrella* Stainton (Lepidoptera : Phyllocnistidae) in Haryana. *Indian Journal of Ecology,* **11** (1) : 146-153, Ludhiana.

LAL, K.B., 1952 : Insect pests of fruit trees grown in the plains of U.P. and their control. *Agriculture and Animal Husbandry India,* **3** (1-3) : 54-84, Lucknow.

LAL, K.B., 1960 : Insect pests of fruit trees grown in the plains. *Agriculture and Animal Husbandry, U.P.,* **11** (4) : 30-45, Lucknow.

LATIF, A. AND K.L. AHMED, 1955 : Control of citrus psylla with diazinon emulsion. *Punjab Fruit Journal,* **19** (72-73) : 42-43, Lahore.

LATIF, A. AND MOHAMMAD YUNUS, 1950 : Observations on the relationship between citrus leaf-miner (*Phyllocnistis citrella Stn.*) and citrus canker (*Pseudomonas citri* Hasse). *Indian Journal of Entomology,* **12** (1) : 27-29, New Delhi.

LATIF, A. AND CH. MOHAMMAD YUNUS, 1951 : Food plants of citrus leaf-miner in *Bulletin of Entomological Research, London,* **42** (2) : 311-316, London.

LEES, A.H., 1916 : Some observations on the egg of *Pyrilla mali. Annals of Applied Biology,* **11** : 251-257, Cambridge.

LEEROY, H. MAXWELL, 1906 : Indian Insect Pests, 315 pp. Superintendent, Government printing, India, Calcutta.

LEFROY, H. MAXWELL, 1907 : The more important insects injurious to Indian Agriculture. *Memoirs of the Department of Agriculture, India,* **1** (2) : 152 pp., Calcutta.

LEFROY, H. MAXWELL AND F.M. HOWLETT, 1909 : Indian Insect Life, 788 pp., Thacker Spink and Co., Calcutta (reprinted 1971, Today & Tomorrow, Printers and Publishers, New Delhi).

LINDGREEN, D.L., 1941 : Factors influencing the results of fumigation on the California red scale. *Hilgardia,* **13** : 491-511, Oakland.

LINNAEUS, CAROLUS, A., 1758 : Systema Naturae, edition decima. Reformate, **1** : 453 pp. Laurentil, Salvii, Holmige.

LU, B.Q., X.H. HUANGF, Y. XIAO, Y.G. XHANG AND C.B. LI, 1995: Studies on the control of *Panonychus citri* Mcg. by mixed spraying of various chemical fertilizers and surface active agents. *China Citrus,* **24** (2) : 35-37.

LUH, NIEN-TSIN, 1936 : A survey on the falling off of citrus fruits due to insect pests in Hangyuen during, 1935. *Entomology and Phytopathology,* **4** (6) : 102-107, Hangchow.

LUSHINGTON, A.W., 1910 : The genus *Citrus. Indian Forestor,* **36**: 323-353, Dehradun.

MACKIE, D.B., 1928 : Progress in citrus whitefly suppression in the Northern California areas. *Monthly Bulletin California Department of Agriculture,* **16** : 677-686, Sacramento, Cal.

MACKIE, D.B., 1931 : The citrus whitefly in California. *Monthly Bulletin, California Department of Agriculture,* **20** : 599-612, Sacramento, Cal.

MANALAE, SAN JAUN, J., 1924 : *Prays citri* Millière, a rind insect pest of Philippines orange. *Philippine Agriculture,* **12** (8) : 339-348, Los Banôs.

MANGAT, B.S., 1966 : Biology and control of citrus psylla (*Diaphorina citri* Kuw.). *Plant Protection Bulletin,* **18** (3) : 18-20, New Delhi.

MANGAT, B.S. AND SARDAR SINGH, 1960 : Biology and control of citrus psylla (*Diaphorina citri* Kuwayama) in the Punjab. *Proceedings 47th Indian Science Congress, Bombay, III* : 353-354, Calcutta.

MAINI, M., 1994 : Recovery of the indigenous *Coccidoxenoides peregrinus* and the exotic *Leptomastix dactylopii* in lemon and acid-lime orchards. *Biocontrol Science and Technology,* **4** : 49-52, Oxfordshire (U.K).

MANI, M., 1995 : Comparative development progeny production and sex ratio of the exotic parasitoid, *Leptomastrix dactylopii* How. (Hymenoptera : Encyrtidae) on *Planococcus lilacinus and Planococcus citri* (Homoptera: Pseudococcidae). *Entomon,* **20** : 23-26, Thiruvananthapuram (Trivandrum).

MANI, M. AND A. KRISHNAMOORTHY, 1990 : Predation of *Mallada boninensis* on *Ferrisia virgata, Planococcus citri* and *P. lilacious. Journal of Biology Control,* **4** : 122-123, Bangalore.

MANI, M. AND A. KRISHNAMOORTHY, 1991 : Contact toxicity of synthetic pyrethroids to some parasitoids and predators of mealybugs. *Indian Journal of Plant Protection,* **19** : 93-95, Hyderabad.

MANI, M. AND A. KRISHNAMOORTHY, 1996 a : Biocontrol of mealybugs on fruit crops. *Indian Institute of Horticulture Research, Folder No. 68,* 6 pp., Bangalore.

MANI, M. AND A. KRISHNAMOORTHY, 1996 b : Discovery of the coccinellid predator *Chilocores circumdatus* on the green scale *Coccus viridis. Entomon,* Thiruvananthapuram (Trivandrum).

MANI, M., A. KRISHNAMOORTHY AND M. SRINIVASA RAO, 1993 : Toxicity of different pesticides to the exotic parasitoid, *Leptomastix dactylapii* How. *Indian Journal of Plant Protection,* **21** : 98-99, Hyderabad.

MANI, M., S.N. SUSHIL AND A. KRISHNAMOORTHY, 1955 : Influence of some selective pesticides on the longevity and progeny production of *Leptomastix dactlopii* How. a parasitoid of citrus mealybug, *Planococcus citri* (Risso). *Pest Management in Horticultural Ecosystems,* **1** (2) : 81-86, Bangalore.

MANI, M.S., 1939 : Description of new and records of some known Chalcidoids and other Hymenopterous parasites from India. *Indian Journal of Entomology,* **I** : 69-99, New Delhi.

MANI, M.S., 1941 : Studies on Indian parasitic Hymenoptera — I. *Indian Journal of Entomology,* **3** : 34-35, New Delhi.

MANI, M.S., 1945 : Introduction to Entomology, 356 pp. Agra University Publication, Agra.

MANI, M.S., 1973 : General Entomology, 597 pp. Oxford & IBH Publishing Co., New Delhi.

MARGABANDU, V., 1933 : Insect pests of orange in Northern circle. *Madras Agricultural Journal,* **21** (2) : 60-68, Coimbatore.

MASKELL, W.M., 1895 : Contribution towards a monograph of Aleyrodidae, a family of Hemiptera-Homoptera. *Transactions of New Zealand Institute,* **28** : 411-449, Wellington.

Mc Coy, C.W., 1978 : Entomopathogens in arthropod pest control prorgrams in citrus. p. 211-223. *In* : Microbial Control of Insect Pests — Future Strategies in Pest Management System, Gainesville, Florida.

Mc Coy, C.W., R.I. Sailor and J.L. Knapp, 1980 : History and establishment of parasite, *Prospattella lahorensis* for the biological control of citrus whitefly in Central Florida. *Proceedings of Florida State Horticultural Society*, **93** : 47-49, Gainesville, Florida.

Mc Donald, R.E., P.D. Greany, P.E. Shaw, W.J. Schroeder, T.T. Hatton and C.W. Wilson, 1988 : Use of Gibberellic acid for Caribbean fruit-fly (*Anastrepha suspensa*) control in grapefruit. p. 1147-1152. *In: Proceedings of the Sixth International Citrus Congress, Tel Aviv, Israel, March 6-11, 1988* : Balaban Publishers, Rehovot (Israel).

Mehra, B.P. and B.N. Shah, 1970 : Bionomics of *Thiacidas postica* Walker (Lepidoptera : Noctuidae), a pest of *Zizyphus mauritiana* Lamarck. *Indian Journal of Entomology*, **32** (2) : 145-151, New Delhi.

Mehta, P.R. and B.K. Verma, 1978 : Plant Protection, 587 pp. Directorate of Extension. Ministry of Food & Agriculture, Community Development and Co-operative, New Delhi.

Melichar, L., 1903 : Homopteren — Fauna von Ceylon : 148, Verlag von felix L. Dames, Berlin.

Menenghini, M., 1948 : (Experiments on transmission of the tristeza disease of citrus by the black orange aphid, *Aphis tavaresi*) O. *Biologico*, **14** : 115-118.

(Van der) Merwe, C.P., 1923 : The citurs psylla (*Spanioza erythrae* Deg.). *Journal of Department of Agriculture Union of South Africa*, **7**: 135-141, Pretoria.

Mingdu, H. and L. Shuxin, 1989 : The damage and economic threshold of citrus miner, *Phyllocnistis citrella* stainton to citrus. In Studies on the Integrated Management of Citrus Pests, 84-89 pp. Academic Books and Periodical Press, Guanzhous, China.

Mishra, S.C. and N.D. Pandey, 1965 : Some observations on the biology of *Papilio demoleus* Linn. (Papilionidae : Lepidoptera). *Labdev Journal of Science and Technology*, **3** (2) : 142-143, Kanpur.

Misra, C.S., 1920 : Index of Indian fruit pests. *Report of Proceedings of 3rd Entomological Meeting, Pusa (Bihar), February 1991*, **3** : 564-595, Calcutta.

Misra, C.S., 1924 a : The citrus whitefly, *Dialeurodes citri* in India and its parasites together with the life history of *Aleurodes citri*. *Report of Proceedings of 5th Entomological Meeting, Pusa (Bihar), February 1923* : 129-135, Calcutta.

Misra, C.S., 1924 b : A list of Coccidae in the Pusa collection. *Report of Proceedings of 4th Entomological Meeting, Pusa (Bihar), February 1923* : 345-351, Calcutta.

Mitra, S.K. and Khongwir, P.C., 1928 : Orange cultivation in Assam. *Bulletin Department of Agriculture, Assam*, **2** : 19, Shillong.

Modawal, C.N., 1940 : The tachinid *Erycia nymphalidephaga* Bar., parasite of *Papilio demoleus* Linn. *Indian Journal of Entomology*, **2** (2) : 242, New Delhi.

Morgan, A.C.F., 1892 : A new genus and species of Aleyrodidae. *Entomologists Monthly Magazine*, **3** : 29, London.

MORRIL, A.W., 1905 : The greenhouse whitefly (*Aleyrodes vaporariorum* West.). *United States Department of Agriculture, Bureau of Entomology, Bulletin No. 57,* 9 pp. Washington, D.C.

MORRIL, A.W., 1908 : Fumigation for the citrus whitefly. *United States Department of Agriculture, Bureau of Entomology, Bulletin No. 76,* 73 pp. Washington, D.C.

MORRIL, A.W. AND E.A. BACK, 1911 : Whiteflies injurious to citrus in Florida. *United States Department of Agriculture, Bureau of Entomology, Bulletin No. 92,* 109 pp. Washington, D.C.

MORSE, J.G. AND O.L. BRAWNER, 1986 : Managing citrus thrips resistance to pesticides. *Citrograph,* **71** (5) : 93, 96-97 & 108, Los Angeles, Cal.

MOULTON, D., 1909 : The orange thrip. *United States Department of Agriculture Bureau of Entomology,* Technical Series, Bulletin No. **12** (7) : 119-122, Washington, D.C.

MOUND, L.A. AND S.H. HALSEY, 1978 : A Systematic Catalogue of Aleyrodidae (Homoptera) with host plant and natural enemy data, 340 pp. British Museum (Natural History), London and John Wiley and Sons, New York.

MUKERJEA, T.D., 1966 : A new chemical for control of scarlet and pink mite. *Two leaves and a bud,* **13** (3) : 94-101, Cinnamara (Assam).

MUKHERJEE, D. AND D.K. NATH, 1971 : *Paralebeda plagifera* (Walker) (Lasiocampidae : Lepidoptera) A new record of citrus pest in West Bengal. *Indian Journal of Entomology,* **33** (1) : 103, New Delhi.

MUKHERJEE, S.K. AND P.K. MAJUMDAR, 1973 : Vegetative propogation of tropical and subtropical fruit crops. *Indian Council of Agriculture, Technical Bulletin No. 45,* New Delhi.

MURTHI, B.K., 1922 : The lime tree borer. *Mysore Agricultural and Experimental Union Journal,* **4** : 69-75, Bangalore.

MUTHUKRISHNAN, T.S., T. SANTHAKUMAR, G. RAJENDRAN, J. CHANDRASEKARAN, R. SANKARANARAYANAN AND P. RAJAGOPALAN, 1975 : Varietal susceptibility of certain root stock to citrus nematode, *Tylenchulus semipenetrans* Cobb, 1913. *South Indian Horticulture,* **23** (3-4) : 91-93, Coimbatore.

NAGARAJANA, N.R., 1967 : A preliminary study of citrus mites in Madras State. *South Indian Horticulture,* **15** (1-2) : 31-35, Coimbatore.

NAIR, M.R.G.K., 1960 : Notes on Crop Pests and Their Control, 72 pp. Alliance Printing Works, Thiruvananthapuram (Trivandrum).

NAIR, M.R.G.K., 1975 : Insect and Mites of Crops in India, 404 pp. Indian Council of Agricultural Research, New Delhi.

NAKAO, S., K. NOHARA AND A. NAGATOMI, 1977 : Studies on pests and their predators at two groves of "Kuroshima-mikan", a native citrus on Nagashima, Kagoshima Pref. *Journal of Applied Entomology and Zoology,* **12** : 334-346.

NAKAO, S., K. NOHARA AND T. ONO, 1972 : Fundamental study on the integrated control of citrus red mite in the summer orange groves. *Mushi,* **46** : 1-29, Fukuoka.

NAKAO, S.K. NOHARA AND T. ONO, 1974 : Experimental study on the integrated control of three important pests on the summer orange goves. *Mushi,* **47** : 81-110, Fukuoka.

NARAYANANA, E.S. AND H.N. BATRA, 1960 : Fruitflies and Their Control, 68 pp. Indian Council of Agricultural Research, New Delhi.

NARAYANAN, E.S. B.R. SUBBA RAO AND R.B. KAUR, 1957 : Notes on the biology of the parasites of the leaf miners of citrus plants. p.396-397. *In: Proceedings 44th Indian Science Congress, Calcutta, III*, Calcutta.

NARAYANAN, K. AND S. JAYARAJ, 1974 a : Effect on *Bacillus thuringiensis* endotoxin on hemolymph cation leaves in the citrus leaf caterpillar, *Papilio demoleus*. *Journal of invertebrate Pathology,* **23** : 125-126, New York.

NARAYANAN, K. AND S. JAYARAJ, 1974 b : Studies on the mode of action of *Bacillus thuringiensis* Berliner in citrus leaf caterpillar, *Papilio demoleus, Indian Journal of Experimental Biology,* New Delhi.

NARAYANAN, K. AND S. JAYARAJ, 1974 c : Observations on the pathogenicity of *Serratia marsescens* Bizio for certain lepidopteran insects. *Madras Agricultural Journal,* **61** : 92-95, Coimbatore.

NARAYANAN, K. AND S. JAYARAJ, 1976 : Anorexient effect of *Bacillus thuringiensis* Berliner on citrus caterpillar, *Papilio demoleus. Madras Agricultural Journal,* **63** (2) : 135-137, Coimbatore.

NARAINI, T.K., H.S. SAHAMBI AND B.L. CHOPRA, 1965 : Occurence of triteza virus in Northern India. *Indian Phytopathology,* **18** : 220, New Delhi.

NATH, DINA, 1932 : The lemon buterfly and its control. *Seasonal Notes. Punjab Agriculture Department,* **9** (1) : 35-36, Lahore.

NATH, DINA, 1940 : Better spraying, better fruits. *Punjab Fruit Journal,* **4** (15) : 821-822, Lahore.

NATH, DINA, 1941 a : Black citrus fly and its control. *Seasonal Notes. Punjab Agriculture Department,* **19** (1) : 53-55, Lahore.

NATH, DINA, 1941 b : Citrus psylla situation and what the citrus growers should do. *Punjab Fruit Journal,* **5** (17) : 896-897, Lahore.

NATH, DINA, 1970 : Occurrence of the slug caterpillar, *Narosa propolia* Hampson (Cochilididae, Lepidoptera) on *Citrus* spp. in West Bengal. *Indian Journal of Entomology,* **32** (2): 177, New Delhi.

NATH, D.K., 1972 : *Callantra minax* (Enderlein) (Tephrytidae: Diptera), a new record of Ceratitinid fruitfly on orange fruits (*Citrus reticulata* Blanco) in India. *Indian Journal of Entomology,* **34** (3) : 246, New Delhi.

NATH, D.K., 1973 : Insect transmission of sooty mould (*Capnodium sp.*) to orange orchards at Darjeeling district, West Bengal. *Science and Culture,* **39** (6) : 262-263, Calcutta.

NATH, D.K., AND A.C. BASU, 1969 : Observations on the behaviour of orange trunk borer, *Anoplophora versteegi* (Rits.) (Lamiidae : Coleoptera) in Darjeeling hills, West Bengal. *Indian Journal of Agricultural Sciences,* **39** : 713-717, New Delhi.

NAYAR, R.K., T.N. ANANTHAKRISHNAN AND B.V. DAVID, 1976 (Third reprint 1982) : General and Applied Entomology, 589 pp., Tata McGraw-Hill Publishing Co. Ltd., New Delhi.

NECHOLS, J.R. AND T.F. SEIBERT, 1985 : Biological control of the spherical mealy bug, *Nipaecoccus vastator* (Homoptera: Pseudococcidae) assessment by ant exclusion. *Environmental Entomology,* **14** : 45-47, Lanham, Maryland.

NEL, J.J.C., L.DE LANGE AND H. VAN ARK, 1979 : Resistance of citrus red scale, *Aonidiella aurantii* (Mask.) to insecticides. *Journal of Entomological Society of South Africa,* **42** : 275-281, Pretoria.

NEWMANN, L.J., BA.O. CONNOR AND H.G. ANDERWARTHA, 1929 : Wax-scale

(*Ceroplastes ceriferus* Anderson). *Journal of Department of Agriculture for Western Australia,* (2) **6** (4) : 516-526, Perth.

NIJJAR, G.S., 1977 : Citrus. Punjab Agriculture University, 50-62 pp. Ludhiana.

NIJJAR, G.S., 1977 : Fruit Breeding in India, 219 pp. Oxford & IBH Publishing Co., New Delhi.

NOBLE, E.R. AND G.D. NOBLE, 1965 : Parasitology, 724 pp. Lea and Febiger, Philadelphia.

NOHARA, K., 1970 : Study of the biological and chemical control of citrus pests, with special reference to the integrated control of *Unaspis yanonensi* (Kuwana) and *Panonychus citi* (McGregor). *Special Bulletin Yamaguchi Agricultural Research Station,* **23** : 255-280.

NOHARA, K., 1988 : The effect of pesticides on citrus fauna in Japan. p.1113-1119. *In: Procedings of Sixth International Citrus Congress, Tel Aviv, Israel, March 6-11, 1988*, Balaban Publishers, Rehovot, (Israel).

OMOTO, C., T.J. DENNIHY, C.W. MCCOY, S.E. CRANE AND J.W. LONG, 1995 : Management of citrus rust mite (Acari : Eriophydae) resistance to dicofol in Florida citrus. *Journal of Economic Entomology*, **88** (5) : 1120-1128, Lanham, Maryland.

PAGDEN, H.T., 1931 : Two citrus fruit borers. *Malay States Department Science Series*, 7 : 29pp., Kuala Lumpur.

PAL, S.K., 1975 : Notes on the incidence and control of the citrus whitefly, *Dialeurodes citri* Ashmead (Aleurodidae: Homoptera) in Himachal Pradesh. *Indian Journal of Agricultural Research,* **9** (2) : 91-93, Karnal.

PANDEY, N.D., 1958 : Life history of leaf miner. *Kanpur Agricultural College Journal,* **17** (1-2) : 30-32, Kanpur.

PANDEY, N.D. AND Y.D. PANDEY, 1964 : Bionomics of *Phyllocnistis citrella* Stn. (Lepidoptera : Gracillariidae). *Indian Journal of Entomology,* **26** (4) : 417-422, New Delhi.

PANDEY, N.D. AND T.R. SUKHANI, 1965 : *Achoea janata* Linnaeus the citrus fruit sucking moth (Lepidoptera : Noctuidae), *Agra University Journal of Research (Science),* **14** (2) : 23-24, Agra.

PANDEY, N.D., T.R. SUKHANI AND R.L. GUPTA, 1966 : Bionomics of *Achaea janata* Linnaeus (Lepidoptera : Noctuidae). *Labdev Journal of Science and Technology,* **6** (2) : 127-128, Kanpur.

PARANJPYE, H.P., 1937 : The cultivation of oranges and allied fruits in Bombay Presidency. *Bombay Department of Agriculture, Bulletin No. 19,* Pune.

PATCH, EDITH M., 1914 : Maine aphids of the rose family. *Maine Agricultural Experimental Station, Bulletin No. 133* : 253-280, Maine.

PATEL, D.P., 1966 : Indian Field Crops : 203-227, Atul Prakashan.

PATEL, G.A. AND H.K., PATEL, 1952 : Control of termites and pests of fruit trees. *Proceedings of Zoological Society of Bengal,* **5** : 133-140, Calcutta.

PATEL, G.A. AND G.M. TALGERI, 1956 : Fruit crops. p.131-157. *In*: Crop Pests and How to Fight Them. Directorate of Publicity, Government of Maharashtra, Mumbai (Bombay).

PATEL, N.C., V.M. VALAND AND J.R. PATEL, 1994 : Effect of weather factors on activity of citrus leaf miner (*Phyllocnistis citrella* infesting lime

(*Citrus aurantifolia*). *Indian Journal of Agricultural Sience,* **64** (2) : 132-134, New Delhi.

PATEL, V.C., H.K. PATEL AND R.C. PATEL, 1964 : Effect of various insecticides on the bark borer (*Indarbela* sp.) of the road side trees in Gujarat. *Indian Journal of Entomology,* **26** (1): 103-106, New Delhi.

PATIL, A.V., N.P. KAHARE AND B.R. RAUT, 1972 : Screening the genetic stock of different varieties of citrus root-stock against citrus leaf miner and citrus psylla. *Pesticides,* **6** (9) : 48-52, Mumbai (Bombay).

PEAL, H.W., 1903 : Contribution towards a monograph of the oriental Aleurodidae. *Journal of Asiatic Society of Bengal,* **63** : 61-98, Calcutta.

PERRAJU, A., 1960 : Preliminary observations in the trials for chemical control of tobacco and citrus nematode root galls. *Andhra Agricultural Journal,* **7** : 33-38, Bapatla.

PELEG, B.A., 1988 : The effect of a new phenoxy juvenile hormone analog on California red scale and Florida wax scale. p.1141-1146. *In: Proceedings of the Sixth International Citrus Congress, Tel Aviv (Israel), March 6-11, 1988* : Balaban Publishers, Rehovot (Israel).

PERACCHI, A.L., 1971 : Dois aleurodideos pragas de *Citrus* no Brasil (Homoptera Aleyrodidae). *Archos Mus. nac Rio de Jenerio,* **54** : 145-151, Rio de Jenerio.

PHILLIPS, P.A. AND G.P. WALKER, 1997 : Increase in flower and young fruit abscission caused by citrus bud mite (Acari : Eriophyidae) feeding in the axillary buds of lemon. *Journal of Economic Entomology,* **10** (5) : 1273-1282, Lanham, Maryland.

PRADHAN, S., 1960 : Insect Pests of Crops, 208 pp., revised by M.G. Jotwani, 1983, 198 pp., National Book Trust, India, New Delhi.

PRADHAN, S., M.G. JOTWANI AND P. SARUP, 1950 : Comparative toxicity of some important insecticides to the adults of psylla, *Diaphorina citri* (Psyllidae : Homoptera) a pest of citrus. *Indian Journal of Horticulture,* **16** (4) : 252-254, Bangalore.

PRASAD, S.N., 1973 : Bionomics of the citrus blossom midge. *Cecidologia indica,* **8** (1-2) : 64 pp. Allahabad.

PRASAD, V., 1974 : A Catalogue of Mites of India, 320 pp. Indira Publishing House, Ludhiana.

PRASAD, V.G., 1992 : Thrusts in citrus pest management. *Pesticides Information,* **18** (3) : 14-17, New Delhi.

PRASAD, V.G., DHAMO K. BUTANI AND M.M. SHARMA, 1981 : Insect pests of fruits, their incidence and management. *Proceedins All India Workshop on Pest management — Recommendations and future guidelines* : 71-88, Entomological Society of India (Udaipur), New Delhi.

PREM CHAND, 1995 : Agricultural and Forest Pests and Their Management, 374 pp. Oxford & IBH Publishing Co. Pvt. Ltd., New Delhi.

PRIESNER, H. AND M. HOSNY, 1934 : Contribution to a knowledge of the whiteflies (Aleurodidae) of Egypt-III. *Ministry of Agriculture, Egypt, Technical Scientific Service, Bulletin No. 145,* Cairo.

PRINSLOO, G.L., 1982 : Two new South African species of *Timberlakia* Mercet (Hymenoptera : Encytidae), parasitic in mealybug on citrus. *Journal of Entomological Society of South Africa,* **45** : 221-225, Pretoria.

PRUTHI, HEM SINGH, 1925 : The morphology of the male genitalia in Rhynchota. *Transanctions of the Entomological Society of London,* **73** : 127-254, London.

PRUTHI, HEM SINGH, 1944 : Spread of the "fluted scale" *Icerya purchase* Maskell : to the Bombay Presidency. *Indian Journal of Entomology*, **6** (1-2) : 163, New Delhi.

PRUTHI, H.S., 1950 : A foreign insect menace to Indian citrus industry checked. *Indian Farming*, **11** : 5-6, New Delhi.

PRUTHI, HEM SINGH, 1969 : Textbook on Agricultural Entomology, 977 pp. Indian Council of Agricultural Research, New Delhi.

PRUTHI, HEM SINGH AND H.N. BATRA, 1938 : A preliminary annotated list of fruit pests of the N.W.F.P. *Miscellaneous Bulletin Imperial Council of Agricultural Research, India, No. 19,* 22 pp., New Delhi.

PRUTHI, HEM SINGH AND H.N. BATRA, 1960 : A preliminary annotated list of fruit pests of North-west India. *Indian Council of Agricultural Research, Bulletin No. 80,* 113 pp., New Delhi.

PRUTHI, HEM SINGH AND M.S. MANI, 1942 : Distribution, hosts, habits etc. of the Indian Serphoidea and Bethyloidea. *Memoirs, Indian Museum,* **13** : 405-444, Calcutta.

PRUTHI, HEM SINGH AND M.S. MANI, 1945 : Our knowledge of the insect and mite pests of citrus in India and their control. *Imperial Council of Agricultural Research, Monograph No. 16,* 42 pp., New Delhi.

PRUTHI, HEM SINGH AND L.N. NIGAM, 1939 : The bionomics, life-history and control of the grasshopper, *Poecilocerus pictus* (Fab.) a new pest of cultivated crops in North India. *Indian Journal of Agricultural Sciences,* **9** (4) : 626-641, New Delhi.

PURI, S.N. AND K.M. SINGH, 1977 : Persistence of monocrotophos in citrus as influenced by various methods of application. *Indian Journal of Entomology,* **39** (1) : 46-52, New Delhi.

QUAINTANCE, A.L., 1899 : New or little known Aleurodidae. *Canadian Entomologist,* **31** : 1-4, Ontario.

QUAINTANCE, A.L., 1900 : Contribution towards a monograph of the American Aleurodidae. *United States Department of Agriculture, Bureau of Entomology, (Technical series), Bulletin No. 8,* 64 pp., Washington, D.C.

QUAINTANCE, A.L., 1903 : New Oriental Aleurodidae. *Canadian Entromologist,* **35** : 61-64, Ontario.

QUAINTANCE, A.L., 1907 : The more important Aleurodidae infesting economic plants with description of new species infesting the orange. *United States Department of Agriculture, Bureau of Entomology (Technical Series), Bulletin No. 12* (5) : 89-94, Washington, D.C.

QUAINTANCE, A.L., 1909 : A new genus of Aleurodidae, with remarks on *Aleyrodes nubifera* Berger and *Aleurodes citri* Riley and Howard. *United States Department of Agriculture, Bureau of Entomology (Technical series), Bulletin No. 27,* 93 pp., Washington, D.C.

QUAINTANCE, A.L. AND A.C. BAKER, 1913 : Classification of Aleyrodidæ, Part I. *United States Department of Agriculture, Bureau of Entomology (Technical Series), Bulletin No.* **27**, 93 pp., Washington, D.C.

QUAINTANCE, A.L. AND A.C. BAKER, 1914 : Classification of Aleyrodidae, Part II. *United States Department of Agriculture, Bureau of Entomolgy (Technical Series), Bulletin No.* **27**, (2) : 94-109, Washington, D.C.

QUAINTANCE, A.L. AND A.C. BAKER, 1916 : Alheurodidae or whiteflies attacking the orange with description of three new species of economic importance. *Journal of Agricultural Research*, **6** : 459-466, Washington, D.C.

QUAINTANCE, A.L. AND A.C. BAKER, 1917 : A contribution to our knowledge of the whiteflies of the subfamily Aleurodinae (Alyerodidae). *Proceedings of the United States National Mueum*, **51** : 335-445, Washington, D.C.

QUAYLE, HENRY J., 1914 : Citrus fruit insects in Mediterranian countries. *United States Department of Agriculture, Bulletin No. 134* : 7-9 & 20, Washington, D.C.

QUAYLE, HENRY J., 1938 : Insects of Citrus and Other Subtropical Fruits, 583 pp., Comstock Publishing Co., Inc. Ithaca, New York.

RAE, D.J., D.M. WATSON, W.G. LIANG, B.L. TAN, M.LI., M.D. HUANG, Y. DING, J.J. XIONG, D.P. DU, J. TANG AND G.A.C. BEATTIE, 1996 : Comparison of petroleum spray oils, abamectin, cartap, and methomyl for the control of citrus leafminer (Lepidoptera : Gracillariidae) in Southern China *Journal of Economic Entomology*, **89** (2) : 493-500, Lanham, Maryland.

RAHMAN, K.A., 1938 : Citrus psylla and its control. *Punjab Fruit Journal*, **2** (8) : .218-221, Lahore.

RAHMAN, K.A., 1940 a : Insect pests of citrus in the Punjab — *Insect Pest Number, Agricultura! College Magazine, Punjab*, **7** (5-7) : 43 pp., Faislabad (Lyallpur).

RAHMAN, K.A., 1940 b : Important insect pests of fruit trees in the Punjab and their control. *Punjab Fruit Journal*, **4** (1-3) : 715-718, Lahore.

RAHMAN, K.A., 1941 a : Preliminary observations on the use of nicotine sulphate against citrus psylla. *Proceedings of 28th Indian Sciene Congress, Banaras, III* : 192, Calcutta.

RAHMAN, K.A., 1941 b : Biology of the citrus leaf minor. *Proceedings of 28th Indian Science Congress, Banaras, III* : 193, Calcutta.

RAHMAN, KHAN A., 1944 : Insect pests of fruit trees. *Indian Farming*, **5** (10) : 463-466, New Delhi.

RAHMAN, KHAN, A., 1946 : Insect pests of fruit trees. *Punjab Fruit Journal*, **10** (40) : 141-144, Lahore.

RAHMAN, KHAN A. AND A.R. ANSARI, 1945 a : Insect pests of citrus plants in the Punjab and their control. *Punjab Fruit Journal*, **9** (33 & 34) : 118-121, Lahore.

RAHMAN, KHAN A. AND A.R. ANSARI, 1945 b : The citrus red scale in the Punjab, *Aonidiella aurantii* (Mask.), *Punjab Farmer*, **1** (1) : 29-30, Lahore.

RAHMAN, KHAN A. AND M. ABDUL LATIF, 1944 : Description, bionomics and control of the giant mealybug, *Drosicha stebbingi* Green (Hemiptera: Coccidae). *Bulletin of Entomological Research*, **35** : 197-209, London.

RAHMAN, KHAN A. AND A.N. SAPRA, 1940 : Mites of the family *Tetranychidae* from Lyallpur with description of new species. *Proceedings of Indian Academy of Sciences*, **B-11**: 177-179, Bangalore.

RAHMAN, KHAN A. AND A.N. SAPRA, 1940 : Mites of the family *Tetranychidae* from Lyallpur with description of new species. *Proceedings of Indian Academy of Sciences*, **B-11**: 177-179, Bangalore.

RAHMAN, KHAN A. AND G.S. SOHI, 1940 : Important insect pests of fruit trees in the Punjab. *Punjab Fruit Journal*, **4** (13) : 715-718, Lahore.

RAHMAN, KHAN A. AND D.N. TANDON, 1942 : Black citrus fly (*Aleurocanthus woglumi Ash.*) and its control. *Proceeding of 29th Indian Science Congress Baroda, III* : 176, Calcutta.

RAHMAN, KHAN A. AND M. YUNUS, 1945 : The citrus leaf miner. *Indian Farming,* **6** (5) : 21-22, New Delhi.

RAI, B.K., 1964 : Relative toxicity of some pesticides to *Eutetranychus banksi. Indian Oilseeds Journal,* **8** (4) : 304, Hyderabad.

RANJHEN, P.L., 1949 : On the morphology of the immature stages of *Dacus* (*Stameta*) *cucurbitae* Coq. (the melon fruit fly) with notes on its biology, *Indian Journal of Entomology,* **11** (1) : 83-199, New Delhi.

RAM DASS AND USHA SARABHAI, 1979 : Some observations on *Corcyra cephalonica* Stainton and *Tribolium castaneum* Herbst attacking combs of yellow wasp, *Polistes herbraceus* Fabr. *Bulletin of Entomology,* **20** (1-2) : 159-161, New Delhi.

RAI, B.K., H.C. JOSHI, Y.K. RATHOD, S.M. DUTTA AND V.K.R. SHINDE, 1969 : Studies on the bionomics and control of whitegrubs (*Holotricha consanguinea* Bl,) in Lalsot district Jaipur (Rajasthan). *Indian Journal of Entomology,* **31** (2) : .135-142, New Delhi.

RAIZADA, USHA, 1965 : Life-history of *Scirtothrips dorsalis* Hood with detailed external morphology of the immature stages, *Bulletin Entomology,* Loyolla College, **6** : 30 Chennai (Madras).

RAKSHPAL, R., 1945 : Citrus sucking moth and their control. *Indian Farming,* **6** (10) : 441-443, New Delhi.

RAMCHANDRA RAO Y. AND M.C. CHERIAN, 1944 : The fluted scale *Icerya purchasi* Mask. as a pest of wattle in South India and its control by biological method. *Madras Agriculture Journal,* **32** : 20, Coimbatore.

RAMCHANDRACHARI, C. AND V. PADMANABHAN, 1960 a : An assessment of incidence of different fruit moths on sweet oranges in Cuddapah district. *Andhra Agricultural Journal,* **7** (3) : 121-126, Bapatla.

RAMCHANDRACHARI, C. AND V. PADMANABHAN, 1960 b : A method of attracting and capturing fruit sucking moth, *Achoea janata. Andhra Agricultural Journal,* **7** (4) : 197-198, Bapatla.

RAMACHANDRACHARI, C., V. PADMANABHAN AND C. KRISHNAMURTHY, 1960 : Citrus leaf miner, (*Phyllocnistis citrella* Staint.) and its control in Cuddapah district. *Andhra Agricultural Journal,* **8** (5) : 234-238, Bapatla.

RAMACHANDRAN, S., 1951 : Plant Protection work in Madras during 1948-51. *Plant Protection Bulletin,* **3** : 61-68, New Delhi.

RAMACHANDRAN, S., 1952 : A note on the identity of the cerambycid borer of oranges in South India. *Indian Journal of Entomology,* **14** (3) : 214-215, New Delhi.

RAMCHANDRAN, S., 1954 : Insect pests of oranges in Madras State, *Bulletin Madras Agricultural Department,* 16 pp., Chennai (Madras).

RAMAKRISHNAN, N. AND S. KUMAR, 1976 : Biological control of insects by pathogens and nematodes. *Pesticides Annual*: 32-47, Mumbai (Bombay).

RANDHAWA, G.S., 1974 : Horticulture — Importance of pest control. *Pesticides Annual* : 85-87. Mumbai (Bombay).

RANDHAWA, S.S., 1967 : How Japan cultivates citrus? *Indian Horticulture,* **12** (1) : 23-27, New Delhi.

RANGARAO, P.V., P.L. NARASIMHA RAO AND V. TIRUMALA RAO, 1957: Association of pinkish brown blotch of citrus fruits with an eriophid mite (*Phyllocoptes olerorius* Ashmead). *Indian Journal of Entomology*, **19** (2) : 146-147, New Delhi.

RANGASWAMY, G. AND K. RAMAMOORTHY, 1964 : Preliminary studies on the pathogenicity of *Bacillus thuringiensis* on some crop pests. *Current Science*, **33** : 222-223, Bangalore.

RAO, D. SESHAGIRI, 1972 : A Handbook of Plant Protection, 841 pp. S.V. Rangaswamy & Co. Pvt. Ltd., Bangalore.

RAO, K.P. AND CH. B.G. TILAK, 1993 : Effect of botanical pesticides in the control of citrus butterfly, *Papilio demoleus* L. *Neem Newsletter*, **10** (1-2) : 5, New Delhi.

RAO, V.P., 1943 : *Icerya purchasi* Maskell in South India. *Indian Journal of Entomology*, **5** (1-2) : 246-247, New Delhi.

RAO, V.P., 1950 : Iceryine scale insects recorded from the Orient, *Indian Journal of Entomology*, **12** (2) : 132, New Delhi.

RASMY, A.H., 1977 : Biology of citrus brown mite, *Eutetranychus orientalis* (Klein) as affected by some citrus species. *Acarologia*, **19** (2) : 222-224.

RATAUL, HARBANS SINGH, 1957 : A note on endrin against citrus psylla (*Diphornia citri* Kuw.) (Hemiptera : Psyllidae). *Indian Journal of Horticulture*, **14** (3) : 185, Bangalore.

REDDY, D. BAP, 1962 : Your Plant Protection Guide, 80 pp. Directorate of Extension, Government of India, New Delhi.

REDDY, D. BAP., 1968 : Plant Protection in India, 454 pp. Allied Publishers, Calcutta.

REINKING, A.O., 1921 : Notes on Coccids and Aleurodides on various hosts in Indo-China and Siam. *Philippines Agriculturist*, **9** (6-7) : 185, Los Banos, Manila.

RILEY, C.V. AND L.O. HOWARD, 1892 : The orange-leaf Aleurodes. *Insect Life*, **4** : 274, Washington, D.C.

RILEY, C.V. AND L.O. HOWARD, 1893 : The orange Aleurodes (*Aleurodes citri* n. sp.) *Insect Life*, **5** : 219-226. Washington, D.C.

RICHARDS, O.W. AND R.G. DAVIES, 1977 : Imms' General Textbook of Entomology, 1354 pp. Chapman and Hall, London.

ROSE, M., 1988 : Twenty years of biological control success on *Citrus* in the United States. p. 1153-1161. *In: Proceedings of the Sixth International Citrus Congress, Tel Aviv, Israel, March 6-11, 1988*, Balaban Publishers, Rehovot (Israel).

ROSE, M. AND P. DE BACH, 1981 : Citrus Whitefly parasites established in California. *California Agriculture*, **35** : 21-23, Berkeley, California.

ROSE, M., P. DE BEACH AND J. WOOLLEY, 1981 : Potential new citrus pest : Japanese bayberry whitefly. *California Agriculture*, **35** : 23-24, Berkeley, California.

ROSEN, D., 1967 : The hymenopterous parasites and hyperparasites of soft scales on citrus in Israel. *Beiträge zur Entomologic*, **17** : 255-283, Berlin.

RU, N. AND R.I. SALLER, 1979 : Colonization of citrus whitefly parasite *Prospaltella lahorensis* in Gainesville, Florida. *Florida Entomologist*, **62** : 59-65, Gainesville, Florida.

RUSSELL, LOUISE M., 1962 : Citrus black-fly. *FAO Plant Protection Bulletin*, **10**

: 36-38, Rome.

RUSSELL, LOUISE M., 1963 : Hosts and distribution of five species of *Trialeurodes* (Homoptera : Aleyrodidae). *Annals of Entomological Society of America, College Park,* **56** : 149-153, Madison.

SADANA, G.L. AND R. JOSHI, 1976 : Host range of the mite, *Brevipalpus californicus* (Banks) infesting citrus. *Science and Culture,* **42** (6) : 106-107, Calcutta.

SADANA, G.L. AND V. KANTA, 1971 : Predators of the citrus mite, *Eutetranychus orientalis* (Klein) in India. *Science and Culture,* **37** (11) : 530, Calcutta.

SADANA, G.L. AND V. KANTA, 1972 a : Susceptibility of different varieties of citrus to citrus mite, *Eutetranychus orientalis* (Klein). *Science and Culture,* **38** (12) : 525-526, Calcutta.

SADANA, G.L. AND V. KANTA, 1972 b : Studies on the range of citrus mite, *Eutetranychus orientalis* (Klein). *Science and Culture,* **38** (12) : 528-530, Calcutta.

SAINI, B.S., G. SINGH AND O.S. BINDRA, 1968 : Studies on the host range of Red spider mite, *Tetranychus telarius L.* (Tetranychidae : Acarina).p. 619. *In: Proceedings of 56th Indian Science Congress, Bombay, III,* Calcutta.

SAINI, B.S., G. SINGH AND O.S. BINDRA, 1971 : Laboratory studies on the host range of red spider mite, *Tetranychus telarius* Linnaeus (Tetranychidae : Acarina). *Journal of Research, PAU.,* **7** (4) : 487-490, Ludhiana.

SAINI, M.L. AND P.D. SHARMA, 1970 : The relative efficacy of insecticides against lemon butterfly. *Journal of Research, PAU,* **7** (2) : 203-207, Ludhiana.

SAJJAN, S.S., 1966 : Punjab survey on vegetable and fruit pests. *Indian Horticulture,* **10** (3) : 33, New Delhi.

SALEM, S.A., AND F.N. ZAKI, 1985 : Economic threshold level of the citrus wax scale *Ceroplastes floridensis* Comstock on citrus Egypt. *Bull. Soc. ent. Egypt,* **65** : 333-344, Cairo.

SAMPSON, W.W. AND E.A. DREWS, 1941 : Fauna Maxicana IV. A review of the Aleyrodidae of Mexico (Insecta : Homoptera). *Anales de le Esula nacienal de Ciencias biologicas,* **2** : 143-198, Mexico.

SAMPSON, W.W. AND E.A. DREWS, 1956 : Keys to the genera of Aleyrodidae and notes on certain genera (Homoptera : Aleyrodidae). *Annals and Magazine of Natural History,* **9** (12) : 689-697, London.

SAMUEL, C.K., 1940 : Plant hosts of two important aphids, *Myzus persicae* (Sulzer) and *Aphis maidis* (Fitch) at Pusa. *Indian Journal of Entomology,* **2** (2) : 244-245, New Delhi.

SAN JHAN, J.M., 1934 : *Prays citri* Mill. a rind pest of Philippines oranges *Philippines Agriculturist,* **12** : 339-348, Los Banos, Manila.

SANDU, G.S., G.C. VERMA AND O.S. BINDRA, 1968 : Relative efficacy of some systemic insecticides against the adults of citrus psylla, *Diaphorina citri* Kuw. (Hemiptera : Psyllidae). *Indian Journal of Horticulture,* **25** (3 & 4) : 179-180, Bangalore.

SAPRA, A.N., 1940 : Notes on the mite, *Tenuipalpus* sp. *Indian Journal of Entomology,* **2** : 244, New Delhi.

SARDAR SINGH AND G.S. SOHI, 1957 : A note on insect pests of economic importance in the Punjab. *Government Agricultural College Magazine*, **4** (3) : 19-34, Ludhiana.

SARDAR SINGH, G.S. SONI AND S.S. BAINS, 1958 : Control measures against some insect pests of economic importance in the Punjab. *Government Agricultural College Magazine*, **5** (2) : 12-30, Ludhiana.

SATHIYANANDAM, V.K.R., GOWDER, R. BETTAI, I. IRULAPPAN AND A.S. MATHUR, 1971 : A note on the occcurrence of Cerambycid root borer on citrus. *South Indian Horticultural*, **19** (1-4) : 93-95, Coimbatore.

SATPATHI, C.R. T.K. BHUSAN, R.S. SINGH AND A. PRAMANIK, 1994 : Relationship between larval length of citrus leaf miner, *Phyllocnistis citrella* Stainton (Phyllocnistidae: Lepidoptera) and the leaf miner. *Environment and Ecology*, **12** (1) : 216-217, Kalyani, West Bengal.

SAXENA, D.K. AND R.R. RAWAT, 1968 : Chemical control of *Drosicha mangiferae* (Green), a serious pest of *Citrus* sp. in Madhya Pradesh. *Indian Journal of Entomology*, **30** (1) : 66-68, New Delhi.

SAXENA, P.K. AND H.K. CHHABRA, 1972 : Control of nematodes attacking fruit-trees. *Agriculture and Agro-Industries Journal*, **5** (11) : 33-38, Mumbai (Bombay).

SAXENA, R.C., 1987 : Thrips of economic importance and their management. p. 141-168. *In* : *Recent Advances in Entomology*, Gopal Prakashan, Kanpur.

SCHAFFER, B., J.E. PENA, A.M. CULLS AND A. HUNSBERGER, 1997: Citrus leafminer (Lepideptera : Gracillariidae) in lime ; Assessment of leaf damage and effects on photosynthesis. *Crop Protection*, **16** (4) : 337-343. Elsevier Science Ltd., Great Britain.

SCHROEDER, W.J., 1988 : Entomogenous nematodes for root weevil control in citurs. p.1223-1226. *In: Proceedings of the Sixth International Citrus Congress Tel Aviv., Israel, March 6-11, 1988*, Balaban publishers, Rehovot (Israel).

SEKHAR, P.S. AND SUSEELA SEKHAR, 1964 : Investigations on thrips occurring on arabica coffea, *Coffea arabica* Linn. - 1. Bionomics. *Indian Coffee*, **28** (8) : 173-179, Bangalore.

SEN, A.C. AND D. PRASAD, 1956 : Biology and control of mango mealybug (*Drosicha mangiferae* Green). *Indian Journal of Entomology*, **18** (2) : 127-140, New Delhi.

SEN, P.C., 1921 : The large brown cricket, *Brachytrypes portentosus* Licht. *Bengal Agricultural Journal*, **1** (4) : 111-112, Dacca.

SENGUPTA, GOKUL CHANDRA AND BASANTA KUMAR BEHURA, 1957 : Annotated list of crop pests in the state of Orissa. *Memoirs Entomological Society of India No. 5*, 44 pp., New Delhi.

SETHI, S.L., 1965 : Insecticides for the control of citrus caterpillar, *Papilio demoleus* Linn. *Journal of Research*, PAU, **2** (3) : 205-207, Ludhiana.

SETHI, S.L., 1967 a : Control of Texas citrus mite, *Eutetranychus banksii* in India with new pesticides. *Journal of Economic Entomology*, **60** : 180-181, Madison.

SETHI, S.L., 1967 b : Insecticides for the control of citrus psylla in India. *Journal of Economic Entomology*, **60** : 270-271, Madison.

SETHI, S.L. AND J.S. JAWANDA, 1964 a : Control of citrus pests of arid irrigated regions of the Punjab. *Punjab Horticulture Journal*, **4** (2) : 67-

69, Patiala.

SETHI, S.L. AND J.S. JAWANDA, 1964 b : Watch out for the new pest on citrus. *Progressive Farming, PAU,* **1** (2) : 23-24, Ludhiana.

SETHI, S.L. AND J.S. JAWANDA, 1965 : Aphids relish citrus too. *Progressive Farming,* **1** (5) : 7, Ludhiana.

SETHI, S.L., J.S. JAWANDA AND A.S. ATWAL, 1964 : Insect pests of citrus in the Punjab. VII-Biology and control of citrus mite, *Paratetranchus citri* (Acarina: Tetranychus). *Punjab Horticulture Journal,* **4** (3-4) 142-144, Patiala.

SHAHA, R., 1946 : An effective and inexpensive method for the control of stem borers in fruit trees with special reference to Santra trees in C.P. and Behar. *Current Science,* **15** : 135, Bangalore.

SHAPIRO, J.P. AND R.R. GOTTWALD, 1995 : Resistance of eight cultivar of citrus root stock to larval root weevil, *Diaprepes abbreviatus* L. *Journal of Economic Entomology,* **88** (1) : 148-154, Lanham, Maryland.

SHARGA, U.S., 1948 : Biological control of some hemipterous insect pests of crops in India. *Current Science,* **17** : 302-303, Bangalore.

SHARMA, D.R. AND R.C. BATRA, 1987 a : Seasonal history and chemical control of citrus leaf-roller in Punjab. *Punjab Horticulture Journal,* **27** (1 & 2) : 76-79, Patiala.

SHARMA, D.R. AND R.C. BATRA, 1987 b : Preliminary studies on the development of life tables of citrus leaf roller, *Psorosticha zizyphi* Stainton. *Research and Development Report,* **4** (2) : 145-149, Ludhiana.

SHARMA, D.R. AND R.C. BATRA, 1988 a : Seasonal variations in the duration of development stages of citrus leaf-roller. *Journal of Insect Science,* **1** (3) : 96-98, Ludhiana.

SHARMA, D.R. AND R.C. BATRA, 1988 b : Effect of temperature on survival of different developmental stages of citrus leaf-roller. *Indian Journal of Ecology,* **15** (2) : 196-198, Ludhiana.

SHARMA, D.R. AND R.C. BATRA, 1989 a : Parasitoids of citrus leaf-roller and influence of abiotic factors on the parasitisation. *Journal of Insect Science,* **2** (2) : 162-164, Ludhiana.

SHARMA, R.D. AND R.C. BATRA, 1989 b : Influence of abiotic and biotic factors on population of citurs leaf-roller. *Journal of Research, Punjab Agricultural University,* **20** (2) : 222-226, Ludhiana.

SHARMA, R.D. AND R.C. BATRA, 1992 : Role of abiotic factors on fecundity, egg viability and preference of egg laying of citrus leaf-roller. *Indian Journal of Ecology,* **19** (2) : 215-218, Ludhiana.

SHARMA, H.N. AND B.N. SINGH, 1940 : Insect pests of fruit trees in Kumaon and their control. *U.P. Department of Agriculture, Bulletin No. 21,* Allahabad.

SHARMA, P.L. AND O.P. BHALLA, 1964 : A survey study of insect pests of economic importance in Himachal Pradesh. *Indian Journal of Entomology,* **26** : 318-331, New Delhi.

SHARMA, R.P. AND O.S. SRIVASTAVA, 1970 : Studies on the chemical control of *Papilio demoleus* Linn. on *Citrus reticulata*. *Madras Agricultural Journal,* **57** (5) : 300-302, Coimbatore.

SHARNAGAT, B.K., D.B. KUCHANWAR AND M.G. HARDAS, 1981 : Control of citrus aleyrodid puape with newer insecticides, *Nagpur Agricultural College Magazine,* **54-55** : 57-62, Nagpur.

SHIRAKI, T., 1934 : Insect pests of citrus trees in Formosa. I & II. *Journal of Society of Tropical Agriculture,* **6** : 20-36 and 187-194, Taiwan.

SIDDIQI, Z.A., 1952: Insect pests of fruit trees in Kumaon and their control. *Agriculture and Animal Husbandry,* U.P., **3** : 24-27, Lucknow.

SIDHU, D.S. AND R.C. BATRA, 1992 : Influence of abiotic factors on the population of citrus whitefly. *Punjab Horticultural Journal,* **32** (1-4) : 34-41, Patiala.

SILVESTRI, F., 1927 a : Contribuzione alla conoscenza degli Aleurodidae (Insecta : Hemiptera) viventi su Citrus in Estremo Oriente e dei loro parassiti. *Bollottino del laboratoria de Zologia generale, agraria della R. Scuola superiore d'agricoltura,* **21** : 1-160, Protici.

SILVESTRI, F., 1927 b : Sobre la mosca prieta (on *Aleurocanthus woglumi).* *Bollettino Formento,* **6** : 16-18, Sanjose, Costa Rica.

SINGH, D., 1969 : Citrus industry in India today, *Indian Horticulture,* **13** (2) : 16-22, New Delhi.

SINGH, D.S., P. SIRCAR AND S. DHINGRA, 1995 : Relative efficacy and persistence of residual toxicity of important systemic insecticides against citrus psylla, *Diaphorina citri. Indian Journal of Entomology,* **57** (1) : 26-32, New Delhi.

SINGH, HARCHARAN AND GURMEL S. SANDHU, 1974 : *Bibliography of Agricultural Zoology-Entomology Work Conducted in the Punjab,* 114 pp., Punjab Agricultural University, Ludhiana.

SINGH, H.H., 1978 a : Studies on neuro secretion in the lemon butterfly, *Papilio demoleus,* L. (Lepidoptera). III-Cyclic activity in median neuro secretory cells and intracerebral axonal pathways during post embryonic development. *Proceedings of 65th Indian Science Congress III, Abstract No. 2220* : 228, Calcutta.

SINGH, H.H., 1978 b : Histomorphological studies on the corpora cardiaca-allata complex of the adult lemon butterfly, *Papilio demoleus* L. (Lepidoptera). *Proceedings of 65th Indian Science Congress, III, Abstract No. 221* : 289, Calcutta.

SINGH, H.P., 1996 : Citrus. p. 433-444. *In* : 50 Years of Crop Science Research in India. (Eds.) R.S. Paroda and K.L. Chadha : Indian Council of Agricultural Research, New Delhi.

SINGH, KARAM (LAMBA), 1931 : A contribution towards our knowledge of Aleyrodidae (whiteflies) of India. *Memoirs, Department of Agriculture, India, Entomological Series, 12,* 98 pp. Calcutta.

SINGH, L., B. SINGH, A.A. KHAN AND A. RAM, 1940 : Citrus fruits, *Punjab Fruit Journal,* **4** : 648-661, Lahore.

SINGH, RANJIT, 1969 : Fruits, 213 pp. National Book Trust Ltd., New Delhi.

SINGH, S., 1961 : How to save fruit trees from insects? *Kheti Bari,* **12** (7) : 26-29, New Delhi.

SINGH, S. AND B.S. SAINI, 1956 : New acaricides for the control of vegetable mites. *Indian Journal of Horticulture,* **13** : 30-25, Bangalore.

SINGH, SARDAR AND G.S. SOHI, 1957 : A list of insect pests of economic importance in the Punjab. *Government Agricultural College Magazine,* **4** (3) : 19-34, Ludhiana.

SINGH, SARDAR, G.S. SOHI AND S.S. BAINS, 1958 : Control measures against some insect pests of economic importance in the Punjab. *Government Agricultural College Magazine,* **5** (2) : 12-30, Ludhiana.

SINGH SHAM, S. KRISHNAMURTHI AND S.L. KATYAL, 1967 : Fruit Culture in India, 451 pp. Indian Council of Agricultural Research, New Delhi.

SINGH, S.P., 1980 : Biological control of insect pests of citrus. p.135-137 *In: Proceedings of the Third All India Workshop on Biological Control of Crop Pests and Weeds, Punjab Agricultural University, Ludhiana, October 27-30,* 1980, Ludhiana.

SINGH, S.P., 1991 a : Potential of two commercial formulations of *Bacillus thuringiensis* against citrus butterfly caterpillar, *Papilio demoleus* Linnaeus (Lepidoptera : Papilioinidae). *Indian Journal of Horticulture,* **48** (2) : 135-138, New Delhi.

SINGH, S.P., 1991 b : Natural enemies of *Papilio* spp. *Indian Journal of Horticulture,* **48** (3) : 237-242, New Delhi.

SINGH, S.P., 1993 : Species composition and diapause in citrus butterflies. *Journal of Insect Science,* **6** (1) : 48-52, Ludhiana.

SINGH, S.P., B.S. BHUMANNAVAR AND N.S. RAO, 1982 : Evaluation of different pesticides against red spider mite *Tetranychus fijiensis* Hirst on Italian lemon. *Pestology,* **6** (6) : 15-16, Mumbai (Bombay).

SINGH, S.P. AND N.S. RAO, 1976 : Evaluation of some insecticides against the black citrus aphid, *Toxoptera aurantii* (B. de F.). *Pesticides,* **10** (11) : 21-22, Mumbai (Bombay).

SINGH, S.P. AND N.S. RAO, 1977 a : Effectiveness of different contact insecticides against soft green scale, *Coccus viridis* Green (Coccidae : Homoptera) on citrus. *Pesticides,* **11** (3) : 33-36, Mumbai (Bombay).

SINGH, S.P. AND N.S. RAO, 1977 b : Effectiveness of Agricultural Spray Oil (E-9267) against *Coccus viridis* (Green) and its phytotoxic effect on Coorg mandarin leaves. *Indian Journal of Plant Protection,* **5** (2) : 144-147, Hyderabad.

SINGH, S.P. AND N.S. RAO, 1977 c : Chemical control of citrus leaf-miner, *Phyllocnistis citrella* Stainton with various systematic insecticides. *International Symposium on Citriculture, Bangalore, Abstracts* : 33-34 and *Proceedings International Citrus Symposium* : 352-355, Horticultural Society of India, Bangalore.

SINGH, S.P. AND S.N. RAO, 1978 a : Relative susceptibility of different species/varieties of citrus to leaf-miner, *Phyllocnistis citrella* Stainton. *Abstract No. 130, International Horticultural Congress, Sydney, Australia and Proceedings International Society of Citriculture* : 174-177.

SINGH, S.P. AND S.N. RAO, 1978 b : Comparative efficacy and relative residual toxicity of some insecticides to the citrus butterfly caterpillar, *Papilio demoleus* Linnaeus (Papilionidae : Lepidoptera). *Entomon,* **3** (1) : 51-56, Thiruvananthapuram (Trivandrum).

SINGH, S.P. AND S.N. RAO, 1978 c : Occurrence of *Aphis citricola* van der Goot in India. *Science and Culture,* **44** (7): 330-331, Calcutta.

SINGH, S.P. AND S.N. RAO, 1978 d : Further studies on the control of green scale, *Coccus viridis* (Green) (Coccidae: Homoptera). *Pesticides,* **12** (7) : 40-41, Mumbai (Bombay).

SINGH, S.P. AND S.N. RAO, 1978 e : Chemical control of soft green scale, *Coccus viridis* (Green) (Coccidae : Homoptera) on Coorg mandarin. *South India Horticulture,* **26** (1) : 28-30, Coimbatore.

SINGH, S.P. AND S.N. RAO, 1979 a : Preliminary studies on the control of *Coccus viridis* with soil application of certain granulated systemic insecticides on potted Coorg mandarin. *Indian Journal of Plant Protection,*

7 (1) : 11-14, Hyderabad.

SINGH. S.P. AND S.N. RAO, 1979 b : Field evaluation of fish oil insecticidal resin soap against soft green scale, *Coccus viridis* (Green) (Coccidae : Homoptera) on citrus. *Indian Journal of Plant Protection,* **7** (2) : 208-210, Hyderabad.

SINGH, S.P., S.N. RAO AND B.S. BHUMANNAVAR, 1983 : Chemical control of oriental red mite, *Eutetranychus orientalis* (Klein) on Coorg mandarin. *Pesticides,* **17** (12) : 27-29, and *Pesticides,* **20** (8) : 51-53, Mumbai (Bombay).

SINGH, S.P., S.N. RAO AND K.K. KUMAR, 1977 : Evaluation of various acaricides against *Eutetranychus* sp. on Coorg mandarin. *International Symposium on Citriculture, Bangalore, Abstracts* : 36 and *Proceedings International Citrus Symposium* : 370-372, Horticultural Society of India, Bangalore.

SINGH, S.P., S.N. RAO AND K.K. KUMAR, 1978 : Field evaluation of some insecticides for controlling soft brown scale, *Saissetia coffeae* (Walker) (Coccidae : Homoptera) on Coorg mandarin. *Pestology,* **2** (2) : 22-24, Mumbai, Bombay.

SINGH, S.P., S.N. RAO AND K.K. KUMAR, 1980 a : Evaluation of certain insecticides against soft brown scale, *Saissetia coffeae* (Walker) (Coccidae : Homoptera) on Coorg mandarin. *Pesticides,* **14** (8) : 35-36, Mumbai (Bombay).

SINGH, S.P., S.N. RAO AND K.K. KUMAR, 1980 b : Studies on the combination of insecticides for the control of soft green scale, *Coccus viridis* (Green) (Coccidae : Homoptera) on Coorg mandarin. *Pesticides*, **14** (10) : 12-15, Mumbai (Bombay).

SINGH, S.P., S.N. RAO AND K.K. KUMAR, 1983 : Studies on the borer pests of citrus, *Indian Journal of Entomology,* **45** (3): 286-294, New Delhi.

SINGH, S.P., S.N. RAO, K.K. KUMAR AND B.S. BHUMANNAVAR, 1983: Field evaluation of citrus germplasm for the incidence of the oriental red mite, *Eutetranychus orientalis* (Klein). *Indian Journal of Plant Protection,* **11** (2) : 140-142, Hyderabad.

SINGH S.P., S.N. RAO, K.K. KUMAR AND B.S. BHUMANNAVAR, 1988: Field evaluation of citrus germplasm against the citrus leaf-miner, *Phyllocnistis citrella* Stainton. *Indian Journal of Entomology,* **50** (1) : 69-75, New Delhi.

SINGH, UMRAO, A.M. WADHWANI AND B.M. JOHRI, 1983 : Dictionary of Economic Plants of India, 228 pp., Indian Council of Agricultural Research, New Delhi.

SINHA, M.K., R.C. BATRA AND D.K. UPPAL, 1972 : Role of citrus leaf miner (*Phyllocnistis citrella* Stainton), on the prevalence and severity of citrus canker [*Xanthomonas citri* (Hasse) Dowson]. *Madras Agriculture Journal,* **59** : 240-245, Coimbatore.

SMITH, H.S., 1948 : Biological control of insect pests. In : The Citrus Industry, **2**, Chapter 12, University of California Press, Berkeley.

SMITH, K.M., 1931 : A Textbook of Agricultural Entomology, 41-42, Cambridge.

SOHI, G.S. AND G.C. VERMA, 1969 : Studies on the soil application of systemic insecticides for the control of leaf miner. *Indian Journal of Entomology.* **31** (1) : 59-63, New Delhi.

SOHI, G.S. AND M.S. SANDHU, 1968 : Relationship between citrus leaf miner (*Phyllocnistis citrella* Stainton) injury and citrus canker (*Xanthomonas citri* (Hasse) Dowson (— incidence on citrus leaves). *Journal of Research, Punjab Agricuttural University,* **5** (1) : 66-69, Ludhiana.

SOHI, G.S. AND G.C. VERMA, 1965 : Feeding habits of *Phyllocnistis citrella* Stainton in relation to the anatomical structure of the leaf. *Indian Journal of Entomology,* **27** (4): 483-485, New Delhi.

SONTAKAY, K.R., 1943 : Lemon butterfly and its control. *Indian Farming,* **4** : 456-457, New Delhi.

SONTAKAY, K.R., 1945 : The bark borer of orange. *Indian Farming,* **6** : 74-75, New Delhi.

SONTAKEY, K.R. AND R.L. GUPTA, 1945 a : Biology and control of *Indarbela quadrinotata* a pest of orange. p. 106. *In: Proceedings of 32nd Indian Science Congress Nagpur, III,* Calcutta.

SONTAKAY, K.R. AND R.L. GUPTA, 1945 b : Seasonal incidence of *Phyllocnistis* citrella sta., the citrus leaf miner. *Indian Farming,* **1** : 106, New Delhi.

SRIVASTAVA, A.S., 1957 : Entomological investigations on citrus. *Agriculture Research in Uttar Pradesh,* **2** : 126, Allahabad.

SRIVASTAVA, A.S., 1961 : Research Memoir — A Review. Suprintendent, Printing and Stationary, Lucknow.

SRIVASTAVA, A.S., 1964 : Pests of fruit trees. Entomological research during last ten years. Research Memoir, Government of Uttar Pradesh, **3** : 66-67, Lucknow.

SRIVASTAVA, A.S. AND M.S. SIDDIQUI, 1961 : A note on control of citrus insect, pests, *Proceedings of the National Academy of Sciences, India : 13th Annual session (Abstracts)* : 31-34, Allahabad.

SRIVASTAVA, K.C., 1968 : Citrus diseases and pests. *Pesticides,* **2** (7) : 32, Mumbai (Bombay).

SRIVASTAVA, K.C. AND C.K. NANJAPPA, 1969 : Means of check citrus decline in the South. *Indian Horticulture,* **13** (2) : 33-35, New Delhi.

SRIVASTAVA, K.P., 1957 : Lemon butterfly of citrus. *Proceedings of the National Academy of Sciences, India,* **B-27** (3) : 113-128, Allahabad.

SRIVASTAVA, K.P., 1961 : Studies on lemon butterfly, *Papilio demoleus* L. (Lepidoptera), *Indian Journal of Entomology,* **23** (3) : 202-213, New Delhi.

SRIVASTAVA, O.S., 1972 : Chemical control of bark eating caterpillar, *Indarbela quadrinotata* Walker in guava trees. *Indian Journal of Agricultural Sciences,* **42** (9) : 847-848, New Delhi.

SRIVASTAVA, P.D., 1955 : Studies on the feeding habits and certain aspects of the digestive physiology of different stages of *Papilio demoleus* Linn. *Proceedings of the National Academy of Sciences,* **25-B** (3-4) : 53-62, Allahabad.

STAINTON, H.T., 1856 : Description of three species of Indian micro-lepidoptera. *Transactions of Entomological Society, London (New series),* **3** : 301-304, London.

STEARNS CHARLES, R. JR. J.T. GRIEFITHS, W.L. THOMPSON AND E.J. DESZYCK, 1951 : Progress report on concentrated sprays on citrus in Florida. *Proceedings of the Florida State Horticultural Society,* **64** : 64-66, Florida.

SUBRAMANYAM, V.K., 1949 : Attempts at the introduction of *Cryptochaetum iceryae* Will. into India. *Indian Journal of Entomology,* **11** (1) : 61-70,

New Delhi.

SUBRAMANYAM, V.K., 1950 : *Homalotylus flaminius* (Dalman) — A parasite of *Rodolia* grubs predatory on the fluted scale, *Icerya purchasi* Maskell. *Indian Journal of Entomology*, **12** (1) : 103-106, New Delhi.

SUBRAMANIAM, V.K., 1954 : Control of the fluted scale in peninsular India, *Indian Journal of Entomology*, **16** (4) : 391-415, New Delhi.

SUBRAMANIAM, V.K., 1955 : Control of the fluted scale in Peninsular India-IV. Account of control work in the different states. *Indian Journal of Entomology*, **17** (1) : 103-120, New Delhi.

SULZER, J.H., 1776 : Abgekurzte Geschichte der insekten nach dem Linnaeischen system, Erdter Theil, 1776 : xxviii + 274 pp.

SUSAINATHAN, B., 1923 : Fruit moth. *Report of the Proceedings of 5th Entomological Meeting, Pusa (Bihar), February 1922* : 23-27, Calcutta.

SUSAINATHAN, B., 1924 : The fruit moths problem in Northern Circara. *Agriculture Journal of India*, **19** : 402-404, Pusa (Bihar).

SWIRSKI, E., D. BLUMBERG, M. WYSOKI AND Y. ISHAR, 1987 : Biological control of the Japanese bayberry whitefly, *Parabemisia myricae* (Kuwana) (Homoptera: Aleyrodidae) in Israel. *Israel Journal of Entomology*, **21** : 11-18, Bet Dagon, Israel.

SWIRSKI, E., D. BLUMBERG, M. WYSOKI AND Y. ISHAR, 1988 : Phenology and biological control of the Japanese bayberry whitefly, *Parabemisia myrieae* on citrus in Israel. p.1165-1168. *In: Proceedings of the Sixth International Citrus Congress, Tel Aviv, Israel, March 6-11, 1988*, Balaban Publishers, Rehovot (Israel).

SWIRSKI, E., Y. ISHAR, M. WYSOKI AND D. BLUMBERG, 1986 : Overwintering of the Japanese bayberry whitefly, *Parabemisia myricae* in Israel. *Phytoparasitica*, **14** : 281-286, Rehovot, Israel.

TAKAHASHI, R., 1932 : Aleyrodidae of Formosa, part I. *Report Department of Agriculture, Government Research Institute, Formosa*, **59**, 57 pp. Taioku, Tokyo.

TAKAHASHI, R., 1940 : Some foreign Aleyrodidae (Hemiptera): II — a few species attacking citrus in Australia (*Aleyroplatus citri sp.* n., on lemon in North South Wales). *Transactions of Natural History Society Formosa*, **30** : 381-382, Formosa, Taiwan.

TAKAHASHI, R., 1942 : Some foreign Aleurodidae (Homoptera)-V. Species from Thailand and Indo-China. *Transactions of Natural History Society, Formosa*, **32** : 168-175, Formosa, Taiwan.

TALHOUK, ABDUL MONIM S., 1969 : Insects and mites injurious to crops in middle eastern countries. *Mono-graphien zur angew Entomologie* No. **21**, 239 pp. Verlag Paul Parey, Berlin.

TANAKA, T., 1937 a : Citrus fruits of India. *Allahabad Farmer*, **11** : 2, Allahabad.

TANAKA, T., 1937 b : Further revision of Rutaceae — Aurantioideae of India and Ceylon. *Journal of Indian Botanical Society*, **16** : 227-240.

TANDON, D.N., 1941 : Citrus psylla situation and what citrus growers should do. *Punjab Fruit Journal*, **5** : 896-897, Lahore.

TANDON, P.L., 1986 : Intra-tree spatial distribution of eggs and egg colonies of *Aleurocanthus woglumi*, Ashby (Homoptrea: Aleurodidae)on lime. *Entomon*, **13** (1) : 1-8, Thiruvananthapuram (Trivandrum).

TANDON, P.L., 1988 : Insect pests of tropical fruit trees and their management. p. 278-293. *In* : Tree Protection (Eds. Y.K. Gupta and N.K. Sharma). Society of Tree Scientists, Solan.

TANDON, P.L., 1993 : Insect and mite pests of tropical fruits. p.1527-1555. *In* : *Advances in Horticulture*, Vol. **3** (Ed. K.L. Chadha and O.P. Pareek). Malhotra Publishing House, New Delhi-110064.

TANDON, P.L., 1993 : Problems and prospects of insect pest management in fruit crops. p.373-416. *In* : Recent Trends in Pest Management (Eds., G.S. Dhaliwal and Ramesh Arora) : Commonwealth Publishers, New Delhi.

TANDON, P.L. BECHE LAL AND R.P. SRIVASTAVA, 1978 : New records of additional hosts of mango mealy bug, *Drosicha. magniferae* Green (Margaroidae : Homoptera). *Indian Journal of Horticulture,* **35** (3) : 281-282, Bangalore.

TANDON, P.L. AND G.K. VEERESH, 1986 a : Appropriate transformation for the population counts of citrus green scale, *Coccus viridis* (Green) Coccidae : Homoptera). *Insect Science Application,* **8** (2) : 255-257, Nairobi (Kenya).

TANDON, P.L. AND G.K. VEERESH, 1986 b : Inter-tree spatial distribution of citrus green scale, *Coccus viridis* (Green) (Coccidae : Homoptera). *International Journal of Tropical Agriculture,* **6** (3-4) : 270-275, Hisar.

TANIGOSHI, L.K. AND H.J. GRIEFFITHS, 1982 : A new look at biological control of citrus thrips. *Citrograph,* **67** : 157-158, Los Angeles, California.

TARGE, A. AND L. DEPORTES, 1953 : L'aleurode des agrumes *Dialeurodes citri* Ash. dans les Alpes Maritimes. Premiers resultats d'experimentations de traitements. *Phytoma,* **6** (44) : 9-15, Paris.

TAGORE, A.L. DEPORTES AND R. JOUBERT, 1954 : (*Dialeurodes citri*) et les traitemens des argumes dens les Alpes Maritimes. *Phytoma,* **7** (59) : 28-32, Paris.

TEOTIA, T.P.S. AND S. CHAUDHRI, 1966 : Some observations on life-history of *Euproctis fraterna* Moore (Lepidoptera : Lymantriidae) on castor. *Labdev Journal of Science and Technology,* **4** (1) : 45-47, Kanpur.

THAMMI RAJU, N.B., K. RAMASUBBA REDDY, S. VENKATESWARA RAO, K. LAKSHMINARAYANA AND B.H. KRISHNA MURTHY RAO, 1977 : Efficacy of different insecticides for the control of citrus leaf miner *Phyllocnistis citrella* Stainton in Andhra Pradesh. *International Symposium on Citriculture, Bangalore, Abstracts* : 34-35, Horticultural Society of India, Bangalore.

THAPA, RESHAM B., FANINDRA P. NEUPANE AND DHAMO K. BUTANI, 1986 : Insect pests of citrus in Nepal and their control. *Pestology,* **10** (4) : 24-27, Mumbai (Bombay).

THAPER, A.R., 1961 : Horticulture in the Hill Region of North India: 139-142, Directorate of Extension, Ministry of Food & Agriculture, New Delhi.

THOMPSON, W.L., 1939 : Cultural practices and their influence upon citrus pests. *Journal of Economic Entomology,* **32** : 782-789, Madison.

THOMPSON, W.R., 1950 : A Catalogue of the Parasites and Praedators of Insect Pests. Section I — Parasite Host Catalogue, Part 3 — Parasites of the Hemiptera, 149 pp. Ottawa, Ontario.

THOMPSON, W.R. AND F.J. SIMMONDS, 1964 : A Catalogue of Parasites and Predators of Insect Pests. Section III — Predator Host Catalogue, 204 pp. Farnham, Royal, Bucks (UK).

TIRUMALA RAO, V., 1956 : Stray notes on some crop pest outbreaks of South India. *Indian Journal of Entomology,* **18** (2) : 123-126, New Delhi.

TIRUMALA RAO, V. AND LEELA A. DAVID, 1958 : The biological control of coccid pests in South India by the use of the beetle, *Cryptolemus mentrouzieri* Mulls. *Indian Journal of Agricultural Sciences,* **28** (4) : 545-552, New Delhi.

TIRUMALA RAO, V., LEELA A. DAVID AND K.R. MOHAN RAO, 1954 : Attempt at the utilisation of *Chilocerus nigritus* Fabricius (Coleoptera : Coccinellidae) in the Madras state. *Indian Journal of Entomology,* **16** (3) : 205-209, New Delhi.

TIRUMALA RAO, V., P. GOVIND RAO AND P.V. RANGA RAO, 1954 : Guide showing Methods to Control Crop Pests and Diseases. Andhra Pradesh, Department of Agriculture and Fisheries, 141 pp.

TIRUMALA RAO V., N. RAGHAVA RAO, A. PERRAJU AND B.R.K. PARAMAHAMSA, 1954 : Crop pests—Beware of minor pests becoming serious. *Andhra Agricultural Journal,* **1** (6) : 344-348, Bapatla.

TOMKINS, A.R., G. FALLAS, C. THOMSON AND D.J. WILSON, 1994 : A new insect growth regulator, CGA, with promising activity against scale insects. *New Zealand Plant Protection Society,* Wellington.

TREHAN, K.N., 1944 : Distribution of whitefly in the Punjab. *Indian Farming,* **5** : 514-515, New Delhi.

TREHAN, K.N., 1957 : Brief notes on crop pests and their control in the Punjab. *Journal of Bombay Natural History Society,* **54** (3) : 581-626, Mumbai (Bombay).

TREHAN, K.N. AND DHAMO K. BUTANI, 1960 : Bibliography of Aleyrodidae. *Beitrage zur Entomologie,* **10** (3-4) : 333-388, Berlin.

TREHAN, K.N. AND S.V. PINGALE, 1946 : Annotated list of crop pests in Bombay Province. *Journal of Bombay Natural History Society,* **47** : 139-153, Mumbai (Bombay).

TREHAN, K.N., G.M. TALGERI AND S.R. DHARAESHWAR, 1949 : Control of red ants (*Oecophylla smaragdina* Fab.). *Proceeding Indian Academy of Science,* **B-30** : 338-342, Bangalore.

TSUCHIYA, M., K. FURUTASHI AND S. MUSVI, 1995 : Control of yellow tea thrips (*Scirtothrips dorsalis* Hood) by reflective sheet in satsuma mandrin (*Citrus unshin*) orchard. *Japanese Journal Applied Entomoloty and Zoology* **39** (1) : 219-225, Toshima, Tokyo.

USMAN, S. AND M. PUTTARUDRIAH, 1955 : A list of the insects of Mysore including mites. *Department of Agriculture, Mysore State (Entomological series) Bulletin No. 16,* 194 pp. Bangalore.

VACANTE, V., A. NUCIFORA AND A.G. TROPEA GARZIA, 1988 : Citrus mites in the Mediterranean area. p.1325-1334. *In: Proceedings of the Sixth International Citrus Congress Tel Aviv, Israel, March 6-11, 1988,* Balaban Publishers, Rohovet (Israel).

VARADARAJAN, B.S. AND C.K. SUBRAMANIAN, 1953 : Mandarin orange in Coorg. *Indian Journal of Horticulture,* **10** : 89-97, Bangalore.

VARMA, P.M., D.G. RAO AND R.S. VASUDEVA, 1960 : Additional vectors of Tristeza disease of citrus in India. *Current Science,* **29** : 359, Bangalore.

VASUDEVA, R.S. AND S.P. KAPOOR, 1952 : Outbreaks and new records :

India. Citrus decline in Bombay State. *FAO, Plant Protection Bulletin,* **6** : 91, Rome.

VASUDEVA, R.S., P.M. VERMA AND D.G. RAO, 1959 : Transmission of citrus decline virus by *Toxoptera citricidus* Kirk. in India. *Current Science,* **28** : 418-419, Bangalore.

VENKATAKRISHNIAH, N.S., 1957 : Canker disease of sour lime and its control. *Journal of Mysore Horticultural Society,* **2** : 40-44, Bangalore.

VENKATARAMAN, T.V. AND RAMESH CHANDRE, 1967 : Experiments on the possible use of *Bacillus thuringiensis* Berliner in the control of crop pests. II. Subsceptibility of some lepidopterous pests to *B. thuringiensis. Proceedings of Indian Academy of Sciences,* **B-66** : 231-236, Bangalore.

VERMA, A.N., J.S. BALAN AND S.M. SINGHVI, 1970 : Relative efficacy of different insecticides for the control of *Ak* grasshopper, *Poecilocerus pictus* F. on castor. *Telhan Patrika,* **2** (2) : 1-5, Hyderabad.

VERMA, A.N. AND A.D. KHURANA, 1974 : Further new hosts records of *Indarbela* sp. (Lepidoptera: Metarbelidae). *Haryana Agricultural University, Journal of Research,* **4** (3) : 253-254, Hisar.

VERMA, A.N., A.D. KHURANA AND R. SINGH, 1974 : Chemical control of bark eating caterpillar, *Indarbela quadrinotata* (Walker) infesting pomegranate. *Indian Journal of Entomology,* **36** (3) : 249, New Delhi.

VERMA, A.N., R. SINGH AND A.D. KHURANA, 1974 : New host records of *Indarbela quadrinotata* (Walker) (Lepidoptera). *Indian Journal of Entomology,* **36** (3) : 247, New Delhi.

VERMA, G.C. AND G.S. SOHI, 1967 : Studies on the chemical control of citrus leaf miner, *Phyllocnistis citrella* Stainton (Gracilariidae: Lepidoptera). *Journal of Research, PAU,* **4** (2) : 227-232, Ludhiana.

VERMA, P.M., D.G. RAO AND S.P. CAPOOR, 1965 : Transmission of Tristeza virus by *Aphis craccivora* and *Dactynotus jaceae* L. *Indian Journal of Entomology,* **27** : 67-71, New Delhi.

VEVAI, E.J., 1969 : Know your crop : its pest problems and control — Citrus. *Pesticides,* **3** (8) : 31-37, Mumbai (Bombay).

VEVAI, E.J., 1971 a : Know your crop, its pest problems and control — 33 : Banana : Plantain. *Pesticides,* **5** (6) : 37-56, Mumbai (Bombay).

VEVAI, E.J., 1971 b : Know your crop, its pest problems and control, Minor tropical fruits. *Pesticides,* **5** (11) : 31-34, Mumbai (Bombay).

VIGGIANI, G., 1984 : Side effect of pesticides on beneficial arthropods in citrus orchards. *CILB. WPRS. Bulletin No. 7* : 59-74.

VIGGIANI, G., 1988 : Citrus pests in the Mediterranean basin. p.1067-1073, *In: Proceedings of the Sixth International Citrus Congress, Tel Aviv, Israel, March 6-11, 1988,* Balaban publishers, Rehovot (Israel).

VISALAKSHI, A., NASEEMA BEEVI, T. PREM KUMAR AND M. R.G.K. NAIR, 1980 : Biology of *Leptoglossus australis* (Fabr.) (Coreidae : Hemiptera) a pest of snake gourd. *Entomon,* **5** (1) : 77-79, Trivandrum.

VOUTE, A.D., 1935 a : *Crytorrhynchus gravis* F. und die Ursachen : sciner Massenvermeehrung in Java. *Archieves Neerlandaiares de Zoologie,* **2** : 112, Den Helder, The Netherlands.

VOUTE, A.D., 1935 b : Die plagen van de djeroekeultuur in Nederlandschindie. *Mededelingen van het Institute voor Plantenziekten,* **86** : 65, Buitenzorg.

Vyas, H.N., 1994 : Pest complex of *citrus sp. Crop Research Journal*, **7** (1) : 168-169, Hisar.

Wadhi, S.R. and H.N. Batra, 1964 : Pests of tropical and subtropical fruit trees. p. 227-260. *In* : Entomology in India. Silver Jublee Number, Entomological Society of India, New Delhi.

Ware, A., 1994 : The biology and control of citrus leaf miner. *Citrus Journal*, **4** (4) : 26-28.

Wat Anabe, M.A., C. Yashii and R.C. Siloto, 1994: Parasitism of the scale insect *Selenaspidus* (Morgon) (Hemiptera: Diaspididae) on citrus in the regions of Jaquriuma and Limerra. *Revista de Agricultura Piracicaba*, **69** (2): 193-200.

Watson, J.R. and E.W. Berger, 1932 : Citrus insects and their control. *University of Florida, Agricultural Experiment Station, Bulletin No. 67* : 63-64, Gainesville, Florida.

Westwood, J.O., 1840 : An Introduction to the Modern Classification of insects. **2** : 442-443, London.

Wheaton, T.A., C.C. Childers, L.W. Timmer, L.W. Duncun and S. Nikdel, 1985 : Effects of Aldicarb on yield, fruit quality and the condition of Florida citrus. *Proceedings of Florida State Horticultural Society*, **98** : 6-10, Gainesville, Florida.

Woglum, R.S., 1913 : Report of a trip to India and the Orient in search of the natural enemies of the citrus whitefly. *USDA Bureau of Entomology, Bulletin No. 120* : 52-53, Washington, D.C.

Woglum, R.S., 1919 : Recent results in the fumigation of citrus trees with liquid hydocyanic acid. *Journal of Economic Entomology*, **12** : 117-123, Madison.

Woglum, R.S., C.C. Plummer and J.G. Shaw, 1949 : Insecticides for citrus blackfly. *Citrograph* : **34** : 146, Los Angeles, California.

Woodworth, C.W., 1901 : Notes on respiration of *Aleurodes citri. Canadian Entomologist*, **33** : 175-176, Ontario, Toranto.

Woodworth, C.W., 1902 : The red spiders of citrus trees. *California Agricultural Experimental Station, Bulletin No. 145*, California.

Wu, T., 1995 : Integrated control of *Phyllocnistis citrella, Panonychus citri and Phyllowptruta ollivora* with periodic releases of *Mallada basalis* and pesticide application. *Chinese Journal of Entomology*, **15** (2) : 13-123.

Xu, C.F., Y.H. and C. Ke, 1994 : A study on the biology and control of the citrus psylla, *Acta Phytophylacica Sinica*, **21** (1) : 53-56, Beijing, China.

Yadava, C.P.S., 1969 : Combating fruit sucking moths. *Indian Horticulture*, **13** (3) : 32-34, New Delhi.

Yang, C.K. and F.S.Li, 1984 : Nine new species and a new genus of psyllids from Yannon province. *Entomotaxonomia*, **41** : 251-260, Beijing, China.

Yang, Y., J.C. Allen, J.L. Knapp and P.A. Stansly, 1995a: Frequency distribution of citrus rust mite (Acari : Eriophyidae) damage on fruit in 'Hamlin' orange trees. *Environmental Entomology*, **24** (5) : 1018-1023, Lanham, Maryland.

Yang, Y., J.C. Allen, J.L. Knapp and P.A. Stansly, 1995b: Relationship between population density of citrus rust mite (Acari : Eriophyidae) and damage to 'Hamlim' orange fruit. *Environmental Entomology*, **24** (5) : 1024-1031, Lanham, Maryland.

YOUNG, B., 1942 : Whiteflies attacking citrus in Szechwam. *Sinensia*, **13** : 95-101, Peipeh.

YOTHERS, W.W., 1919 : The wooly whitefly in Florida citrus groves. *United States Department of Agriculture, Farmers' Bulletin No. 1011*, 14 pp. Washington, D.C.

YOTHERS, W.W., 1930 : The citrus rust mite and its control. *United States Department of Agriculture, Technical Bulletin No. 176*, Washington, D.C.

ZACHAR, F., 1928 : Die spinnmilken der Himbeere [The spinning mite of raspberry] — *Narchrichenble Bl. deuts Pfl SchDienst*, **8** (11) : 103-104.

ZANG, X.N. AND Z. SHANHVAN, 1995 : Investigation of timing spray insecticides against citrus leaf miner (*Phyllocnistis citrella*). *Journal South China Agriculture University* **16** (1) : 44-49.

ZEHAVI, A. AND D. ROSEN, 1987 : Population trends in the spirea aphid, *Aphis citricola* van der Goot, in a citrus grove in Israel. *Journal of Applied Entomology*, **104** : 271-277; Berlin.

ZHANG, A.C., C.O. LEARY AND W. QUARLES, 1994 : Chinesa IPM for citrus leaf miner (*Phyllocnistis citrella*) *IPM Practiner*, **16** (8) : 10-13.

ZHANG, X.W. AND D.X. GU, 1994 : Experimental release of *Aphytis melinus* to control *Aonidiella aurantii* in a citrus orchands in Guangdong, *Chinese Journal Biological control*, **10** (3) : 103-105, Beijing, China.

ZHAU, C.C. J.K. ZOU AND S.D. HUANG, 1994 : The effect of Agaratun inter planting in hilly citrus orchards in Humcum province on citrus mite and insect populations. *Acta Phytophylacica Sinica*, **21** (1) : 57-61, Beijing, China.

3

MAJOR PATHOGENIC DISEASES OF CITRUS IN INDIA AND THEIR MANAGEMENT

Y.S. AHLAWAT* AND R.P. PANT*

Citrus is cultivated at 40⁰ of either side of equater encompassing tropical to sub-tropical climate. North-Eastern part of India is considered to be the natural home of citrus from where a number of citrus species/varieties have their origin and later taken to different parts of the world.Presently citrus has become an important fruit crop in the world trade for fresh fruits, its processed products and by-products with 77.50 million tonnes total world production. The approximate contribution to citrus products in the world is: Brazil (19.11%), USA (17.50%), China (8.87%), Spain (6.67%), Mexico(5.57%), India (4.77%), Italy (3.54%), Egypt (3.18%), and Argentina (2.54%). Thus these top citrus producers contribute more than 70% to the world citrus production (FAO, 1996).

In India, citrus is grown in 4.4 lakh ha with a production of 37.0 lakh tonnes. The important commercial citrus fruits in India are the mandarins (*Citrus reticulata* Blanco), sweet orange (*Citrus sinensis* Osbeck) and acid lime (*Citrus aurantifolia* Swingle). Commonly grown commercial cultivars of mandarin are Coorg, Nagpur, Khasi and Darjeeling orange. The state of Andhra Pradesh, Maharastra, Karnataka and Punjab contribute each more than 10% towards total production. In addition to biotic stresses,inadequate nutrition, erratic irrigation schedules, cultural practices like inefficient weed control and use of incompatible inter crop especially during pre-bearing years are also some of the

* Division of Plant Pathology, Indian Agricultural Research Institute, New Delhi-110012

bottlenecks accounting for overall low average productivity of 7-8 tonnes/ha which is far below the average productivity of 25-30 tonnes/ha obtained in the frontline citrus growing countries.

With increasing emphasis on increased production alongwith improved fruit quality compatible at international standards and subsequently keeping an eye on global markets for exports in comming years warrants a thorough stock taking of constraints associated with citrus production in India and identify the future thrust areas to be tackled in a systematic manner.In order to establish the citrus industry of the country on a sound scientific footing it is necessary to develop agro-techniques which improve productivity as well as quality of the produce.

The productivity of the citrus fruits in India is comparatively low owing to many biotic stresses of which some fungi, bacteria, viruses, viroids and phytoplasma play a very significant role. Although more than 30 virus diseases, two phytoplasma diseases, one spiroplasma disease, three viroid diseases, 11 fungal diseases, 3 bacterial diseases and two nematode diseases are known to occur in citrus throughout the world alongwith some diseases of uncertain etiology like citrus blight etc.(Table-1a and 1b). Some of these diseases are of high economic significance as they can debilitate or wipe out the whole citrus industry if not managed in time. The role of these pathogens has been well established in declining tree health and production.

Among virus diseases, tristeza, ringspot, mosaic, satsuma dwarf, tatter leaf, and psorosis are globally important. Citrus stubborn disease caused by *Spiroplasma citri* is a limiting factor in citrus production wherever it occurs. Among phytoplasma diseases, witche's broom of lime in Omen and India and rubbery wood in India are the major diseases. Citrus exocortis and cachexia are two most important widely distributed viroid diseases of citrus.In addition to these pathogens, citrus greening disease which was attributed to be caused by Phytoplasma is now known to be caused by a bacterium, *Liberobacter asiaticum* in Asia and *L. africanum* in South Africa. This disease wiped out the citrus industry in several countries of South East Asia. Among fungal diseases,*Phytophthora*

Table 1a. Current status of virus and virus-like diseases of citrus

Disease	Causal agent	Transmission	Distribution	Reference
A. Diseases of known etiology				
A.1 Virus diseases				
1. Psorosis/Ringspot	* Two flexuous and filamentous Viruses: ** Linear-spirovirus *** Ophiovirus	G, MI, D	World wide, imp. in Florida, California, Argentina, Brazil, Texas, Israel, Spain, Vietnam, Central America, Australia, U.S.A., France, Italy, Iran	Swingle and Webber (1896), Fawcett (1932), *Bouhida (1984), Roistacher (1993), ** Derrick *et al.* (1993), *** Milne (1996)
2. Blind pocket	* Sobemovirus	G	California, Florida, Chile and Mediterrean basin	Fawcett and Lee (1926), *Navas-castillo *et al.* (1995)
3. Crinkly leaf	* Polyhedral virus	G, S, MI	California, Australia, India	Fawcett and Lee (1926), *Yot-Dauthy and Bove (1968), Ahlawat and Sardar (1976)
4. Leprosis	* Bacilliform-particles like Rhabdovirus	G, I, MI,	Florida, South America, Brazil	Knorr (1968), Lovisolo *et al.* (1996)
5. Infectious variegation	* Polyhedral particles ** Ilarvirus	G, MI	California, Sicily, Florida, India, Algeria, Argentina, Israel, Italy, Uruguay, U.S.A.	Fawcett and Klotz (1939), *Yot-Dauthy & Bove (1968), **Garnsey (1974), Yora *et al.* (1977)
6. Tristeza	* Flexuous rod ** Closterovirus	G, MI, I	World wide distribution nearly in all citrus growing countries	Moreira (1942) *Kitajima *et al.* (1964) ** Gonsalves *et al.* (1978), Vasudeva and Capoor (1958), Chakraborty *et al.* (1993)

*, **, *** : Causal agent and corresponding reference

Table 1a. Continued

7. Satsuma dwarf	Isometic virus, *Serologically related to como- and nepo-viruses	G, MI	Widespread in Japan, China, Korea, Turkey	Yamada and Sawamura (1952), *Iwanami and Leki (1995)
8. Vein enation and woody gall	* Spherical virus	G, S, I	California, S. Africa, Peru, Japan, Florida, India, China, New Zealand, Nepal, Phillippines, Australia, Spain	Wallace and Drake (1953), *Iwanami *et al.* (1992), Mali *et al.* (1976b)
9. Citrus mosaic	* Spherical virus	G, S, I	Japan,	Ishigai and Jinno (1958), *Tanaka and Imada (1976)
10.Citrus yellow mosaic	* Badnavirus	G, M, I	India	Dakshinamurty and Reddy (1975), **Ahlawat *et al.* (1996)
11.Tatter leaf	* Flexuous rod, Capillovirus	G	California, Nepal, China, Japan, Taiwan, Korea, S. Africa, U.S.A., New Zealand	Wallace and Drake (1962), *Kawai *et al.* (1995)
12.Rumple of lemon	*Virus ds RNA (?)	G	Florida, Italy, Turkey	Knorr *et al.* (1963) *Davino *et al.* (1995)
13.Citrus leaf rugose	**Ilarvirus*	G, MI	Florida, Argentina	Miyakawa *et al.* (1977), *Garnsey (1975)
14. Indian citrus ringspot	*Capillovirus	G, D, MI	Wide spread in India	Ahlawat (1989), *Byadgi and Ahlawat (1995) Pant *et al.* (1997)

15. Yellow vein clearing of lemon	*Filamentous particles **Capillovirus	G	Pakistan	*Catara *et al.* (1993) **Grimaldi & Catara (1996)
A.2 Viroid diseases				
1. Xyloporosis	*Citrus viroid (s)	G, S	Wide spread in Mediterreanean countries, Brazil, Argentina, South Africa, Florida, Texas, Philippines, U.S.A., India	Reichert and Perlberger (1934), *Semancik & Duran-Vila (1991), Nagpal (1959)
2. Exocortis	* Viroid, 371 nts and its ** sequence variants 370-375 nts	G, D, CPT	Wide spread in over 44 countries	Fawcett and Klotz (1948), *Semancik & Weathers (1972), **Visvader & Symon (1985), Patil and Warke (1968) Ramachandran *et al.* (1993)
3. Cachexia	*Citrus viroid 299 nt	G	Wide spread in Mediterranean countries : Brazil, Argentina, S. Africa, Florida, Texas, California, Mexico, India, Nepal, Venezuela	Childs (1950), *Semancik (1986), Semancik *et al.* (1988), Levy and Hadidi (1993)
4. Gummy bark	*Citrus viroids	G	Egypt, Saudi Arabia, Sudan, Turky, S. Africa	Nour-Eldin (1956) *Onelge *et al.* (1996)
5. Gum pocket	*Viroid	G, SSI	S. Africa, Argentina Australia	Schwarz and McClean (1969), *Marais *et al.* (1996)
6. **Yellow corky vein**	*Viroid	G. MI	India	*Rustem Ali (1998) Reddy *et al.* (1974)

A. 3 Diseases caused by mollicutes				
1. Stubborn	*Spiroplasma citri*	G, D, I	California, Mediterranean countries, Middle east, North Africa, Western U.S.A.	Fawcett (1946) *Igwegbe & Calavan (1970)
2. Rubbery wood	*Phytoplasma	G	India	Ahlawat & Chenulu (1985), Ahlawat (1987)
3. Witches broom of lime	**Phytoplasma aurantifolia*	G, D	Omen, United Arab Emirates	Bove *et al.* (1988) *Bove *et al.* (1995)
A. 4 Diseases caused by Fastidious bacteria				
1. Greening	*(1) *Liberobacter asiaticum* (2) *L. africanum*	G, I G, I	(1) India, Nepal, Sri Lanka, Vietnam, Cambodia, Malaysia, Indonesia, Philippines, Taiwan, China and Saudi Arabia (2) S. Africa, Zimbabwe and Yemen Both (1) and (2) are present in Mauritius islands	Oberholzer (1947), *Jagoueix *et al.* (1995) Bove *et al.* (1993), Varma *et al.* (1993)
2. Variegated chlorosis	**Xylella fastidiosa, bacterium*	B, I	Brazil	Lee *et al.* (1991) *Garnier *et al.* (1993)
3. Citrus shoot yellowing	* Citrus shoot yellowing organism (CSYO)	G	China	*Dewang *et al.* (1995)

B. Diseases of uncertain etiology				
1. Concave gum	Virus-like (?)	G	California, India, Japan, Vietnam	Fawcett and Lee (1926) Mali (1979)
2. Impietratura	Virus-like	G	Mediterranean basin, Iran, Venezuela, S. Africa, India	Ruggieri (1961) Ahlawat *et al.* (1984)
3. Yellow vein	Virus (?)	G	California	Weathers (1957)
4. Bud-union crease	Virus-like (?)	G	Palestine, Florida, California, Egypt, S. Africa, Brazil, India, Spain, Texas, Israel, U.S.A., Japan	Grimm *et al.* (1955) Bakhshi and Dhillon (1964)
5. Leaf curl	Virus (?)	G	Florida, Brazil, India	Salibe (1959) Mali *et al.* (1976a)
6. Tarocco pit	Virus (?)	B	Sicily	Russo and Klotz (1963)
7. Cristacortis	*Virus (?) **Viroid (?)	G	Mediterranean basin	*Vogel and Bove (1964) **Vogel and Bove (1987)
8. Multiple sprouting	Phytoplasma (?)	TW	Rhodesia, S. Africa, India	Searle (1969) Ahlawat and Chenulu (1985)
9. Blastomania	Phytoplasma (?)	B	India	Mali *et al.* (1975)
10. Leathery leaf	Virus (?)	G, MI, I	India	Ahlawat (1975)
11. Leaf yellow mid vein	Viroid (?)	G, MI	India	Sharma and Pandey (1983)
12. Citrus chlorotic dwarf	Virus (?)	I, SSI	Turkey	Kersting *et al.* (1996)

G = Grafting, CPT = Contaminated pruning tools, MI = Mechanical inoculation, SSI = Stem slash inoculation, D = Dodder, I = Insect vector, B = Budding, S = Seed, TW = Top working.

Table 1b : Important Fungal, and Bacterial Diseases prevalent in citrus nurseries and orchards

Disease	Casual organism
FUNGAL	
Damping off/Foot rot and Gummosis/Root rot/Twig blight/ Twig dieback	*Phytophthora spp. Sclerotinia sclerotiorum, Diplodia natalensis, Fusarium solani, Stemphylium sp.*
Anthracnose/Withertip	*Colletotrichum gloeosporioides, Gloeosporium limetticola*
Powdery mildew	*Acrosporium tingitaninum (Syn: Oidium tingitaninum)*
Leaf spot	*Alternaria citri*
Botrytis blight	*Botrytis cinerea*
Leaf and fruit spot	*Cercospora angolensis*
Pink disease and thread blight	*Corticium salmonicolor*
Dry root rot	*Macrophomina phaseoli, Fusarium sp. Diplodia natalensis*
Premature fruit drop	*Colletotrichum gloeosporioides, Alternaria, Botrytis sp.*
Scab	*Elsinoe fawcettii, Sphaceloma fawcettii*
BACTERIAL	
Canker	*Xanthomonas axonopodis* pv *citri*
Leaf and twig blast	*Pseudomonas syringae*
Citrus variegated chlorosis	*Xylella fastidiosa*

induced diseases, powdery mildews, wither tip and twig blight are of high significance whereas, citrus canker is the major bacterial disease in all citrus growing countries. The major diseases of citrus in India are summarized in the following paragraphs.In addition to the orchard diseases, post-harvest diseases (Table-2) are also important and will be described later in this chapter. The brief discription of important diseases is given below.

VIRUS DISEASES

1. **TRISTEZA**: Between 1840 and 1900 destruction of sweet orange seedlings in Spain, Australia and South Africa by *Phytophthora* spp. resulted in wider use of sour orange as the rootstock to avoid *Phytophthora* induced diseases. It was soon realised that the sweet orange trees budded on sour orange were not profitable as many of them died and this problem was thought to be due to incompatibility (Bas-Joseph *et al.*,1981). It was also postulated that the sweet orange top produced some substance that was toxic to sour orange roots. Moreira in 1942 named the disorder as "Tristeza" to the so called 'incompatibility'. Later, Webber, 1943 reported that tristeza was a vector transmissible disease possibly caused by a virus. In 1946, Fawcett and Wallace demonstrated the bud transmission of the tristeza virus. Eventually the virus was transmitted by brown citrus aphid, *Toxoptera citricidus* and virus- vector relationship were studied.

Tristeza virus had wiped out the citrus industry in many countries before it was actually established and managed. For example, in Argentina and Brazil where the citrus industry expended after first world war, tristeza destroyed about 30 million trees. Similar situations were also reported from Spain, Japan and United States. The estimates indicate that tristeza destroyed about a million trees in India (Ahlawat,1997).

In India, Brown 1920 recorded the failure of malta sweet orange on sour orange rootstock, providing evidence of flourishing tristeza disease in India but was first reported in 1958 from Maharashtra state by Vasudeva and Capoor.

Table 2 : Post harvest Diseases of Citrus

Disease	Casual organism	Preferred host (Citrus fruit)
FUNGAL		
Alternaria rot*	*Alternaria alternata, A.citri*	orange, grapefuit,
Anthraenose*	*Colletotrichum, gloeosporioides (Glomerella cingulata)*	mandarin
Black mould rot*	*Aspergillus niger*	lemon, orange
Blue mould*	*Penicillium italicum*	orange, lemon, mandarin
Green mould*	*Penicillium digitatum*	orange, lemon, mandarin
Stem end rot*	*Diplodia natalensis (Syn: Botryodiplodia theobromae) Diaporthe citri (Phomopsis citri)*	mandarin, orange, grapefruit
Grey mould rot	*Botrytis cinerea*	lemon, mandarin
Sour rot	*Endomyces geotrichum, (Geotrichum candidum)*	lemon, grapefruit
Fusarium rot	*F. moniliforme, F. oxysporum F. solani*	orange, mandarin, grapefruit
Black spot	*Guignardia citricarpa (Phyllosticta citricarpa)*	Valencia orange
Brown rot*	*Phytophthora spp.*	Majority citrus fruits
Cottony rot	*Sclerotinia minor S. sclerotiorum*	lemon, grapefruit
Greasy spotrind blotch	*Mycosphaerella citri (Stenella citri-grisea)*	grapefruit
Melanose	*Diaporthe citri, (Phomopsis citri)*	orange orange, sweet
Scab	*Elsinoe fawcettii, Sphaceloma fawcettiisour*	orange
Septoria spot	*Septoria sp.*	lemon
Trichoderma rot	*Trichoderma viride*	grapefruit, mandarin, orange
BACTERIAL		
Black pit	*Pseudomonus syringae pv. syringae*	lemon
Canker*	*Xanthomonas axonopodis pv. citri*	acid line

* Serious disease, prevalent in most citrus growing countries

Later, its transmission was demonstrated by *T.citricidus* (Vasudeva *et al.*, 1959) *Aphis gossypii, Myzus persicae* (Verma *et al.*, 1960), *A.craccivora and Dactynotus jaceae* (Verma *et al.*, 1965).The virus was reported to be non-persistently transmitted from citrus to citrus.

Tristeza virus occurs almost in all citrus growing regions in the world and has been reported in the form of various strains. However, destructive strains are known to occur only in places wherever the vector *T.citricidus* is present and active.

There are no apparent symptoms of the disease on leaves of most *Citrus* spp.but pitting in the stem and decline of trees are indication of the presence of tristeza infection. Schneider (1954) showed that the seive tubes of Tristeza affected trees become necrotic and degenerated below the bud union resulting in the decline of such trees.

Kitazima *et al.*, (1964) showed that thread like virus particles were constantly associated with tristeza infection. Bar-joseph *et al.* (1970) for the first time partially purified citrus tristeza virus (CTV) which lead to the development of serological methods for CTV detection. Kagzi lime *(C.aurantifolia)* is commonly used as a indicator host for biological detection of the virus and trifoliate orange can be used to filter tristeza from mixed infection of other viruses already present in field trees (Tanaka *et al.*, 1971). A major breakthrough in CTV research was the production of virus specific polyclonal and monoclonal antibodies against CTV for successful detection in enzyme-linked immunosorbent assay (ELISA) (Bar-joseph *et al.*, 1979) and the use of dsRNA technology (Dodds and Bar-Joseph, 1983) which later helped in identification of CTV-free planting material.

Tristeza virus infects nearly all Citrus species and citrus relatives and hybrids. The only non-rutaceous host of CTV is *Passiflora* sp. In general, mandarins are normally tolerant to CTV infection but some strain of CTV can even infect certain mandarins such as coorg mandarin in India. Tristeza virus was apparently originated in Asia where it might have been existed for centuries but was unrecognised bacause many of the commonly propagated citrus cultivars in Asian countries are tolerant to CTV and propagated by seed. Citrus was introduced to new world as seed and since CTV was not seed transmitted, the seedling trees were free from this virus. Eventually, vegetative propagation started to maintain best horticultural traits. Citrus when moved as vegetative propagation material nationally or internationally, CTV definitely moved through such materials resulting in wider distribution of the disease. The epidemics were also experienced in countries where sour orange was used as rootstock and virus vectors, *T.citricidus* was present and active.

CTV is more damaging to Indian sweet oranges than to other cultivars (Ahlawat *et al.*,1995). Kagzi lime is used as an indicator host for CTV detection, but the diagnostic vein-flecking symptoms on this host depend on temperature.No foliar symptoms are developed at temperature higher than 27°C and hence its biological indexing on large scale without controlled condition is difficult.

Characteristic symptoms of Tristeza on kagzi lime develop as vein flecking of leaves (C-3.8), leaf cupping and stem pitting on inoculated plants (C-3.9). CTV is a highly flexuous filamentous virus of 10-12 nm in width and 2000 nm in length (B-3.11). The virus is located only in phloem tissue and is a member of closterovirus group. CTV has single stranded positive sense RNA of 20,000 nucleotides as its genome and a single type coat protein with Mr 26,000. The CTV coat protein gene (CPG) was selectively identified in PCR and was cloned (Manjunath *èt al.*,1993).The coding region of CPG was expressed in *Escherichia coli* (Migula) Castellani and Chalmers. In addition to the biological indexing on kagzi lime and virus specific antibodies, diagnostic reagents based on CTV- coat protein and genomic RNA have been developed for quick and reliable indexing of CTV. Strain specific monoclonal antibodies (MAbs) such as CTV-MCA 13 are being used to identify protective mild strains for the management of CTV by cross-protection against severe and disastrous strains.

Electron and light microscopy have also been used to identify CTV infection either by directly observing the virions in EM or detection of CTV-specific inclusions in cells both by EM and light microscope. Indexing by serological methods to analyse more samples at a time is now possible by newer

C-3.8 Leaves of Kagzi lime showing flecking of veins due to tristeza virus infection.

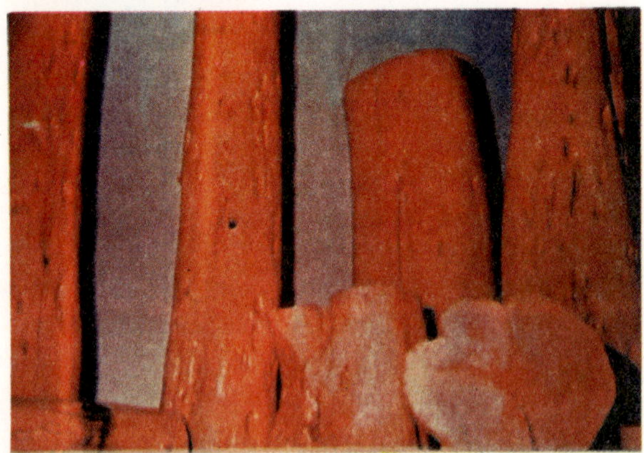

C-3. 9 Citrus wood showing pitting due to citrus tristeza disease.

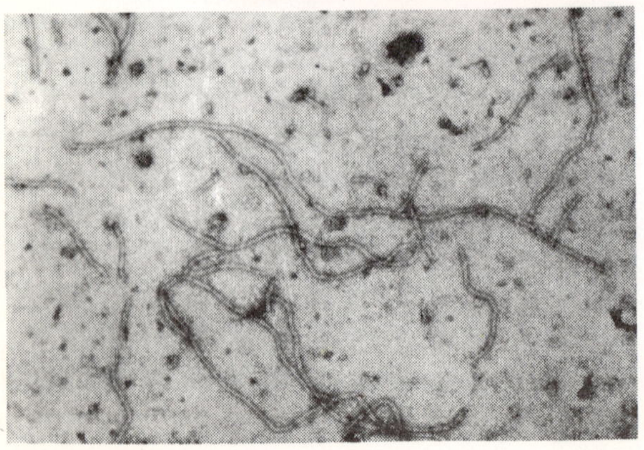

B-3.11 Flexuous virus particles of citrus triteza virus (80,000X)

Table-3. Selective transmission of tristeza virus strains by aphid species.

Aphid species	Source fed upon	Strain transmitted
Toxoptera citricida	Mild(Tm)	Mild(Tm)
	Severe(Tt)	Severe(Tt)
	Mixture	Mild (Tm)
	(Tm+Tt)	Severe (Tt)
Aphis gossypii	Mild	Mild
	Severe	Severe
	Mixture	Mild
	Tm+Tt	Severe
Aphis craccivora	Mild	Mild
	Severe	Severe
	Mixture	Mild
	(Tm+Tt)	Severe
Myzus persicae	Severe	——
	Mixture	——
	(Tm+Tt)	——
Dactynotus jaceae	Mild	Mild
	Severe	——
	Mixture	——
	(Tm+Tt)	
Toxoptera aurantii	Mild	Mild
	Severe	——
	Mixture	——
	(Tm+Tt)	

(After Capoor and Rao, 1967)

Table:4. Indexing of CTV cross-protection experiments in Bangalore and Tirupati.

Trees inoculated with	No.of trees	DAS-ELISA with MAbs* 3DF1 Bangalore	Tirupati	MCA-13 Bangalore	Tirupati
Mild strain	6	0.78	1.06	1.14	1.01
Severe strain	6	0.92	1.05	1.19	1.39
Mild+Severe strain	6	0.80	0.85	1.16	0.81
Uninoculated control	6	0.52	0.00	1.07	0.00
Healthy control (glasshouse)	-	0.00	0.00	0.04	0.00
Buffer	-	0.00	0.00	0.00	0.00
Control T-36 (severe strain)	-	1.13	1.04	1.26	1.35
Control T30 (mild strain)	-	1.26	0.92	0.00	0.00

*ELISA values are the average of two replications. Zero was adjusted with buffer.

techniques like ELISA (Ahlawat and Raychaudhary, 1988: Chakraborty *et al.*, 1993). For more specific detection of CTV, techniques like nucleic acid hybridization and polymerase chain reaction (PCR) are being developed in India. Monoclonal antibody MCA-13 is being used to descriminate strains in CTV. Protective strains are being used for immunization the nursery plants to manage the disease in the field by cross protection in several countries.

Quarantine restrictions and bud-wood certification are the main methods for virus management. However, these programmes will be successful only if they are supplemented with eradication of infected trees. Decline of valuable material even on sour orange rootstock can be prevented by inarch grafting of the tree with virus-tolerant seedlings. Since CTV is known to occur in variable strains, identification and use of a mild strain for immunization will help to manage the disease through cross protection technology. For successful implementation of this technology, selection of a suitable mild strain is essential which can be achieved by using differential host (Balaraman and Ramakrishnan, 1977) or by differential transmission (Table- 3) (Capoor and Rao, 1967) or by serological techniques using monoclonal antibodies (Chakraborty *et al.*, 1993). CTV is being managed successfully in Florida and Brazil by cross protection. However, in India such programmes could not be successful possibly bacause of the selection of the right protective strain. The analysis of immunized trees showed the infection of severe strain when tested with MCA-13 antibodies (Table-4) (Ahlawat, 1997).

2. PSOROSIS

Psorosis was the first disease of citrus proved to be graft transmissible (Fawcett, 1934, 1938). This discovery led to the first eradication programme for citrus diseases in the United States of America. Psorosis as originally described by Swingle and Webber (1896) is a disease associated with typical bark scaling in trunks and limbs of sweet orange, mandarin and grapefruit. In addition to bark scaling, certain leaf symptoms like vein clearing, oak-leaf pattern and rings were also associated with the disease (Fawcett, 1934). Wallace 1945 reported shock reaction and vein clearing of young leaf as diagnostic symptoms of psorosis disease.

Fawcett and Klotz, (1938), proposed two types of psorosis, psorosis A and psorosis B. Both had similar bark lesion symptoms but differed in young leaf symptoms. Psorosis A developed vein clearing on young leaves whereas B type developed oak-leaf symptoms on sweet orange upon inoculation. However, the young leaf symptoms are short lived. In fruits only B type causes discolored, circular to semi circular rings or groves.

Psorosis is diagnosed in the field by typical bark scaling symptoms. The major susceptible varieties of sweet orange, mandarin and grapefruit show bark scaling but sour orange, sour lemon, pummelo and rough lemon usually do not show such symptoms.

In addition to bud transmission, some isolates of Psorosis particularly the ones causing ringspot symptoms are transmitted mechanically to *Chenopodium quinoa* (Garnsey and Timmer, 1988)

Psorosis disease is also seed transmitted in seeds collected from fruits of infected carrizo citrange (15-31%) from Florida (Childs and Johnson, 1966), Spain, Navarro *et al.* (1980) from Salto region of Uruguay, Campiglia *et al.* (1976). Vogel and Bove (1980) reported pollen transmission of a psorosis isolate from Corsica. Timmer and Garnsey (1980) showed natural spread of necrotic ringspot, probably a psorosis B type isolate in nucellar virus free grapefruit in Texas. Pujol and Benatena (1965) reported presence of psorosis in seedling trees in Argentina and considered it either due to seed or vector transmission.

Derrick *et al.* (1988) reported a 48KDa capsid protein from several psorosis and ringspot isolates from Florida. However, da Graca *et al.* (1991) and Derrick *et al.* (1988, 1991) from California, Garcia *et al.* (1991) from Argentina, Navas Castillo *et al.* (1991) from Spain reported 48-50 KDa capsid protein as top and bottom component. Bauhida (1984) on the contrary had shown 29 KDa capsid protein from psorosis infected citrus trees in California. So far various worker have shown different capsid protein in psorosis isolates suggesting the variability in the pathogen. A filamentous virus is associated with psorosis.

Derrick *et al.*(1991) reported cross protection tests among various isolates of psorosis and concluded that psorosis-B is a severe form of psorosis-A and the ringspot is also a similar disease, if not identical to psorosis-B. This view was further supported by recent finding that the isolates of ringspot and psorosis are either identical or strains of the same virus (Roistacher, 1993). Thus ringspot and psorosis must be considered as synonymus.

3. CITRUS RINGSPOT

A ringspot disease of citrus was first described by Wallace and Drake (1968). Later it was found that ringspot shares properties with psorosis (Roistacher, 1993). In India, a disease with ringspot symptoms was first reported as a strain of psorosis-A disease (Ahlawat, 1989) but he observed that field trees in India were devoid of bark lesions, a characteristic symptoms of all psorosis strains. Another major difference was in characteristic flecking or vein clearing of young leaves of inoculated sweet orange plants which unlike psorosis persisted till the leaves became mature rather than for a short while, a characteristic of most psorosis strains. These observations suggested that the disease in India was apparently different from other known psorosis diseases.

The incidence of ringspot disease was observed up to 100% in most of the kinnow mandarin orchards in North India especially in Punjab. The incidence in sweet orange ranged from 20-50% in Maharashtra, Andhra Pradesh and Karnataka. The incidence of the disease was also recorded on different rootstocks with kinnow as a scion variety and it was found that the disease incidence was more on sohsarkar *(C. karna* Raf.)(56%) and Karna khatta (54.7%) as compared to Troyer citrange (7.3%) in a root stock trial at the Indian Agricultural Research Institute, New Delhi. The mature leaves of affected trees or inoculated plants showed distinct ringspot symptoms without bark scaling. (B-3.12)

The yield loss (number of fruits) in 7 to 10 yr.old kinnow orchards varied from 20.54 to 98.38%. The health of the affected trees deteriorated year after year and finally became totally unproductive.

Except bud transmission, no other mode of natural spread of the disease could be established so far (Byadgi and Ahlawat, 1995). However, citrus ring spot virions were trapped in Immuno Electron microscopy (IEM) from the pollens of ringspot affected kinnow trees, indicating the possibility of its transmission through pollen (Ahlawat, 1997). Recently, the virus was mechanically transmitted from citrus to herbaceous hosts like *Chenopodium quinoa* and *Phaseolus vulgaris* (Pant *et al.*,1997)

Citrus ringspot virus (CRSV) infects most of the commercial citrus cultivars and rootstocks used in India causing variable symptoms (Table-5) (Byadgi and Ahlawat, 1995). The virus associated with citrus ringspot disease in India (CRSV-I) was purified from citrus and *P.vulgaris* and based on virion morphology it was tentatively identified as a member of capillovirus group. However,the virus associated with most of the citrus ringspot diseases reported elsewhere has distinct morphology and designated as spirovirus based on its helical morphology (Derrick *et al.*,1993). Later, it was established that citrus ringspot virus and citrus psorosis associated viruses are members of the new genus 'Ophiovirus'. The virion morphology of the Indian citrus ringspot virus (CRSV-I) is totally distinct from ophioviruses and hence this virus has been tentatively named as Indian citrus ring spot capillovirus. However, CRSV-I did not react with antisera of capilloviruses, apple stem grooving virus, citrus tatter leaf virus, lilac chlorotic leaf spot virus, apple chlorotic leaf spot virus and potato virus-T further indicating the complex taxonomy of CRSV-I.

The virions of CRSV-I capillovirus are filamentous with a size of 640x15nm (B-3.12) but in purified preparations some very thin virus-like particles measuring 690x9nm and others in the form of tubule 2250x40nm were also observed (Byadgi and Ahlawat, 1995) which are yet to be characterized. CRSV-I has a single stranded RNA as its genome with a single polypeptide of 29 KDa as its coat protein (Byadgi *et al.*,1993; Ahlawat *et al.*,1995). The virus specific antiserum was prepared and used in ELISA for the detection of CRSV-I in planting material (Ahlawat *et al.*, 1995).

Table 5 : Host range of the Indian CRSV in family Rutaceae

Host	No. of plants infected out of 10 inoculated	Symptom observed
Citrus sinensis cvs.		
Mosambi	8	VC,LF,RS,OLP
Sathgudi	7	CS,RS,LF
Malta	7	LF,CS,RS
C. reticulata cvs.		
Darjeeling roange	7	CS
Nagpur orange	6	LF,NS,RS,WS
Kinnow mandaris	7	CS,NS,RS
C. aurantium		
Sour orange	7	VC,LF,WS,LC
C. paradisi		
Grapefruit	4	LF,CS
C. limettiodes		
Sweet lime	7	CS,VC
C. jambhiri		
Rough lemon	5	CS, VC
C. decumana	5	CS
C. limon cv. Galgal	3	VC
C. aurantifolia cvs.		
Kagzi lime	5	CS,LF
Kagzi kalan	6	CS,RS
C. medica Etrog citron	3	VC,WS

VC = Vein clearing; CS = Chlorotic spot
OLP = Oak leaf pattern; RS = Ring spot
LF = Leaf flecking; NS = Necrotic spot
WS = Water soaking; LC = Leaf curling

The disease can be easily recognised in field trees by its characteristic ringspot symptoms. Affected trees show conspicuous rings on mature leaves which may be one to several per leaf (C-3.8). Most of the CRSV affected trees show decline or die-back symptoms and the quantity and quality of fruits is greatly affected. The glasshouse inoculated plants of sweet orange and mandarin show shock reaction after 2-3 months of inoculation. Young leaves of mosambi sweet orange plants also develop vein clearing symptoms on young leaves which persist till the maturation of the leaves. The virions of CRSV-I both from field and glasshouse infected plants can be detected in electron microscope in leaf dip preparations with negative staining. The virus can also be detected in

ELISA and IEM using homologous antibodies.

C-10 Ringspot symptoms on kinnow mandarin leaves due to ringspot virus.

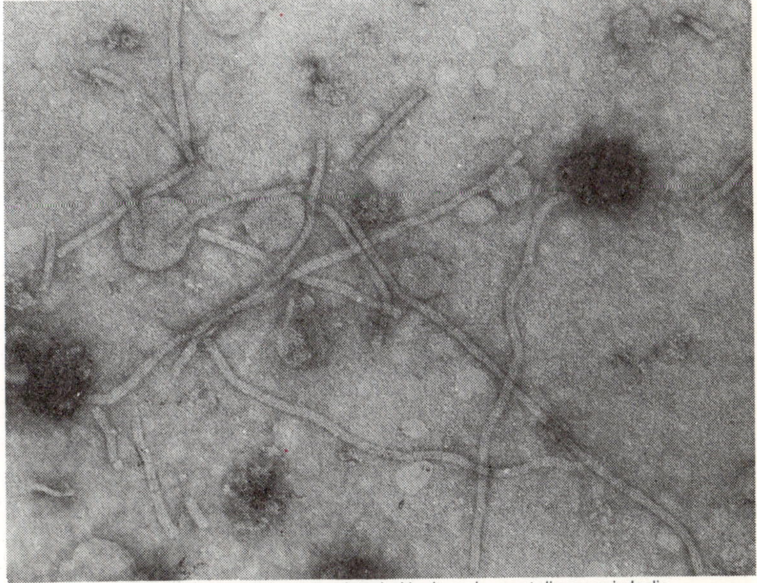

B-3.12 Filamentous particles associated with citrus ringspot diseases in India.

The optimum conditions determined for PTA-ELISA were: 1:20 (w/v) extract of infected tissue in PBST,(pH-7.4), 1μg/ml of purified IgG in carbonate buffer (pH-9.6), 1:1000 dilution of goat anti-rabbit conjugated with alkaline phosphatase and 0.6mg/ml of the substrate, para-nitrophenyl phosphate (PNPP). The virus could also be detected in IEM using 1:50 dilution of homologous antiserum. In limited tests on CRSV-I, the virus was also detected in dot immuno binding assay (DIBA) and western blotting. In IEM tests, CRSV-I did not react with other filamentous viruses *viz.*, citrus ringspot virus- Florida isolate (CRSV-F), Apple chlorotic leafspot virus (ACLSV), Citrus tristeza virus-Florida isolate (CTV-F), Garlic mosaic virus (GMV),Garlic latent virus (GLV), Shallot latent virus (SLV),Henbane mosaic virus (HBMV),Papaya ringspot virus (PRSV), Potato virus Y (PVY),Sweet potato feathery mottle virus (SpFMV), Potato virus X (PVX) and Tobacco mosaic virus (TMV) indicating that Indian citrus ringspot virus is serologically distinct from these filamentous viruses. Recent studies on CRSV-I have shown this virus to be distinct from known viruses and requires new taxonomic position (R.G.Milne- personal communication).

4. CITRUS MOSAIC

A mosaic disease of citrus was reported from Andhra pradesh by Murti and Reddy (1975) on sathgudi sweet orange and on khasi mandarins from North-eastern states (Ahlawat *et al.*,1985). The incidence of the disease was recorded to an extent of 5.84%. Recent surveys, however, showed the incidence of the mosaic disease from 10 to 70% in citrus orchards and nurseries in Andhra Pradesh. The disease can cause yield reduction of fruits upto 77% and the juice was 10% less with 1.3% less ascorbic acid having more acidity.Rao and Narasimhan (1974) reported decrease in chlrophyll a,b and IAA in mosaic affected leaves. The mosaic disease from Andhra Pradesh was reported to be transmitted by aphid,*Toxoptera citricida* and that of from North eastern region by aphids, *Myzus persicae* and *Aphis craccivora*. However, no further information is available on these two mosaic diseases.In view of the occurrence of the mosaic disease only in India and its wide spread and the losses caused, it was studied in detail by the authors.

Unlike earlier reports, citrus mosaic disease could not be transmitted by aphid vectors but was transmitted by a mealybug, *Planococcus citri*. It was transmitted to various citrus species by graft (Table-6) and to sweet orange, galgal and sugarcane by mechanical inoculations. In samples, collected from mosaic affected citrus trees from Andhra Pradesh, a mixed infection of three viruses, CRSV-I, CTV and mosaic was often observed (B-3.13). However, sap inoculation separated the mosaic virus from mixed infections.

Table-6. Host range of Citrus mosaic virus in the sub-family Aurantioideae.

Plant species	No. plant infected out of 10 inoculated	Symptoms
C.limonia	9	CS,VB,MM
C.volkameriana	8	CS,VB,M
C.jambhiri	8	VF,M,CS
C.sinensis cvs.		
sathgudi	6	M,CS
mosambi	10	M,CS
chini	10	CS,M
C. reticulata cv.Nagpur orange	6	VB.M
C. limettiodes	2	MM
C. aurantifolia	0	-
C. grandis	10	M,SC
C. paradisi cv.*duncun*	4	VF, M
C. medica	2	VF,M
C. aurantium	10	CS,SM
C. mitis	10	CR,SL,M
C. decumana	10	M,CS,SL

CS = Chlorotic Spot VF = Vein flecking
VB = Vein banding M = Mosaic
MM = Mild mosaic SL = Smalling of leaves

In purified and leaf dip preparations of mosaic affected leaf samples, bacilliform virions measuring 130x30nm (B-3.14) were constantly observed (Pant and Ahlawat, 1997). This bacilliform virus reacted specifically in ISEM tests to antisera of seven badnaviruses, Sugarcane mosaic bacilliform virus, commelina yellow mottle virus, banana streak virus, cacao swollen shoot virus, Diascorea bacilliform virus, canna yellow mottle virus and Kalanchoe top spotting virus but not

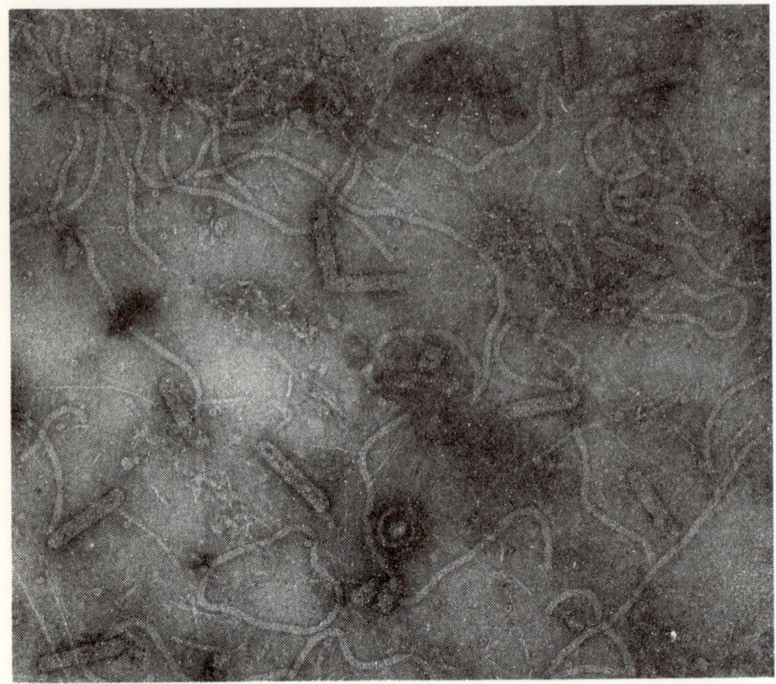

B-3.13 Mixed infection of tristeza, ringspot and mosaic viruses isolated from field tree.

to rice tungro bacilliform virus. Based on its bacilliform morphology and serological reactivity, the citrus mosaic bacilliform virus was identified as a member of badnavirus group.Its badnavirus nature was further confirmed by amplifying its genomic nucleic acid in PCR using degenarated primers located on consenses sequences of badnaviruses.(Ahlawat *et al.*,1996). These experiments confirmed that citrus mosaic disease is caused by a badnavirus (CMBV) which is the first record of a dsDNA virus in citrus.

SDS-PAGE and western blot analysis of purified virions showed one major band of 32KDa and two minor bands of 22 and 19 KDa. The 32 KDa is the major capsid protein and two minor bands are degraded product of 32 KDa. The nucleic acid extracted from purified virions showed two bands of 21Kb and 6.6Kb in non-denaturing agarose gels. The faster migrating band of 6.6 kb is linear form of viral DNA while

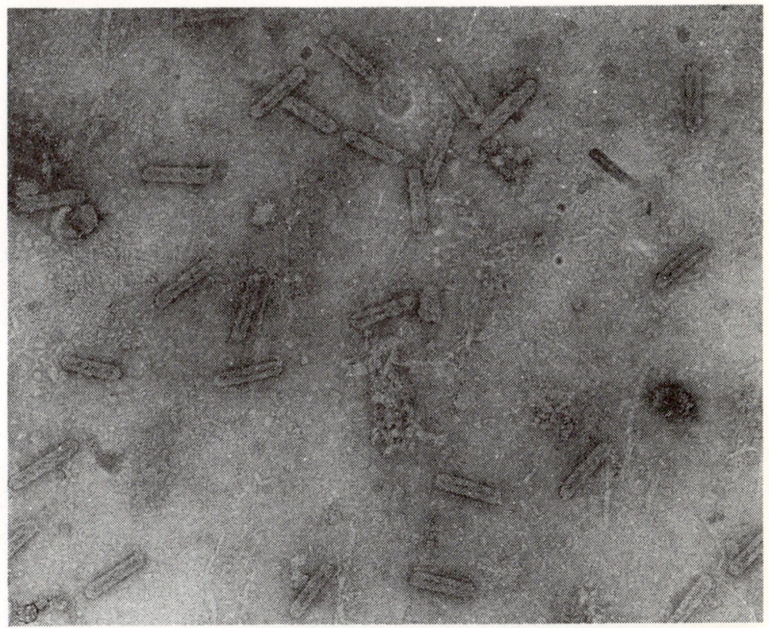

B-3.14 Citrus mosaic bacilliform virus purified from sweet orange (100,000X)

slow migrating band of 21Kb may be circular form of DNA. Electrophoresis of viral DNA digested with S1 nuclease gave two faster migrating bands of approximately 5.0 and 2.2Kb,besides 6.6 Kb linear DNA band. These two bands arose due to the presence of two discontinuities on the viral DNA. The presence of two discontinuities on the genome is hall mark of reverse-transcription process in pararetroviruses to which badnaviruses belong. The viral DNA was cloned at Hind III site in pUC18 vector and transformed into competent *E. coli* strain NM522 cells. The recombinent colonies were screened and finally two virus-specific clones of 2.0 and 4.0Kb were identified. The cloned fragments were radiolabelled and used as probes to detect the virus in field trees by dot-blot hybridization (Table-7, B-3.15). The virus was also detected in mealybug vector using this probe. The 2.0 Kb probe can detect the mosaic virus in DNA-DNA hybridization in even 25ng of the total DNA or 95pg of viral DNA or 25 ng of tissue per spot (Reddy, 1997).

Table- 7. Detection of citrus mosaic virus by dot blot hybridization.

Orchard number	Samples tested	Samples showing symptoms	Samples positive by dot blot
1	10	0	10
2	11	4	8
3	18	4	15
4	23	3	9
5	10	2	5
6	9	2	2
7	7	0	7
8	5	0	1
9	2	0	1
NURSERY(KODUR)	18	4	15
TOTAL	107	19	66

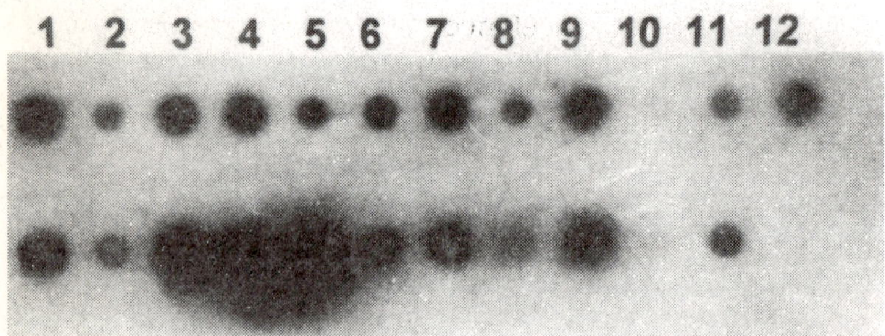

B-3.15 Detection of citrus mosaic badnavirus by dot-blot hybridization.

A mosaic disease of citrus has been reported from Japan (Ishigi and Jinno, 1958) and two from India. To avoid confusion, it was decided in the 13th IOCV conference held in China in 1995 that the name of the disease be changed as citrus yellow mosaic based on its characteristic yellow mosaic symptoms, which was accepted.

5.SATSUMA DWARF

Satsuma dwarf virus (SDV) first discribed in Japan (Yamada and Sawamura, 1952) is now widely distributed. It moved to other countries through the movement of infected budwood. Affected trees remain stunted and yield is greatly reduced. The methods of its natural spread are not known although the disease has covered large area in Japan.

The disease is graft and mechanically transmissible to a wide range of host plants in rutaceous and non-rutaceous plant species. The virus is reported to be seed transmitted in bush bean seeds but not in seeds of other hosts. Natural spread apparently by a soil borne vector has been suspected in many areas of Japan but not established. The most prominent symptoms in satsuma mandarins are stunting, narrow boat shaped or spoon shaped leaves. In other cultivars the symptoms remain mild or absent (Koizumi *et at.*, 1988).

Symptoms of satsuma disease in herbaceous hosts include local lesion in *Chenopodium quinoa*, sesame and cow pea. Systemic vein clearing, leaf mottling or necrosis occurs in *Physalis floridana*, sesame, cowpea and bush bean. The virus particles are isometric having size of approximately 26nm in diameter and contain two RNA species. It is serologically related to natsudaidai dwarf virus (NDV), citrus mosaic virus (CiMV) and naval infectious mottling virus (NiMV), (Usugi and Saito, 1979).

6. CITRUS TATTER LEAF

This disease was first discovered in the United States in 1962 by Wallace and Drake as a latent infection in trees of Meyer lemon, a cultivar introduced from China (Miyakawa, 1980). Later the disease was also reported from Japan on some citrus cultivars originated from China. Investigations in China revealed its wide distribution in cultivars propagated on trifoliate orange or trifoliate orange hybrids. The tatter leaf affected trees are usually stunted with chlorotic foliage and have a pronounced, virus induced bud-union incompatibility. Citrange stunt was originally described as a distinct virus but recent studies have shown that tatter leaf and citrange stunt symptoms are probably caused by the same virus.

The causal virus can infect many citrus species/cultivars and hybrids, sweet orange, sour orange, grape fruit, mandarin and lemon are the latent hosts. Chlorotic leaf symptoms are produced in Mexican lime, Citrus excelsa, citrange and other trifoliate hybrids. Leaves and stem may get deformed due to this virus in some citrus hosts like citrange. Trifoliate orange is immune or highly resistant to this virus.

Tatter leaf virus is mechanically transmitted to many herbaceous non-citrus hosts. It causes necrotic local lesion and variable systemic necrosis in cowpea and bean. Chlorotic local lesions and a systemic chlorotic mottle was produced in *Chenopodium quinoa.* Periwinkle and petunia are symptomless hosts. Reinoculation of citrus from herbaceous hosts has reproduced tatter leaf symptoms in *C.excelsa* and citrange stunt symptoms in citrange.

The citrus tatter leaf virus (CTLV) is filamentous measuring 600-650 nm in length and 13nm in width and belongs to capillovirus group serologically related to apple stem grooving virus (ASGV). SDS-PAGE analysis of coat protein revealed a single band of 27KDa. The virus has a single species of RNA of Mr 2.83×10^6 same as in case of ASGV (Nisho *et al.*, 1989; Takahashi, 1990). Recently complete CTLV-RNA has been sequenced (Ohira *et al.*, 1995). It has 6496 nt long excluding the 3'terminal poly A tract, and contains two putative overlapping open reading frames. The genome structure is homologous to that of ASGV.

Polyclonal antibodies (PAbs) and monoclonal antibodies (MAbs) have been used to detect CTLV in ELISA tests in Japan (Kawai *et al.*, 1996).

7.INFECTIOUS VARIEGATION AND CRINKLY LEAF

Citrus variegation described as a transmisssible virus- like disease in 1939 by Fawcett and Klotz was initially considered a member of the psorosis disease complex but later proved to be different from psorosis family as the virus belongs to Ilarvirus group. Citrus variegation virus (CVV) infects most citrus species and cultivars. Lemon, sour orange, etrog citron and grapefruit usually develop chlorotic leaf symptoms. The affected trees showed stunting and distortion of fruits. It is also mechanically transmitted to cowpea and *Crotalaria spectabilis* and produced variable chlorotic or necrotic local lesions. The citrus variegation ilarvirus has four nucleoprotein components and the particles are spherical in size varying from 26-32nm. It is serologically related to citrus leaf rugose virus and several other ilarviruses.

Citrus crinkly leaf virus (CCLV) is a mild variant of CVV. Doughty and Bove (1965) suggested CIVV and CCLV are the strains of the same virus. Martelli (1968) showed cross protection between CCLV and CIVV. Uyeda and Mink (1983) demonstrated selological relationship between CVV and four other ilarviruses. Davino and Garnsey 1984 develop antiserum of CVV and used in ELISA for detection of CVV isolates.

Citrus crinkly leaf virus is usually transmitted by graft inoculation but is also transmitted by mechanical inoculations to cowpea and bush bean. There are some unconfirmed reports of seed transmission. In India, both the diseases have been reported on citrus cultivars from North-east region and Maharashtra (Ahlawat and Sardar 1976; Yora *et al.*, 1977).

8. VEIN ENATION

Vein enation, an aphid transmitted virus disease of citrus is known to occur in several citrus growing countries, United States of America. South Africa, Australia, Peru, Japan and India. The disease produces severe gall in some rough lemon cultivars (Hooper and Schneider, 1969).

Enations are produced on the leaf veins in some cultivars of Mexican lime. sour orange and rough lemon whereas swelling or galls are developed on the stem of infected Mexican lime which are usually linked with wounds (Wallace and Drake, 1960).

The causal virus is spherical, approximately 25nm in diameter. The virus is transmitted by graft and also by vectors, *Myzus persicae, Toxoptera citricidus and Aphis gossypii.* The disease is not economically important.

9. CITRUS CRISTACORTIS

Unlike citrus tristeza virus infection Vogel and Bove (1964) observed unusual deep pits in both scion and rootstock of some tangelo trees and named the disease as cristacortis. In some grooves upto 75% trees were affected. The disease was transmitted by graft from citrus to citrus and Orlando tangelo was identified as its indicator host. Vogel and Bove (1980) also reported pollen transmission of cristacortis. Troyer citrange, trifoliate orange, citron, lemon and mexican lime did not develop disease symptoms upon inoculation. However,in Italy, cristacortis affected trees showed leaf flecking and different type of pits suggesting a mixed infection in such trees.Vogel and Bove (1972) also reported that it was a distinct disease. Mild, moderate and severe strains are present in cristacortis which can be separated on clementine mandarins.

10. IMPIETRATURA

Impietratura was first described in Israel in 1930 by Reichert and Hellenger as a transmissible disorder of citrus. Ruggieri (1955) named the disease as 'impietratura'means hardening like a stone (Chapot, 1961). The disease is known to occur in Morocco, Lebanon, Turkey, Greece, India, Italy and other mediterranean countries. Ahlawat *et al.* (1984) reported impietratura disease from Darjeeling district of West Bengal in India.

The yield is significantly reduced due to impietratura infection. The fruits become hardened and fall down prematurely. Gum pockets are developed on branches of affected trees which is an additional symptoms for identification of this disease.The symptoms of impietratura can be confused with boron deficiency and certain insect injury and hence transmission tests are necessary to establish the disease.

The disease is graft transmissible and no vector or other natural modes of transmission are known. Leaves of affected trees show leaf flecking and oak leaf pattern which are the diagnostic symptoms (Bar-Joseph and Loebenstein, 1970). Cartia and Catara (1974) found that oranges and grapefruits are most susceptible to impietratura infection than mandarins and lemons but Volkamer lemon can be used as indicator host. Symptoms of the disease generally appear 6 to 8 months after inoculation and hence reliable detection techniques like, ELISA, SSEM, and molecular hybridization are being used to detect the causal pathogen to identify healthy planting material for propagation.

DISEASES CAUSED BY FASTIDIOUS PROKARYOTES
1. GREENING BACTERIUM

The presence of greening disease in India was suggested by Lilian Fraser,a scientist from Australia during her visit to India in 1965. Following this suggestion Dr. S.P.Capoor and his team at the Regional Station of Indian Agricultural Research Institute, Pune, experimentally demonstrated that the greening disease was transmitted by oriental Psyllid, *Diaphorina citri* Kuwayama (Capoor *et al.*,1967) suggesting that possibly a "virus" was involved with the greening disease of citrus as BLOs and MLO's were not known at that time. Occurrence of greening disease was later reported from various parts of the country mostly based on field symptoms and limited transmission tests. Most of these reports were based on Zinc deficiency type symptoms on the leaves of affected trees but leaf mottle,a typical symptom of greening disease has been overlooked by subsequent workers.

Psyllids can acquire CGB during feeding on diseased source (Capoor *et al.*,1967) but the percentage of the transmission is extremely low. After aquisition the BLO passes through intestinal valve and travel to acini in the salivary glands from where it is transferred to the plants. This is normally a long process (1-3weeks) for most of the prokaryotes. Except in few endemic regions for greening like Coorg in Karnataka and Hindupur in Andhra Pradesh,the field spread of CGB was extremly low in spite of the presence of the psyllid vector *(D.citri)* in abundance which suggested possibility of the presence of different biotypes in psyllid vector.

Adult psyllids were unable to acquire CGB from known CGB infected pineapple sweet orange plants after variable acquisition feeding as determined in DNA-DNA hybridization tests (Ahlawat, 1997). It appears that psyllids either acquire and transmit CGB at their nymphal stages or there are biotypes in *D.citri* that alone act as efficient vector. Further comparative DNA-DNA hybridization results revealed that biotypes of Indian psyllids were less efficient vector than that of Malaysian psyllids. Different serotypes of the CGB were also determined with the help of monoclonal antibodies.The MAbs prepared from Pune strain of the CGB also recongnized the Malaysian CGB showing similarities in the pathogen found in two countries but the presence of more number of viruliferous psyllids in nature in Malaysia further confirms the distinct biotypes in the vector, *D.citri.*

Until the reports of Garnier *et al.*(1984 a,b) that the greening pathogen was phloem restricted gram negative bacterium, the organism was considered to be a virus. Although the bacterial nature of the greening organism (GO) was established several years ago but it has not yet been cultured in cell free medium and hence called as bacterium like organism (BLO). Efforts were made all over the world to characterize the BLO associated with greening disease,but no one succeeded till Jagoueix *et al.*(1995) experimentally confirmed that the greening organism was a true bacterium. They classified it as *Liberobacter* genus with two species, *L.asiaticum,*the organism responsible for Asian greening and *L.africanum,* the organism associated with South African greening.These studies were based on 16s rDNA sequences as well as sequences of the rplKAJL-rpoBC operons with two probes of 2.6Kb (Asian greening) and 1.7Kb (South African greening).

Another breakthrough in the study of citrus greening disease was its experimental transmission to periwinkle (*Catharanthus roseus*) by dodder *(Cuscuta spp.)*. In periwinkle, GO multiplies quickly with much higher concentration and equal distribution than in citrus. Therefore it is easy to isolate it from inoculated periwinkle plants for various studies.

In the absence of reliable diagnostic reagents and tools like electron microscope,the actual incidence and distribution of citrus greening bacterium (CGB) in India could not be achieved till a collaborative Indo-French project was developed at IARI, New Delhi (1991-94) with a view to prepare diagnostic reagents and its effective detection by serological and molecular techniques.

Two important reagents,the monoclonal antibodies and CGB specific nucleic acid probes were developed from a Pune strain of bacterium (Bove *et al.*,1993; Varma *et al.*,1993). However, the trees with typical die back symptoms in Nagpur region were without leaf mottle and gave negative results for the presence of CGB in electron microscopy and DNA-DNA hybridization. These results were in contrast to the observations made by Fraser *et al.* in 1966 and hence needs more intensive testing.The nucleic acid probes also detected the bacterium in individual psyllid vector.

During the period of Indo-French project, 51,331 trees from 98 orchards were examined in 8 different states of India (Andhra Pradesh, Delhi, Haryana, Karnataka, Orrisa,Punjab,Rajasthan and Uttar Pradesh) and the presence of CGB was confirmed by electron microscopy, immunofluorescence and DNA-DNA hybridization in leaf samples collected from these places (Table-8).

Among the commercial cultivars, group of sweet oranges (*C.sinensis* (L) Osbeck) is more suceptible to the greening bacterium than the group of lemons (*C.limon* (L.) Burm.f) and limes (*C.aurantifolia* (Christm) Swingle).Trifoliate orange (*Poncirus trifoliata* (L.) Raf.) although known to be tolerant to greening bacterium but it was also found infected with greening in Coorg region.

Field diagnosis of greening disease is very difficult bacause the symptoms on affected trees are often confused with that of zinc and other nutritional deficiencies. Varma *et al.* (1993) tested citrus trees showing variable symptoms with electron microscopy, ELISA and DNA-DNA hybridization and concluded that mottling of the leaves on affected trees was the main foliar symptoms of the greening disease. Unusual colouring of the fruits and aborted seeds were its additional symptoms. Electron microscopy of ultrathin sections is an important technique to detect the bacterium in sieve tubes of affected leaves (B-3.16). No polyclonal antibodies could be developed for this bacterium so far because of its pleomorphic nature and inability to culture in synthetic media. MAbs have however been prepared and used for strain- specific detection (Bove *et al.*, 1993; Varma *et al.*, 1993; Ahlawat *et al.*,1995).The nucleic acid probes detected most bacterial strains of greening present in India. It is, therefore, now possible to detect more strains of greening bacterium in ELISA, immuno-fluorescence and DNA-DNA hybridization (Varma *et al.*, 1993; Ahlawat *et al.*, 1995; Korsten *et al.*, 1993).

Table-8. **Number of leaf samples with and without mottle which tested positive by Electron Microscopy (EM), Dot-Blot hybridization (dbH) or ELISA in various regions of India.**

REGION			Number of leaf sample tested		Sample positiveby EM/sample tested		Sample positive by dbH/sample tested		Sample positive by ELISA/sample tested		Sample positive/ Sample tested	
Nearby town	Cultivars	With +M	Without -M	With +M	Without -M	With +M	Without -M	With +M	Without -M	With +M	Without -M	
A.P.	Tirupati	Sat.sw.or. (1)	3	2	3/3	0/2		NDb	0/2	0/1	3/3	0/1
	Hindupur	Sat. sw.or.	4		3/3		4/4			ND	4/4	
	Kagzi l. (2)	1		1/0		1/0			ND	1/1		
	Delhi	Kinn.m. (3)	3	1	3/3	0/1	4/4	0/1		ND	4/4	0/1
	(IARI)	Mos.sw.or. (4)	2		2/2	0/3	2/2	0/3		ND	2/2	0/3
KARNATAKA	Bangalore	s.or. (5)	1		1/1			ND	1/1		2/1	
	(IIHR)	C. indica (6)	4		1/3		2/2		2/2		4/4	
		R. lem (7)	6		4/4		3/3		1/1		6/6	
		Kagzil.	1		0/0		1/0			ND	1/1	
		lemon	0		0/1		·	ND		ND	1/1	
		undet. (9)	1		1/1		1/1			ND	1/1	
	Kasakatapura	Ch.sw.or. (9)	7		5/5		3/3		0/4		7/7	
	Gonikoppal	Coorg.m. (10)	6		3/3		6/6			ND	6/6	
		und.sw.or. (11)	1		0/1		1/1			ND	1/1	
		R.lem	1		1/1		1/1			ND	1/1	
	Chitalli	Coorg.m.	5	1	4/3	ND	5/5	0/1		ND	5/5	0/1
		C.macrop. (12)	1		1/1		0/1				1/1	
		undet.	2		1/1		1/1			ND		1/1
MAHARASHTRA	Amravati,	Nagpur .(03)	3		3/3			ND	0/3			3/4
ORISSA	Angul	Nagpur m.	3		0/2		0/2			ND	0/2	
		Kagzi l.	1		1/1		1/1			ND	1/1	
	Subalda	Nagpur m.	1		0/1		1/1			ND	1/1	
RAJASTHAN	Jhalawar	Nagpur m.	2	4	2/2	0/2	2/2	0/4		ND	2/2	0

Abbreviations-1,4,9:Satgudi, Mosambi. Chini sweet orange; 3,10,13: Kinnow, Coorg, Nagpur mandarin; 2: Kagzi lime; 5: sour orange; 6: Citrus indica; 7: rough lime; 8: undetermined cultivar; 11: undetermined sweet orange; 12: Citrus macropetra.

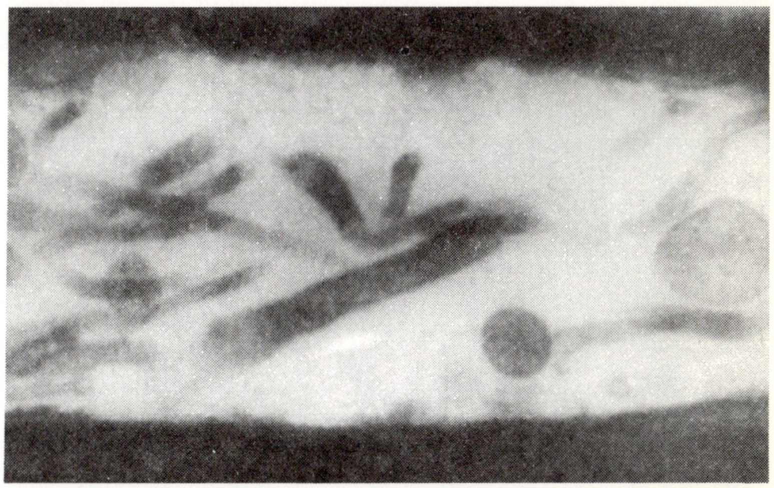

B-3.16 Electron micrograph showing greening bacteriun in the phloem tissue of leaf (see the pleomorphic forms of the organisms)

The prospects of successful breeding of new scion varieties inheriting workable resistance is still remote. However,the disease symptoms can be suppressed by injecting the affected trees with 500 or1000 ppm tetracycline or penicillin under pressure (10Kg^2/cm) (Kapur *et al.*,1992). Regular treatment with these drugs can reduce the incidence and intensity of the greening disease.

2. CITRUS VARIEGATED CHLOROSIS ASSOCIATED BACTERIUM

Citrus variegated chlorosis (CVC) was first reported in Brazil in 1987 (Rossetti *et al.*,1990 and Rossetti, 1993). It is widely distributed in Brazil and is a major constraint in citrus production. CVC is caused by a bacterium *Xylella fastidiosa* (Hartung *et al.*, 1994; Lee *et al.*,1992). It is spreading through vegetative propagation of citrus and appears to have aerial vectors. CVC is more severe where weeds and sharpshooters population are high.

CVC infected trees become non productive and will serve as a source of inoculum for other trees. In CVC affected trees, there is a good co-relation between symptoms and the presence of disease causing bacterium. The bacterium moves very slowly to adjacent branches suggesting the possibility that the disease might be controlled by timely removal of symptomatic limbs. These findings have been used to develop an effective strategy for management of CVC (Baretta *et al.*, 1996).

The bacterium is detected only on symptomatic young and mature leaves and limbs upto 2 cm in diameter by serological tests like dot-blot immunobinding assay and western blotting and PCR (Lee *et al.*, 1992).

MOLLICUTES

These organisms are known to be associated with over 200 plant diseases. They are not cultured in-vitro as yet and hence there is no direct proof for the pathogenicity but indirect evidences as given below are suggestive of their pathogenic nature.

1. Their association with diseased but not with healthy plants.
2. Adverse effect in invaded cells.
3. Disease development following transmission by vector or graft.
4. Disappearance of organisms and remission of disease symptoms following tetracycline treatment.
5. Thermotherapy can inhibit disease development.

Members of the class mollicutes are bounded by a single membrane and do not have cell wall. They are incapable of synthesising cell wall precursors like muramic acid and diaminopalmelic acid. The classification of an organism to this class is based on the following criteria:

- absence of cell wall
- typical "fried egg" appearance of colonies in culture
- filterable through a membrane filter of 450nm diameter pore size.

The class mollicute is divided into 6 families- Mycoplasmataceae, Acholiplasmataceae, Spiroplasmataceae,

Anaeroplasmataceae, Thermoplasmataceae and Entomoplasmataceae.

The diseases caused by mollicutes in citrus are given below:

(i) STUBBORN DISEASE

Stubborn disease was first described in 1915 in California and was considered to be an infectious virus-like disease (Fawcett *et al.*,1944). The disease was later reported from parts of California, N.Africa, Eastern Mediterranean basin and Middle East. The disease is known as little leaf in Israel.

The causal organism of stubborn disease has been established as a culturable, wall free mollicute, *Spiroplasma citri* (Saglio *et al.*,1971, 1972 and Fudl-Allah *et al.*,1971). It is filamentous with spiral morphology and motile. In liquid culture, the spiral filaments are frequently attached to basal bodies, 1-2 μm in diameter. These organisms can be seen in phase contrast or dark field microscope. Fried-egg shaped colonies are developed in semi-solid media.

Spiroplasma citri is transmitted by graft and also by species of leaf hoppers, *Scaphytopius nitridus* and *Circulifer tenellus* in the United States and *C.haematoceps* in Mediterranean region. Recently, Klein and Raccah (1991) reported that species of *C.haematoceps* differ strongly in distribution and host range in Israel than in Mediterranean region.

Spiroplasma citri can infect most citrus species and cultivars. It also infects large number of non-citrus plants. Oranges, grapefruits, mandarins are more susceptible than acid lime, lemons and trifoliate orange. *Spiroplasma citri* infects plant species in 18 dicot families and one monocot family. Natural occurrence of *S.citri* was observed in periwinkle, London rocket, *Brassica* spp. and wild radish. Stubborn disease rarely kills citrus trees but affected trees are often stunted and bushy in appearance. The internodes are shortened and the foliage is dense and abnormally upright. Leaves are cupped and abnormally thick and have variable chlorotic patterns which resemble with nutritional deficiencies. The leaves also show mottling and chlorosis of leaf veins. The stubborn infected trees showed reduction in number of fruits which become lopsided or acorn shaped and thus making them unmarketable.

ELISA and direct culture tests are widely used for detection of this organism from planting material (Kersting and Sengonca, 1992).

(ii). **WITCHES BROOM**

Witches broom disease of lime (WBDL) was first observed in North coastal plain of the Sultanate of Oman near the border of United Arab Emirates (UAE) in 1970. Since then the disease is spreading within Oman and extended all along the coastal plain from UAE to Muscat (Garnier *et al.*, 1991).

A phytoplasma is associated with witche's broom disease of lime. The organism is transmitted to periwinkle (*Catharanthus roseus*) by dodder, *Cuscuta spp.* (Bove *et al.*, 1988). It is graft transmitted to lime, troyer citrange, lemon, rough lemon and trifoliate orange but not to sweet or sour orange (Garnier *et al.*, 1991). The natural host of this phytoplasma in Oman are citron, sweet limetta and Indian and Palestine sweet lime.

Monoclonal antibodies and pathogen specific DNA probes of the phytoplasma have been developed and used for specific detection of the phytoplasma in trees and insect vector (Bove *et al.*, 1993). Among several leaf hoppers *Hishimonus phycitis* reacted positively with phytoplasma-specific probes which was finally proved to be the vector of the disease. *H.phycitis* was also found to feed on lime trees in Sultanate of Oman thus facilitating the spread of witche's broom phytoplasma under natural condition. In India, *H.phycitis* is a vector of the little leaf of brinjal phytoplasma disease (Bindra, 1973). The WBDL-phytoplasma specific monoclonal antibodies and DNA probes did not react with the egg plant little leaf phytoplasma thus showing the two phytoplasma distinct from each other. Recently, witche's broom disease of lime has also been established in India (Ghosh *et al.*, communicated to Plant Disease).

(iii). RUBBERY WOOD

Rubbery wood (RB) disease is known to occur in India for a long time (Ahlawat and Chenulu, 1985). The limbs of affected trees bend downward and are abnormally flexible. The trees become unproductive and may die. The disease is transmitted by graft and dodder, *Cuscuta reflexa*. Phytoplasma bodies were seen only in EM in ultrathin sections of diseased tissue but not in tissue of healthy plants. Phytoplasmas were seen only in sieve tubes of phloem cells (Ahlawat,1987). Rubbery wood has been tested with a panel of MAbs produced against the Indian, Chinese and South African greening bacterium and also with the MAbs produced against witches broom of lime to establish its identity but no positive relationship could be established with these pathogens. The RW phytoplasma is detected either by electron microscopy or through biological indexing on lemon seedlings.

DISEASES CAUSED BY VIROIDS

Citrus trees are known to be naturally infected by five distinct groups of viroid as based on their characteristic of electrophoretic mobility of their nucleic acids,sequence homology and by molecular hybridization as described by Duran-Vila *et al.*,1986, 1988 and also by biological properties in various hosts specially in etrog citron (Puchta *et al.*,1989 and Roistacher, 1977). Semancik and Weather (1972) described the symptoms of exocortis disease on various hosts and a range of symptoms in etrog citron. Levy and Hadidi (1993) described that CVd IIa is 302 nucleotide in length and CVd-IIb is 299 nucleotide in length and differs from CVd-IIa by 3 nucleotide deletions and 2 nucleotides substitutions. CVd-IIa shares more than 99% sequence homology with Japanese hop stunt viroid variants. Some wild forms of citrus exocortis viroid have been reported from Japan but these mild viroids did not show any cross protection with infection by the CEV challenge inoculation (Garnsey *et al.*, 1993). The viroids classified in various categories are CEVd, CVd-I, CVd-II, CVd-IIa, CVd-III and CVd-IV. Viroid associated with exocortis disease fall under CVd-IIa and CVd-IIb is associated with cachexia disease.. Since viroids have been demonstrated to be the causal agent of exocortis and cachexia diseases (Duran-Vila *et al.*,1988a and b). The methods like polyacrylamide gel electrophoresis (PAGE), molecular hybridization and nucleic acid amplification in PCR are widely used for detection and identification of viroids. These method have relevent advantage over conventional biological indexing as they are more specific and reliable.

1. EXOCORTIS

Bark scaling and stunting of infected trees are the main symptoms of citrus exocortis disease as first described in 1948 by Fawcett and Klotz. However, mild to moderate symptoms were interpreted as a reaction of citrus exocortis viroid (CEVd) isolates on indicator plant, Etrog citron. These include epinasty of leaves. browning of petiole, leaf tip and mid- rib (Roistacher *et al.*,1991). The affected trees remain stunted to various degree but CEVd is rarely lethal and fruit quality is not much affected. Trifoliate orange, citranges and Rangpur lime are desirable tristeza tolerant root stocks but are susceptible to CEVd. Mild strains of CEVd have been used to induce tree dwarfing for high density plantations in Australia and Israel but were later replaced by dwarfing root stocks.

Citrus exocortis disease is transmitted by graft and also through mechanical inoculations. The most common method of its natural spread is through contaminated tools employed during orchard operations specially while pruning. Seed and vector transmission have not yet been confirmed experimentally.

CEVd can infect most citrus species and several non-citrus hosts such as tomato (Rutgar) and cucumber (Suyo). The most sensitive citrus species are trifoliate orange, Rangpur lime, some citron specially etrog citron and lemons which develop stem blotching or bark splitting. CEVd infected sweet orange, grapefruit and mandarins are symptomsless but when grafted on sensitive rootstocks characteristic disease symptoms develop.

CEVd is an infectious, circular, ssRNA with 302 nucleotides, which are highly base paired forming a stable rod like structure. Some regions of the CEVd molecule have homologies with potato spindle tuber and chrysanthemum stunt viroids. CEVd is highly resistsnt to heat inactivation and also to chemicals used to inactivate viruses. It may survive for long periods in dry tissue but can be inactivated by hydrolysis or by ribonucleases under appropriate conditions.

In India, Nariani *et al.* (1968) observed symptoms similar to exocortis disease but its etiological agent was not established. Recently, viroids have been identified in citrus species showing exocortis symptoms by return polyacrylamide gel electriphoresis (R-PAGE) in samples collected from various places in India. The R-PAGE analysis revealed the presence of viroids in trees showing (i) typical bark scaling symptoms of exocortis-type disease, which is transmissible to *C. medica* L., (ii) trees without any bark scaling effect but transmissible to suyo cucumber (*Cucumis sativus* L.). However, two different viroid species were associated with two types of symptoms on field trees (Ramachandran *et al.*, 1993). They were identified as citrus viroid IIa (Duran-Vila *et al.*, 1998) and Indian tomato bunchy top viroid (ITBVd) (Mishra *et al.*, 1991). Viroid IIa is a common infection to rootstocks, trifoliate orange (*Poncirus trifoliata* and Rangpur lime (*C. limonia*). However, these two rootstocks are not popular in India and rough lemon (*C. Jambhiri*) is a popular rootstock. Recent investigations have shown that ITBVd can infect the trees even on Rough lemon (*C. Jambhiri*) rootstock without visible symptoms (Ramachandran *et al.*, 1993). Therefore, further evaluation of different scion combinations on rough lemon roots is necessary to establish the damage caused by these pathogens. Presence of these pathogens in Indian citrus may cause concern in the near future as they are symptomless, seed-borne and are transmitted by contaminated tools. Recently, a disease showing yellowing and corking of veins of kagzi lime showed presence of previously unknown viroid (Rustem Ali, 1998).

2. CACHEXIA

Cachexia was reported as a viroid induced disease by Roistacher *et al.*, 1983. It primarily affects mandarins, certain tangelos, alemow and kumquat and commonly found in old line cultivars. The most important symptoms are stunting, chlorosis and tree decline. The most prominent symptoms of cachexia are discolouration and gum impregnation of the bark which can be seen by scrapping of outer bark.The inner bark surface is bumpy with numerous rounded bumpsor pegs which fit into corresponding depressions in the wood. The disease favours warm conditions for better symptoms. Severely affected trees are stunted and chlorotic which eventually decline and die.

The most sensitive host to cachexia is Orlando tangelo but mild reaction also develop on sweet lime, rough lemon and Rangpur lime. The transmission and spread of cachexia is through contaminated tools like exocortis viroid. Semancik *et al.* (1988) proved the viroid nature of cachexia as a low molecular weight RNA of about 299 nucleotides.The nucleic acid hybridization studies indicate that it is a distinct viroid than exocortis and grouped it in CVd-IIa.

DISEASES OF UNCERTAIN ETIOLOGY

Although most of the important diseases of citrus have been studied in detail but the pathogen associated with them have not been characterized in some cases. Among these, citrus blight in Florida (Smith and Reitz, 1977) and citrus declinio in Brazil (Rossetti *et al.*, 1991), citrus dieback in India (Raychaudhuri *et al.*, 1977) are some of the major diseases of concern as they are responsible for heavy losses to citrus industry in their place of origin. The declinio has been demonstrated to be transmitted through root grafts but not by budding (Baretta, *et al.*, 1991). The incidence and severity of the disease is dependent on rootstock used. The cause and mode of its natural spread are not known although some disease specific proteins have been isolated from infected trees which can be used for diagnostic purposes (Derrick *et al.*, 1990). Perhaps the declinio in Brazil and blight in Florida are the diseases of same origin.

Other diseases of uncertain etiology are concave gum (Roistacher, 1963), Algerian navel orange, brittle twig yellow, citrus yellow mottle, fatal yellow disease, gum pocket and gummy pitting, gummy bark,Fovea, Hassaku dwarf, leaf curl, leathery leaf, bud union crease and yellow vein and rumple of lemons (Whiteside *et al.*,1988). These are graft transmissible diseases but the etiology of these diseases has not been established as yet. However, two more diseases namely the leaf rugose (Garnsey and Gonsalves, 1976) caused by an ilarvirus and leprosis and zonate chlorosis (Knorr, 1968) caused by a rhabdovirus are important disorders of sweet orange. Leprosis and zonate chlorosis found in North, Central and South America mainly in Brazil and Argentina. The disease symptoms are developed on foliage, fruit and twigs. Citrus leprosis is transmitted by a mite vector *Brevipalpus phoenicis* in Brazil. Larvae of the mite are more efficient vector than adults. The disease is graft transmitted from citrus to citrus and mechanically to some herbaceous hosts, *Chenopodium amranticolor, C.quinoa,* and *Gomphrena globosa* where local lesions are developed on inoculated leaves (Colariccio *et al.*, 1995). In ultrathin sections of leaves and young bark of sweetorange and cleopatra mandarin bacilliform particles were observed in endoplasmic reticulum. Round structures were occasionally found in the perinuclear space of sweetorange leaf cells. Kitazima *et al.*, 1972 observed bacilliform particles of 100-120x40nm size in tissues of infected sweetorange leaf, some of which appeared to have budded through nuclear membrane. Recently, examination of ultrathin sections of mechanically inoculated citrus revealed particles resembling non-enveloped rhabdovirus measuring 120-130x50-55nm in membranous vesicles of the endoplasmic reticulum (Colariccio *et al.*, 1995 and Lovisolo *et al.*, 1996).

MAJOR FUNGAL AND BACTERIAL DISEASES OF CITRUS

A. Diseases of Nursery and orchards :

Citrus plant is susceptible to number of diseases. List of

important fungal, bacterial and nematode diseases prevalent in citrus nursery and orchards are presented in Table 1b. Among these *Phytophthora* induced diseases and bacterial canker are considered most serious in terms of destructiveness and difficulty or expense in achievening their control.

Phytophthora induced Diseases
Symptoms :

Phytophthora induced symptoms vary from damping-off of seedlings to foot rot and gummosis of scion near ground level, rotting and subsequent decay of fibrous roots and brown rot of citrus fruits. Typical symptoms of damping off results from soil or seed borne fungus penetration of stem just above soil line and causes toppling of seedlings. Fungus can also cause seed rot or pre-emergence rot. However, most serious symptom of *Phytophthora* infection is foot rot and gummosis(Lutz and Menge, 1986). Infection occurs through wounds or natural cracks in the bark of trunk or roots near ground level. Large water soaked patches first develops on the basal portion of stem which may spread further. Infected bark initially remain firm with small cracks through which small or large amounts (depending on citrus variety and weather) of water soluble gums come out. Gummosis symptom sometimes noticed on higher branches also. Infected portion of bark, afterwards shrinks and cracks and becomes sharply deliminated (Feichtenberger, 1990; Davino *et al.*, 1991). Badly affected trees have pale green leaves with yellow veins. If the lesions cease to expand or the fungus dies, the affected area is surrounded by callus tissue. Large trees may be killed but usually only partially girdled and the injury causes decline of canopy. There is decay of fibrous roots of infected plant due to *Phytophthora* infection. Large roots frequently show so called 'frog eye' lesions. *Phytophthora* infection of fruits produce a decay called brown rot (discussed under post-harvest diseases) which results in direct loss of fruit.

Causal Organism:

Most important and widespread *Phytophthora* spp.that attack citrus are *P. parasitica and P. citrophthora. P. parasitica* is more common in sub-tropical areas of the world and causes foot rot, gummosis and root rot but usually does not infect far above the ground. *P. citrophthora* causes gummosis and root rot in Meditterranean climates, where seasonal rainfall occures during the cooler winter months. It attacks aerial parts of the trunk and is most commonly the cause of brown rot. Apart from above mentioned two species others like *P. hibernalis, P. syringae, P. citricola* also occur in limited area (Graham and Timmer, 1992).

Disease cycle and Epidemiology :

The fungus primarily live and reproduce in soil and usually attack susceptible plants at or below the soil line. Sometimes, however, spores of the fungus may be splashed into injured above ground bark or low lying leaves and fruits and cause subsequent infections at these points. The fungus overwinters in infected tissues as chlamydospores, mycelium or oospores. Well aerated and moist soil condition is conducive for germination of chlamydospores. Chlamydospore germinates in response to nutrients from roots to form sporangium that liberates zoospores which finally infects host plant (Tsao, 1969). *P. citrophthora* grows best at 24-28^0 while *P. parasitica* at 28-32^0 C (Lutz and Menge, 1986). Prolonged humid weather with favourable temperature results in disease outbreak.

Disease management :

A. Nursery :

Losses due to *Phytophthora* can be reduced to a greater extent by taking proper sanitary measures in the nursery. Site for nursery should be at some distance from existing orchards. Soil solarization with clear polythene sheet during summer months or fumigation with methyl bromide can completely kill soil borne fungus or nematodes. Seeds should be treated with hot water at 52^{0}C for 10 min. before sowing in nursery to eliminate seed borne fungal infection. Run off water from contaminated areas should not enter into *Phytophthora* free zones. Plant, if infected with *Phytophthora* in nursery should be removed immediately along with surrounding soil which should be followed with soil drenching with systemic fungicides like metalaxyl. (Feichtenberger, 1990; Graham and Timmer, 1992).

B. Orchard :

To control *Phytophthora* disease in orchard following practices should be adopted -

a. Varietal resistance :

Use of resistant rootstocks is the best solution for control of *Phytophthora* diseases. But the problem is some highly resistant rootstocks are either susceptible to other diseases or with poor horticultural performance. Trifoliate orange is almost immune to infection but it grows well in soil with higher pH. Most commercially used rootstocks *viz.*, majority of rough lemon selections, Troyer and Carrizo citrange, Volkamer lemon, Rangpur lime are somewhat tolerant to *Phytophthora* diseases (Graham and Timmer, 1992).

b. Cultural :

i) In areas of high rainfall trees should be budded at 50 cm or so above soil line (Whiteside, 1972).
ii) Injuries to trunk should be avoided which otherwise gives entry to pathogen.
iii) Proper drainage is necessary so that water does not stagnate in field.
iv) Plants with partially damaged foot rot symptom can be recovered by bridge grafting with 2 to 3 resistant rootstock seedlings
v) Canopy budding has been reported very effective to reduce disease (Koller, 1994)
v) *Phytophthora* infected leaves or fruits fallen on ground should be removed and destroyed.

c. Biological :

Vesicular-arbuscular mycorrhizal fungi has been reported highly effective to reduce *Phytophthora* population in citrus (Graham, 1986). This might be due to microbial antagonism or improved host nutrition as a result of phosphorus uptake (Davis and Menge, 1980). Certain strains of *Trichoderma viride* have been identified at IIHR, Bangalore which is effective as biocontrol agent (Sawant *et al.*, 1995).

d. Chemical :

i) Bark tissues from the infected trunk should be removed with knife and followed by wound painting with Bordeaux paste or systemic fungicides (Sawant *et al.*, 1990).
ii) In areas of high foot rot incidence, soil treatment with metalaxyl or regular foliar sprays with fosetyl- Al is a routine recommended practice (Feichtenberger, 1990).
iii) Brown rot of citrus fruits can be controlled by regular preharvesting spraying with fosetyl -AL and metalaxyl (Cohen, 1981).
iv) Foliar sprays with 0.1% phosphorus acid reduces disease in field (Walker, 1988).

Powdery mildew (*Acrosporium tingitaninum*)

All aerial plant parts are susceptible to infection. Characteristic whitish powdery masses are developed on young shoots and leaves. Severe infection leads to leaf and fruit drop and dieback symptom. Sulfur dusting or spraying with Bavistin, and Calixin provides effective control of the disease. (Ram *et al.* 1977; Narayanappa, 1990).

Premature fruit drop

Premature fruit drop may be attributed by number of factors *viz.*, physiological, environmental and pathological. Pathogens involved with this malady are species of *Collectotrichum, Alternaria, Botrytis, Curvularia, Pestalotia* etc. Regular protective spraying with systemic fungicides like benomyl, thiabendazole can reduce fruit drop problem (Timmer and Zitko, 1992).

Anthracnose/wither tip (*Colletotrichum gloeosporioides, Gloeosporium limetticola*)

Leaves, young shoots and tender fruits are generally susceptible to fungal attack. Infected leaves may be distorted or develop localized necrosis. Young shoots with severe infection results in shrivelling of tips. Flower buds when affected fail to develop further, fruits, if developed, results in dropping. Disease can be minimized by proper irrigation, manuring, cultural practices and plant protection measures. Removal of dry twig, application of Bordeaux paste, Zineb (0.75%), 0.2% captafol, lime sulphur or 0.1% Bavistin are suggested control measures (Klotz, 1973; Singh and Agrawala, 1987).

Twig dieback/Twig blight

Number of fungi are associated with the disease. Symptoms caused by *Diplodia natalensis* and *Fusarium spp.* are somewhat similar to anthracnose or Wither tip. Often there are mixed infection by all these fungi. Probably they do more injury in combination than alone (Fawcett, 1936). Other fungi associated with the disease are *Sclerotinia selerotiorum* and *Botrytis cinerea.* The symptoms include wilting, die-back of twig and reddish brown discolouration of dead bark and dusky grey colour of the wood. Trees injured by heat or cold are particularly susceptible with large areas of bark and even larger branches being killed. There may gum exudation from affected branches. Removing infected twigs atleast three inches beyond margin of invasion and disinfection cuts and injuries with copper fungicides are useful.

There are other fungal diseases namely damping off caused by *Rhizoctonia solani, Pythium spp.* (similar to *Phytophthora* induced damping off), scab (*Elsinoe fawcettii*), leaf and fruit spot (*Cercospora sp.*), leafspot (*Alternaria sp.*) etc. which occasionally occur in citrus nursery and orchards.

Citrus Canker

Canker is one of the most widespread and serious disease of citrus. At present, this disease occurs in almost every country in the Pacific region and other parts of the world, except in the Mediterranean countries. In India, citrus canker was recorded in Tamil Nadu, Andhra Pradesh, Karnataka, Rajasthan, Maharashtra, Madhya Pradesh, Assam, U.P. etc. (Balaraman and Purushotman, 1981; Kale *et al.*, 1987; Kishore and Chand, 1987).

Symptoms :

The diseased plants are characterised by the occurrence of conspicuous erumptent lesion that develop on leaves, twigs and fruits. Canker lesions start as pinpoint spots and attain a diameter of 3-10 mm. In leaves, a yellowish halo often surrounds the lesion. Severe infection results in defoliation, die-back and premature fruit drop. Usually canker does not seriously debilitate the tree, but it does cause fruit losses from abscission and nonmarketable quality due to lesions.

Cause :

The disease is caused by the bacterium *Xanthomonas axonopodis* pv. *citri* (Dye *et al.*, 1980; Vauterin *et al.*, 1995). The bacterium is rod shaped, measuring 1.5 - 2 x 0.5 - 0.75 (μm in size, gram negative and forms typical yellowish colonies on nutrient agar media.

Disease cycle and Epidemiology :

The bacterium survives in lesions in leaves, twigs and branches which constitute the main source of inoculum to spread the diseases from season to season. Rao and Hingorani (1963) found that the bacterium survives upto 5 months in the infected leaves. Bacteria are disseminated mostly by wind splashed rains. Disseminations through leaf miner *(Phyllocnistis citrella)* is reported by Nirvan (1961). The long distance dissemination takes place through diseased planting material. Other means of short distance dissemination of the pathogen include contaminated pruning and harvesting tools, equipments and personnel. Infection by the bacteria occurs primarily through stomata, other natural openings and wounds produced during strong winds and by insects.

The disease is most severe in hot, wet cloudy climate, particularly during rainy season. Temperature between 25 -30^{0}C with evenly distributed rains are most suitable for the disease (Ramakrishanan, 1954).

Management :

Control measures include field sanitation, canker free nursery plants, regulation, wind breaks of trees or netting, pruning of diseased summer and autumn shoots, forecasting and chemical sprays. Six or seven sprays of copper are necessary to protect new growth from infection (Kuhara, 1978). Gottwald and Timmer (1995) reported the efficacy of wind breaks in reducing the spread of citrus canker in Argentina. When canker occurred in the United States, the emphasis was on eradication of canker infected plant (Goto, 1992).

In India, recommendations have been made as pruning of affected twigs every year before monsoon and during November-December and 3 to 4 sprays of Bordeaux mixture (1%) in a year (Ramakrishanan, 1954; Patel and Desai, 1970). Two pruning alongwith 4 sprays of 5000 ppm copper oxychloride or 1% bordeaux mixture is reported to be effective against the disease (Kishore and Chand, 1987). Other chemicals effective against the canker are Perenox, ultrasulphur, mixture of sodium arsenate and copper sulphate, Blitox and nickel chloride (Patel and Padhya, 1964 and Ram *et al.*,1972). Six sprays of 1000 ppm streptomycin sulphate along with two prunings reduced the canker in acid lime (Balaraman and Purushotman, 1981). Other effective antibiotics are agrimycin (Sawant *et al.*, 1985), streptocycline in combination with Bordeaux mixture (Krishna and Nema, 1983). Kale *et al.* (1987), in field trials with 7 different chemicals found that the best control of *X.* a.

pv *citri* was given by Paushamycin + Blitox followed by Bordeaux mixture. Singh (1994) at IIHR, Bangalore, found a combination of Blitox 50, Dithane M 45 and streptomycin sulphate gave 68% control of citrus canker. In nursery, application of neem cake solution on the foliage can reduce canker (Reddy and Rao, 1960.)

Kalita *et al.* (1996) isolated a strain of *Bacillus subtilis* from phylloplane of lemon which is an effective antagonist of canker pathogen and can reduce disease upto an extent of 61.9%. Since leaf miner can spread canker bacteria, regular insecticidal spray can control the disease (Goto, 1992).

Resistant Varieties :

Severity of infection varies with the species and varieties of citrus and the prevailling climatic conditions. In India, citrus canker is reported to be relatively more on acid lime and less on mandarin and sweet orange (Ramakrishanan, 1954). Kumquat (*Fortunella spp.*) and Hazara Narangi (*C. microcarpa*) are commonly grown in India for ornamental purpose and these are found resistant to canker. Parsai (1959) suggested that hybridization between Kumquat and Hazara Narangi with the commercial acid lime might give hybrid resistant to citrus canker. *C. latifolia* was also found to be resistant to canker (Kishore and Chand, 1987). Recently ALH-77, an interspecific hybrid lime was reported to be moderately resistant to canker (Prasad *et al.*, 1997).

B. Post-harvest Diseases

Number of fungal and bacterial pathogens are reported to infect citrus fruits after harvest (Table 2). Few of them have the potential for making colossal losses within a very short period. Brief descriptions about symptomatology of important post harvest diseases and their management are given below.

Alternaria rot : Considerable internal rotting (black coloured) takes place in mandarin, grapefruit and sweet oranges, giving rise to common names of black rot or black centre rot. Internal symptom develops before any symptoms appear externally (Singh and Khanna, 1966). The fungus can invade healthy citrus fruit while it is in standing crop through some injury early in the season and thus results in fruit drop (Cohen and Shuali, 1983).

Anthracnose : Fungus usually first infects weakened twigs. During wet or foggy weather fungus spores drip onto fruit where they infect rind leading to the formation of dark brown coloured sunken lesions and rotting. Prolonged rain at blossom time can result in early season infection and premature fruit drop. Alternatively, the infection may remain latent until after harvest and the disease is manifest during transport and storage.

Aspergillus rot : Occurs especially when citrus fruit is stored at fairly high temperature and the fruits are weakend in some way. Infected fruit when cut masses of black powdery spores are seen. However, a soft, water soaked sunken spot may develop in peel (Bhargava, 1972).

Brown rot : Disease (*Phytophthora sp.*) is more prevalent when there is late season rain. Characteristic symptom first develops an greyish-brown spot which later increases but remain firm and leathery. Infected fruits give off a pungent odour (Rao, 1985).

Green Mould and Blue Mould : Early symptoms of both green mould rot (*P. digitatum*) and blue mould rot (*P. italicum*) are development of a soft water soaked area on the peel of the fruit which soon becomes covered with white mould. Coloured spores are formed at the centre of lesion. In green mould there is usually a broad band of white beyond green coloured sporing area, where as in blue mould, white margin is generally not more than 2 mm wide. Frequentlly the two fungi appear together (Barmore and Brown, 1982; Lateef *et al.*, 1994).

Sour rot : The disease (*Geotrichum candidum*) is especially prevalent after warm wet season. Overmature fruits are more susceptible than green or immature fruits. First symptom of infection is water soaked spot in which the affected tissue is extremly soft but not discoloured. Surface of the lesions becomes covered with slimy off-white spores and the fruit tissue beneath is broken down into a sour-smelling watery mass (Kaul *et al.*, 1977).

Stem end rot : It is a serious disease specially in areas having abundant rainfall during growing season. First sign is a slight softening of tissue around the button. Affected rind turns brown and the flesh beneath may darken. In *Botryodiplodia* rot the decay spreads rapidly down the axis of fruit, soon becoming visible at the stylar end. In *Phomopsis* rot the decayed tissue tends to shrink resulting in a ridge of healthy tissue bordering the affected area and in under humid conditions there may be a superficial growth of fine white mould. It is mainly a post-harvest decay occuring in transit and in packing houses (Homma and Yamada, 1969; Ramanjulu and Reddy, 1989).

There are other post-harvest diseases namely greymould rot, melanose, *Trichoderma* rot, scab, *Septoria* spot, Greasy spot rind blotch, cottony rot, *Fusarium* rot Black spot etc. which frequently occur.

Management strategy :

An integrated approach involving application of more than one control measures should be adopted for effective management of post-harvest losses. Sound cultural practices supported by chemical and/or biological methods can minimise crop loss to a greater extent.

Cultural :

i) Pathogenic entry should be stopped by prevention of injury and maintenance of fruit vitality. Harvested fruits should be handled and packed with care.
ii) Fruit harvesting should be done at correct maturity. It is unwise to harvest shortly after rain, since the peel will be turgid and thus more prone to injury and consequent infection.
iii) Fallen fruits should be promptly removed which otherwise may act as a source of inoculum.
iv) Packing boxes should be disinfected periodically with chlorine, quarternary ammonium compound or formaldehyde.
v) Contact between harvested fruit and soil should be

avoided.

vi) Use of vented cartons for transport

vii) Fruits should be stored at low temperature (5-10°C).

Biological :

Bacillus spherieus and *Candida sp.* are reported to be highly effective to control post-harvest diseases caused by *Alternaria citri and Penecillium italicum* (Sharma, 1993). Use of naturally occurring, nonantibiotic producing yeast in combination with chemical (low concentration of TBZ or calcium salts) or physical (heat or UV light) treatment has proved highly useful to control blue and green mould disease in Israel (Chalutz and Droby, 1995).

Chemical :

i) Immersion of fruits in a growth regulating chemical serves to maintain the buttons in a fresh green state thereby delaying fungal invasion of fruit as in case of *Alternaria* rot (Cohen and Shuali, 1983).

ii) Pre-harvest spray with systemic fungicides like benomyl, carbendazim, bengimidazole etc. reduce post-harvest loss. To minimise problems of fungicide resistance, two or more chemically unrelated fungicides should be used. (Ramanjulu and Reddy, 1989; Davino *et al.*, 1991).

iii) After washing, fruits should be coated with wax containing fungicides like benomyl, Carbendazim, thiabendazole etc. Fungicide treatment should be carried out within 24 hours of harvest, before any fungi penetrate too deeply (Gardner *et al.*, 1986).

iv) Following washing and fungicide treatment degreening should be as brief as possible.

v) Post-harvest decay is reduced if citrus fruit is given a pre-harvest application of ethephon which promotes uniform ripening, allowing the crop to be picked at prime maturity (Barmore and Brown, 1978).

vi) Recently, it has been reported that leaf extract of *Ageratum conyzoides* exhibits toxicity against *P. italicum* (Dixit *et al.*, 1995).

FUTURE THRUST

For growing healthy citrus trees, it is essential to have mandatory budwood certification programmes. To start with such a programme it is essential to develop basic information on virus and virus-like pathogens, to prepare diagnostic reagents and to evolve techniques for quick detection of pathogens from the planting material. However, the high average yield to be enjoyed by citrus growers using scientific technology will not depend only by the use of healthy budwood but better disease tolerant stocks and cultural practices also play a major role in citriculture. But unless the orchards are planted with virus-free nursery stock, none of the potential of the improved practices can be fully realised.

From the outset, research and service phases of the protection programmes must be integrated as much as possible. The technology available in the country should go to the field for evaluation. The reduction from laboratory theory to practice may reveal many inadequancies as well as gap in knowledge and refinement could be made leading to cheaper, more rapid and positive virus detection and production techniques.

It is necessary to generate information on epidemiology of citrus virus diseases which has been inadequately studied.

Location-specific researches on viruses and their vectors are warranted to develop integrated management practices. It is important to develop appropriate models on budwood certification to convince the orchardists about the benefit of healthy budwood.

The future breeding strategies for citrus through biotechonological methods must be initiated with a view to develop resistant cultivars, last but not the least, farmers and public must be made educated about the danger of viruses infecting vegetatively propagating crops like citrus.

SUMMARY

Most of the citrus species have their origin in India where they are being cultivated since time immortal. Citrus has moved from India to other parts of the world and has adopted variable agroclimates of different countries. The total citrus production of the world has approximately been estimated 77.50 million tonnes of which 19.1% is produced only in Brazil. India produces 4.8% of the total production and ranks 6th. Citrus is highly susceptible to both biotic and abiotic stresses but virus and virus-like pathogens are the main biotic agents responsible for poor tree health and reduced yield. Over 30 viral diseases, two each of viroid and phytoplasma diseases, one each of fastidious bacterium (*Liberobacter asiaticum, L.africanum*) and citrus stubborn disease caused by *Spiroplasma citri* (Mollicutes) are the major diseases known to cause considerable losses in citrus production in various parts of the world.

During recent years, new viruses infecting citrus were identified and diagnostic reagents were developed for their quick and reliable detection. Among the several such disorders, five diseases, namely, tristeza, ringspot, mosaic, exocortis and greening were identified as potential pathogenic diseases of the decline complex in India. Citrus is also affected with large number of fungal diseases. Among them *Phytophthora* spp. are very important and causing varying degree of root rot, crown rot, gummosis and brown rot of citrus in India.

Among bacterial diseases citrus canker is most devastating disease of kagzi lime in India and is prevalent almost in all acid lime growing areas. Mandarins were supposed to be tolerant to this disease but now certain strains are observed pathogenic to kinnow plantations in India.

REFERENCES

Ahlawat, Y.S., 1975: Leathery leaf - a new virus disease isolated from greening-affected mandarin tree. *Indian Phytopathology*, 28 : 146 (Abs.).New Delhi.

Ahlawat, Y.S., 1987: Association of mycoplasma-like bodies with citrus rubberywood disease.p.12. *In: 3rd Regional workshop on Mycoplasma*, INSA.(Abs).New Delhi.

Ahlawat, Y.S., 1989: Psorosis: a disease of citrus in India. *Indian Phytopathology*, **42**:21-25,New Delhi.

Ahlawat, Y.S., 1997: Virus, greening bacterium and viroids associated with citrus (*Citrus* species) decline in India. *Indian Journal of Agricultural Sciences*, **67**: 51-57,New Delhi.

Ahlawat, Y.S., A.S. Byadgi, A. Varma and N.K. Chakraborty,1995: Detection of ringspot virus and greenig BLO in citrus and their management. *Detection of Plant Pathogen and their Management*.pp-27-44.Ankur Publishers, New Delhi.

AHLAWAT, Y.S. AND V.V. CHENULU, 1985: Rubberywood- a hitherto unrecorded disease of citrus. *Current Science*, **54**: 580-581,Bangalore.

AHLAWAT, Y.S., V.V.CHENULU, N.K.CHAKRABORTY AND S.M.VISWANATH, 1984: Occurrence of impietratura disease of citrus in India. *Current Science*, **53**:384-385, Bangalore.

AHLAWAT, Y.S., V.V. CHENULU, S.M. VISWANATH AND P.K.PANDEY,1985: Studies on a mosaic disease of citrus. *Current Science*, **54**:873-874, Bangalore.

AHLAWAT, Y.S., R.P.PANT, B.E.L. LOCKHART, M. SRIVASTAVA, N.K. CHAKRABORTY AND A.VARMA,1996: Association of a badnavirus with citrus mosaic disease in India.*Plant Disease*, **80**: 590-592, St.Paul,Minnesota.

AHLAWAT, Y.S., T.K. NARIANI AND K.K. SARDAR, 1979: Leathery leaf-A new virus disease of citrus. *Indian Phytopathology*, **32**: 198-201,New Delhi.

AHLAWAT, Y.S. AND S.P. RAYCHAUDHARY, 1988 : Status of citrus tristeza and dieback diseases in India and their detection, p. 871-879. *In* : *Proceeding of Sixth International Citrus Congress*, (eds.) R. Goren and K. Mendel, Tel Aviv, Israel.

AHLAWAT, Y.S. AND K.K. SARDAR, 1976: Lemon crinkly leaf virus in India *Indian Journal of Horticulture*, **33**:168-171, New Delhi.

BALARAMAN, K. AND R. PURUSHOTMAN, 1981: Control of Citrus canker on acid lime. *South Indian Horticulture*, **29** : 175-177,Coimbatore.

BALARAMAN, K AND K. RAMAKRISHNAN, 1977; Studies on strains and strain interaction in citrus tristeza virus, 62 pp.*Technical Series bulletin* 19, University of Agricultural Sciences, Bangalore.

BAKSHI, J.C. AND J.S. DHILLON, 1964: A report on the decline of sweet orange trees in arid-irrigated regions of Punjab. *Punjab Horticulture Journal*, **4** : 15-22,Ludhiana.

BAR-JOSEPH, M., C.N. ROISTACHER, S.M. GARNSEY AND D.J. GUMPH, 1981: A review of tristeza, an ongoing threat to citriculture. p.419-423 *In*: *Proceeding of International Society of Citriculture*.

BAR-JOSEPH, M., G. LOEBENSTEIN, AND J. COHEN, 1970: Partial purification of virus-like particles associated with the citrus tristeza virus disease. *Phytopathology*, **60**:75-78, St. Paul, Minnesota.

BAR-JOSEPH, M., S.M. GARNSEY, D. GONSALVES, M. MOSKOVITZ, D.E. PURCIFUL, M.F. CLARK AND G. LOEBENSTEIN, 1979: The use of enzyme-linked immunosorbent assay for detection of citrus tristeza virus. *Phytopathology*, **69**:190-194,St.Paul,Minnesota.

BAR-JOSEPH, J.M. AND G. LOEBENSTEIN, 1970: Leaf flecking on indicator seedlings with citrus in Israel. A possible indexing method. *Plant Disease Reporter*, **54**: 643-646, Washington, D.C.

BARETTA, M.J.G., V. ROSSETTI, A.R.R. TEIXEIRA AND O. SEMPIONATO, 1991: Positive diagnostic tests for declinio on plants root-graft inoculated in Brazil, *p.256-260. In: Proceeding of 11th Conference of International Organization of Citrus Virologists. IOCV, Riverside.*

BARETTA, M.J.G., V. RODAS, A. GARCIA JUNIOR AND K.S. DERRICK, 1996: Control of citrus variegated chlorosis by pruning. p.378-379. *In*: *Proceeding of 13th Conference of International Organization of Citrus Virologists, IOCV. Riverside.*

BARMORE, C.R. AND G.E. BROWN, 1982: Spread of *Penicillium digitatum and P. italicum* during contact between citrus fruit. *Phytopathology,* **72** : 116-120,St.Paul, Minnesota.

BARMORE, C.R. AND G.E.BROWN, 1978: Pre-harvest ethephon application reduced anthracnose on Robinson tangerine. *Plant Disease Reporter,* **62** : 541-544,Washington. D.C.

BHARGAVA, S.N. 1972: *Aspergillus* rot on *Citrus aurantifolia* fruits in the market. *Plant Disease Reporter,* **56** : 64-65,Washington. D.C.

BINDRA, O.S. 1973: Cicadellid vectors of plant pathogen. *Final report of the P.L.480 Project No.A7-Ent-22,* Punjab Agricultural University, Ludhiana, India.

BOUHIDA, M. 1984: Association of flexuous particles with citrus psorosis disease. M.S. Thesis in Plant Pathology, University of California, Riverside p. 37.

BOVE, J.M., M. GARNIER, A.M. MJÉNI AND A. KHAYRALLAH, 1988: Witches broom disease of small fruited acid lime trees in Oman. First MLO disease of citrus, p.*307-309. In: Proceeding of 10th Conference of International Organization of Citrus Virologists, 10CV, Riverside.*

BOVE, J.M. M. GARNIER, Y.S. AHLAWAT, N.K. CHAKRABORTY AND A. VARMA, 1993: Detection of the Asian strains of the greening BLO by DNA-DNA hybridization in Indian orchards trees and Malaysian *Diaphorina citri* psyllids.p158-263. *In:Proceeding of 12th Conference of International Organization of Citrus Virologists, IOCV. Riverside.*

BOVE, J.M., L. ZREIK, J. DANET JEAN-LUC BONFILS, A.M.M. MJENI AND M. GARNIER, 1995: Witches'broom disease of lime trees: Monoclonal antibody and DNA probes for the detection of the associated MLO and the identification of a possible vector.p.342-348. *In: Proceeding of 12th Conference of International Organisation of Citrus Virologists, IOCV, Riverside*

BRLANSKY, R.H. C.L. DAVIS, R.F. LEE AND L.W. TIMMER, 1993: Immunogold localization of xylem-inhabiting bacteria affecting citrus in Argentina and Brazil.p. 311-319. *In: Proceeding of 12th Conference of International Organization of Citrus Virologists, IOCV, Riverside.*

BROWN, W.R., 1920.: The orange: a trial of stock at Peshawar, p 7. Bulletin 93, Agricultural Research Institute, Pusa,Bihar.

BYADGI, A.S. AND Y.S. AHLAWAT, 1995: A new viral ringspot disease of citrus (*Citrus* species) in India.*Indian Journal of Agricultural Sciences,* **65**:(10): 763-770,New Delhi.

BYADGI, A.S., Y.S. AHLAWAT, N.K. CHAKRABORTY, A. VARMA, M. SRIVASTAVA AND R.G. MILNE, 1993: Characterization of a filamentous virus associated with citrus ringspot in India.p155-162. *In: Proceeding of 12th Conference of International Organisation of Citrus Virologists, IOCV, Riverside.*

CALAVAN, E.C. AND D.J. GUMPF, 1974: Studies on citrus stubborn disease and its agent.*Coll. Inst. Nat. Sante. Rech. Med.* **33**: 181-186.

CAMPIGLIA, H.G., C.M. SILVERA AND A.A. SALIBE, 1976: Psorosis transmission through seeds of trifoliate orange. p 132-134. *In: Proceeding of 7th Conference of International Organisation of Citrus Virologists, IOCV, Riverside.*

CAPOOR, S.P., D.G. RAO AND S.M. VISHWANATH, 1967:*Diaphorina citri* Kuway, a vector of the greening disease of citrus in India. *Indian Journal of*

Agricultural Sciences,**37**: 572-576, New Delhi.

CAPOOR, S.P. AND D.G. RAO, 1967: Tristeza virus infection of citrus in India. p. 723-736. *In: Proceeding of International Symposium on Subtropical and Tropical Horticulture*, Bangalore.

CARTIA, G. AND A. CATARA, 1974: Studies on impietratura disease. p.123-126. *In: Proceeding of 6th Conference of International Organisation of Citrus Virologists, IOCV, Riverside.*

CATARA, A., A. AZZARO, M. DAVINO, AND G. POLIZZI, 1993: Yellow vein clearing of lemon in Pakistan. *In* : p. 364-367. *Proceeding of 12th Conference of International Organisation of Citrus Virologists, IOCV, Riverside.*

CHAKRABORTY, NK, AHLAWAT, Y.S., A. VARMA, K.J. CHANDRA, S. RAMAPANDU AND S.P. KAPOOR, 1993: Serological reactivity in citrus tristeza virus strains in India. p.109-112. *In: Proceeding of 12th Conference of International Organisation of Citrus Virologists, IOCV, Riverside.*

CHAPOT, H., 1961: Impietratura in Mediterranean countries. p. 177-181. *In: Proceeding of 2nd conference of International Organization of Citrus Virologists, IOCV, Riverside.*

CHALUTZ, E. AND S. DROBY, 1995: Biological control of post-harvest diseases of citrus fruit : a viable alternative to synthetic fungicides. p.,414-418. *In: Post-harvest physiology, pathalogy and technologies for horticultural commodities : recent advances.* (Eds. A. Qubahou, and E1- Otmani, M.).

CHILD, J.F.L., 1950: The cachexia disease of orlando tangelo. *Plant Disease Reporter*, **34**:294-298,Washington D.C.

CHILD, J.F.L. AND R.F. JOHNSON, 1966: Preliminary report of seed transmission of psorosis virus. *Plant Disease Reporter*, **50**:81-83.Washington D.C.

COLARICCIO, A., O. LOVIOSOLO, C.M. CHASGAS, S.R. GALLETI, V. ROSSETTI AND E.W. KITAZIMA, 1995: Mechanical transmission and ultrastructural aspects of citrus leprosis disease. *Fitopathologia Brasileira.* **20:** 208-213, Bologna.

COHEN, E., 1981: Metalaxyl for post-harvest control of brown rot of citrus fruit. *Plant Disease*, **65** : 672-673,St.Paul,Minnesota.

COHEN, E. AND M. SHUALI, 1983: Combined treatment with 2,4-D and thiabendazole drencher, before degreening citrus fruits, to delay drying of buttons and stem end rot development, *Alon Hanotea*, **37** : 669-672.

DA GRACA, J.V., 1991: Citrus greening disease. *Annual Review of Phytopathology.* **29**: 109-136, Palo Alto, California.

DA GRACA, J.V., R.F. LEE, P. MORINO, E.L. CIVEROLO AND K.S. DERRICK, 1991: Comparision of citrus ringspot. psorosis and other virus like agents of citrus. *Plant Disease.* **75**:613,St.Paul,Minnesota.

DAVINO, M. AND S.M. GARNSEY, 1984: Purification, characterization and serology of a mild strain of citrus variegation virus from Florida.p.196-203. *In: Proceeding of 9th Conference of International Organisation of Citrus Virologists, IOCV, Riverside*

DAVINO, M., A. CARUSO, F.D. URSO, R. MARINO, G. TERRANOVA AND A. STARRANTINO, 1995: ds RNA analysis of adam lemon affected by rumple disorder. p. 92 *In: Proceeding of 13th Conference of International Organisation of Citrus Virologists, IOCV, Riverside.* (Abstract)

DAVINO, M., O. GAMBERINI, R. AREDDIA AND S.F. ALDARESI, 1991: Field effectiveness of Fosetyl -AL against citrus foot rot and brown rot. Bulletin *OEPP*, **20** : 133-137, Paris.

DAVIS, R.M. AND J.A. MENGE, 1980: Influence of *Glomus fascilatus* and soil phosphorus on *Phytophthora* root rot of citrus. *Phytopathology*, **70**: 447-450,St.Paul,Minnesota.

DEWANG, LI., T. WEIWEN AND F. HWEICHUNG, 1995: Preliminary studies on the methods of rapid serological detection and diagnosis of the BLO associated with citrus shoot yellowing.p.112 *In:Proceeding of 13th Conference of International Organisation of Citrus Virologists, IOCV, Riverside.*

DERRICK, K.S. R.F. LEE AND B.G. HEWITT, 1993: Spirovirus: A new group of plant virus associated with citrus psorosis and ringspot. p.428-429. *In: Proceeding of 12th Conference of International Organisation of Citrus Virologists, IOCV, Riverside.*

DERRICK, K.S., G.A. BARTHE, B.G. HEWITT AND R.F. LEE, 1993: Serological tests for citrus blight. p.121-126. *In: Proceeding of 12th Conference of International Organisation of Citrus Virologists, IOCV, Riverside.*

DERRICK, K.S., R.F. LEE, R.H. BRLANSKY, L.W. TIMMER, B.G. HEWITT AND G.A. BARTHE, 1990: Proteins associated with citrus blight. *Plant Disease*, **74:** 165-170,St.Paul,Minnesota.

DERRICK, K.S.,R.H. BRLANSKY, R.F. LEE, L.W. TIMMER AND T.K. NGUYEN, 1988: Two components associated with the citrus ringspot virus,p.340-342. *In: Proceeding of 10th Conference of International Organisation of Citrus Virologists, IOCV, Riverside.*

DERRICK, K.S:, R.F. LEE, B.G. HEWITT, G.A. BARTHA AND J.V. DA GRACA, 1991: Characterization of citrus ringspot virus.p.386-390.*In: Proceeding of 11th Conference of International Organisation of Citrus Virologists, IOCV, Riverside.*

DIXIT, S.N., H. CHANDRA, R. TIWARI AND V. DIXIT, 1995: Development of a botanical fungicide against blue mould of mandarins. *Journal of Stored Products Research*.**31**:165-172, Oxford, London.

DODDS, J.A. AND M. BAR-JOSEPH, 1983: Double stranded RNA from plants infected with closteroviruses. *Phytopathology*, **73**: 419-423,St.Paul,Minnesota.

DOUGHTY, D. AND J.M. BOVE, 1965: Experiments on mechanical transmission of citrus limes. p. 250-253. *In: Proceeding of 3rd Conference of International Organisation of Citrus Virologists, IOCV, Riverside.*

DURAN-VILA, N., R. FLORES AND J.S. SIMANCIK, 1986: Characterization of viroid like RNAs associated with citrus exocortis viroid. *Virology*. **150**: 75-84, New York, N.Y.

DURAN-VILA, N., J.A. PINA, J.F. BALLESTER, J. JUAREZ, C.N. ROISTACHER, R. RIVERA-BUSTAMENTA AND J.S. SEMANCIK, 1988: The citrus exocortis disease. A complex of viroid-RNAs, p.152-164. *In: Proceeding of 10th Conference of International Organisation of Citrus Virologists, IOCV, Riverside.*

DURAN-VILA. N., C.N. ROISTACHER, R. RIVERA-BUSTAMANTA, J.S. SEMANCIK, 1988: A definition of citrus viroid group and their relationship to the exocortis disease. *Journal of General Virology*.**69**:3069-3080, London.

DYE, D.W., J.F. BRADBURY, M. GOTO, A.C. HAYWORD, R.A. LELLIOT AND M.N. SCHROTH, 1980: International Standards for naming Pathovars from Phytopathogenic Bacteria and a list of Pathovars Names and pathotype strains. *Review of Plant Pathology,* **53** :153-68, Wallingford, England.

ECKERT, J.W. AND B.L. WILD, 1983: Problems of fungicide resistance in *Penicillium* rot of Citrus fruits. p.525-556. In: *Pest resistance to pesticides* G.P. Georghio and T. Saito.(eds.) Plenum Press, New York.

FAWCETT, H.S., 1932: New angles on treatment of bark diseases of citrus. *California Citrograph.* **17** : 406-408, Los Angeles.

FAWCETT, H.S. 1934. Is psorosis of citrus a virus disease ?. *Phytopathology.* **24**: 659-668,St.Paul,Minnesota.

FAWCETT, H.S. (1936). Citrus diseases and their control Mc Graw Hill Book Co. Inc. New york p. 656.

FAWCETT, H.S. 1938. Transmission of psorosis of citrus. *Phytopathology,* **28**: 669 (Abs).St.Paul, Minnesota.

FAWCETT, H.S. AND H.A. LEE, 1926: Citrus diseases and their control. McGraw-Hill, New York and London. p. 582.

FAWCETT, H.S. AND L.J. KLOTZ, 1938: Types and symptoms of psorosis and psorosis like disease of citrus. *Phytopathology,* **28**:670 (Abs).St.Paul, Minnesota.

FAWCETT, H.S. AND L.J. KLOTZ, 1939:Infectious variegation of citrus. *Phytopathology.* **29**: 911-912.

FAWCETT, H.S. J.C. PERRY AND J.C. JOHNSON,1944: The stubborn disease of citrus. *California Citrograph.* **29**: 146-147, Los Angeles.

FAWCETT, H.S. AND J.M. WALLACE. 1946: Evidence of the virus nature of quick decline.*California Citrograph.* **32:** 50-88-89, Los Angeles.

FAWCETT, H.S. 1946. Stubborn disease of citrus, a virosis. *Phytopathology,* **36** : 675-677,St.Paul,Minnesota.

FAWCETT, H.S. AND L.J. KLOTZ, 1948: Exocortis of trifoliate orange. *Citrus Leaves.* **28**(4),8, St. Paul, Minnesota.

FEICHTENBERGER, E.,1990: Control of *Phytophthora* gummosis of citrus with systemic fungicides in Brazil. *Bulletin OEPP,* **20** : 139-148.

FRASER, L.R., D. SINGH, S.P. CAPOOR AND T.K. NARIANI, 1966: Greening virus, the likely cause of citrus dieback in India. *Plant Protection Bulletin, FAO,* **14**: 127-130,Rome.

FUDL-ALLAHH, A.E.S.A., E.C. CALAVAN AND E.C.K. IGWEGBE, 1972: Culture of a mycoplasma-like organism associated with stubborn disease of citrus. *Phytopathology,* **62:** 729-731,St.Paul,Minnesota.

GARCIA, M.L., E.L. ARRESE, O. GRAU AND A.N. SARACHU, 1991:Citrus psorosis agent behaves as a two component ssRNA virus p. 337-344. *In: Proceeding of 11th Conference of International Organisation of Citrus Virologists, IOCV, Riverside.*

GARDNER, P.D., J.W. ECKERT, J.L. BARITILLE AND M.N. BANCCROFT, 1986: Management strategies for control of *Penicillium* decay in lemon packing houses : economic benefits. *Crop Protection,* **5** : 26-32, Guildford, England.

GARNIER, M., C.J. CHANG, L. ZEREIK, V. ROSSETTI AND J.M. BOVE, 1993 : Citrus variegated chlorosis : Serological detection of *Xylella fastidiosa,* the bacterium associated with the disease, p. 301-305. *In: Proceedings of 12th Conference of International Organisation of Citrus Virologists, IOCV, Riverside.*

GARNIER, M., N. DANEL AND J.M. BOVE, 1984a: Etiology of citrus greening disease. *Annals of Microbiology.* **135A**: 169-179.

Garnier, M., N. Danel and J.M. Bove, 1984b: The greening organism is a gram negative bacterium. p.100-108. *In: Proceeding of 9th Conference of International Organisation of Citrus Virologists, IOCV, Riverside.*

Garnier, M., L. Zreik and J.M. Bove, 1991: Witches broom disease of lime trees in Oman: transmission of a mycoplasma like organism (MLO) to periwinkle and citrus and production of monoclonal antibodies against the MLO, p448-453. *In: Proceeding of 11th Conference of International Organisation of Citrus Virologists, IOCV, Riverside.*

Garnier, M and J.M. Bove, 1993: Citrus greening disease and the greening bacterium. p.212-219. *In: Proceeding of 12th Conference of International Organisation of Citrus Virologists, IOCV, Riverside.*

Garnsey, S.M., 1974: Purification and serology of Florida isolate of citrus variegation virus (F-CVV). p. 169-176. *In: Proceeding of 6th Conference of International Organisation of Citrus Virologists, IOCV, Riverside.*

Garnsey, S.M., 1975: Purification and properties of citrus leaf rugose virus. *Phytopathology.* **65**:50-57,St.Paul,Minnesota.

Garnsey, S.M. and D. Gonsalves, 1976: Citrus leaf rugose, CMI/AAB Descriptions of Plant Viruses, No. 164. Commonwealth Mycological Institute and Associations of Applied Biologists, Kew, Surrey, England.

Garnsey, S.M. R.F. Lee and J.S. Simancik, 1993: Lack of cross protection between citrus exocortis viroid and citrus viroids associated with mild symptoms in etrog citron, p. 187-195. *In: Proceeding of 12th Conference of International Organisation of Citrus Virologists, IOCV, Riverside.*

Garnsey, S.M. and L.W. Timmer, 1988: Local lesion isolate of ringspot virus induces psorosis bark scaling,p.334-339, *In: Proceeding of 10th Conference of International Organisation of Citrus Virologists, IOCV, Riverside.*

Gonsalves, D., D.E. Purcifull and S.M. Garnsey, 1978 : Purification and serology of citrus tristeza virus. *Phylopathology,* **68** : 553-559, St. Paul, Minnesota.

Goto, M., 1992: Citrus canker. p. 170-207. *In:* Plant Diseases of International Importance Vol. III. Diseases of Fruit crops , (Ed. Kumar, J. Chaube, H.S., Singh, U.S. and Mukhopadhyay, A.N.) Prentice -Hall, Englewood Cliffs, N.J.

Gottwald, T.R. and L.W. Timmer, 1995: The efficacy of wind breaks in reducing the spread of citrus canker caused by *Xanthomonas campestris* pv. *citri. Tropical Agriculture,* **72**: 194-201, Manila.

Grimaldi, V. and A. Catara, 1996: Association of a filamentous virus with yellow vein clearing of lemon. p. *343-345. In : Proceeding of 13th Conference of International Organisation of citrus Virologists, IOCV, Riverside.*

Grimm, G.R., T.J. Grant and J.F.L. Childs, 1955: A bud-union abnormality of rough lemon rootstock with sweet orange scions. *Plant Disease Reporter,* **39** : 810-811, Washington D.C.

Grahm, J.H., 1986: Citrus mycorrhizae : Potential benefits and interaction with pathogens. *Horticulture Science.* **21**: 1302-1305,Alexandria, Virginia.

Grahm, J.H. and L.W. Timmer, 1992: *Phytophthora* diseases of citrus. p. 250-269. *In:* Plant Diseases of International Importance Vol. III. Diseases of Fruit crops. (Eds).J. Kumar, H.S. Chaube, U.S. Singh, and A.N.

Mukhopadhyay,) Prentice -Hall, Englewood Cliffs, N.J.

HARTUNG, J.S., J.B. BERETTA, R.H. BRLANSKY, J. SPISSO AND R.F. LEE, 1994: Citrus variegated chlorosis bacterium, axenic culture, pathogenicity and serological relationship with other strains of *Xylella fastidiosa. Phytopathology*, **84**:591-597, St.Paul,Minnesota.

HOMMA, Y. AND S.I. YAMADA, 1969: Mechanisms of infection and development of citrus stem-end rot caused by *Diaporthe citri. Bulletin of the Horticultural Research Station, Japan*, **B9** : 99-115.

HOOPER, G.R. AND H. SCHNEIDER, 1969: The anatomy of tumers induced in citrus by citrus vein-enation virus. *American Journal of Botany*, **56**:238-247, Ames.

ISHIGAI,T. AND M. JINNO, 1958: On citrus mosaic.*Annals of Phytopathological Society of Japan*, **23**:29, Tokyo.

IGWEGBE, E.C.K. AND E.C. CALAVAN, 1970: Occurrence of mycoplasmalike bodies in phloem of stubborn-infected citrus seedlings. *Phytopathology* **60** : 1525-1526, St.Paul,Minnesota.

IWANAMI, T. AND LEKI, H., 1995: Nucleotide sequence of the 3'-terminal region of RNA1 of Satsuma dwarf virus, p. 84. *In : Proceeding of 13th Conference of International Organisation of Citrus Virologists, IOCV, Riverside* (Abstract).

IWANAMI, T., T. KANO AND M. KOIZUMI, 1992 : Spherical virus like particle associated with vein enation of Yuzu (*Citrus yunos* Sieb, Tanakal) *Bulletin Tree Research Station*, **23** : 137-143.

JAGOUX, S., J.M. BOVE AND M. GARNIER, 1995: Techniques for the specific detection of the greening *Liberobacter* species: DNA-DNA hybridization and DNA amplification by PCR.p.384-387. *In: Proceeding of 13th Conference of International Organisation of Citrus Virologists, IOCV, Riverside.*

KALE, K.B., J.G. RAUT AND G.B. OHEKAR, 1987: Efficacy of fungicides and antibiotics against acid lime canker. *Pesticides*, **22** : 26-27,Mumbai.

KALITA, P., L.C. BORA AND K.N. BHAGBATI, 1996: Phyloplane microflora of citrus and their role in management of citrus canker. *Indian Phytopathology*, **49** : 234-237,New Delhi.

KAPUR, S.P., S.K. THIND, S.S. CHEEMA AND S.S. SOHI,1992: Citrus greening in the Punjab states and managament. p.52. *In: Proceeding of 13th Conference of International Organisation of Citrus Virologists, IOCV, Riverside.*

KAUL, J.L., R.L. SHARMA AND T.R. SHANDIIYA, 1977: Efficacy of different chemicals against *Geotrichum candidum* causing sour rot of citrus fruits. *Indian Journal of Mycology and Plant Pathology*, **7** : 78-79,Udaipur.

KAWAI, A., T. TSUKAMATO, S. NAMBA AND T. NISHIO, 1996: Citrus tatter leaf virus: a review of its properties and developments of a serological detection system. p.339-342. *In: Proceedings of 13th Conference of International Organisation of Citrus Virologists IOCV, Riverside.*

KERSTING, U., S. KORKMAZ, A. CINAR, B. ERTUGURAL, N. ONELGE AND S.M. GARNSEY, 1996: Citrus chlorotic dwarf, a new whitefly-transmitted citrus disease in the eastern mediterranean region of Turky.p.220-225. *In : Proceeding of International Organization of Citrus Virologists, IOCV, Riverside.*

KERSTING, U. AND SENGONCA, 1992: Detection of insect vectors of the stubborn disease pathogen, *Spiroplasma citri* Saglio *et al.* in the citrus growing area of South Turkey. *Journal of Applied Entomology*, **113**: 356-364, London.

KISHORE, R. AND R. CHAND, 1987: Studies on germplasm resistance and chemical control of citrus canker. *Indian Journal of Horticulture*, **44** : 126-32,New Delhi.

KITAZIMA, E.W., D.M. SILVA, A.R. OLIVEIRA, G.W. MULLUR AND A.S. COSTA, 1964: Thread like particles associated with tristeza disease of citrus. *Nature*, **201** : 1011-1012,London.

KLOTZ, L.J., 1973: Colour handbook of citrus diseases. University California, Riverside p. 112.

KNORR, L.C., R.W. OLSON AND J.W. KESTERSON, 1963: Rumple of lemons, its effect on fresh fruits, lemonade concentrates and peel oil. *Florida Horticulture Society Proceeding*,**76**:36-41, Deland.

KITAZIMA, E.W., G.W. MULLER, A.S. COSTA AND W. YUKI, 1972: Short rod like particles associated with citrus leprosis. *Virology*, **50**: 254-248, New York, N.Y.

KLEIN, M. AND B. RACCAH, 1991: Separation of two leaf hopper population of the *Circulifer haematoceps* complex on different host plants in Israel. *Phytoparasitica*, **19**: 153-155, Bet Dagan,Israel.

KNORR, L.C., 1968: Studies on etiology of leprosis in citrus. p.332-341. *In*: *Proceeding of 4th Conference of International Organisation of Citrus Virologists, IOCV, Riverside.*

KOIZUMI, M., T. KANO, H. LEKI AND H. MAE, 1988: China laurestine a symtomless carrier of satsuma dwarf virus which aaclerates natural transmission in fields. p.348-352. *In*: *Proceeding of 10th Conference of International Organisation of Citrus Virologists, IOCV, Riverside.*

KOLLAR, O.L. AND E. SOPRANO, 1994: Canopy budding - a method that reduces *Phytophthora* problems on citrus limon. *Fruits* (Paris) **49** : 211-215, Paris.

KORSTEN, L., G.M. SANDERS, H.J. SU, M. GARNIER, J.M. BOVE AND J.M. KOTZE, 1993: Detection of citrus greening-infected citrus in South Africa using a DNA probe and monoclonal antibodies. p.224-232. *In*: *Proceeding of 12th Conference of International Organisation of Citrus Virologists, IOCV, Riverside.*

KRISHNA, A. AND A.G. NEMA, 1983: Evaluation of chemicals for the control of citrus canker. *Indian Journal of Horticulture*, **44**: 126-132,New Delhi.

KUHARA, S., 1978: Present epidemic status and control of citrus canker disease, *Xanthomonas citri* (Hasse) Dow in Japan. *Rev. Plant. Protection Research*,**11** : 132-142.

LATEEF, M.F.A.A., M.B. MAHMOUD AND H.A. FAHMY 1994: Effect of time of of application on the effectiveness of fungicides for the control of blue and green mould of oranges.*Egyptian Journal of Phytopathology*, **22** : 59-73, Cairo.

LEE, R.F., K.S. DERRICK, M.J.G. BERETTA, C.M. CHAGAS AND V. ROSSETTI, 1991: Citrus variegated chlorosis : a new destructive disease of citrus in Brazil. *Citrus Industry*, **72** : 12-13., Bartow, Florida.

LEE, R.F., M.J.G. BERETTA, K.S. DERRICK AND M.E. HOOKER,1992. Development of serological assay for citrus variegated chlorosis. A new disease of citrus in Brazil. *Proceeding Florida State Horticultural Society*.**105**: 32-34, Deland.

LEVY, L. AND A. HADIDI, 1993: Direct nucleotide sequencing of PCR-amplified DNAs of the closely related citrus viroid IIa and IIb (cachexia), p.180-186. *In: Proceeding of 12th Conference of International Organisation of Citrus Virologists, IOCV, Riverside.*

LOVISOLO,O., A. COLARICCIO, C.M. CHAGAS, V. ROSSETTI, E.W. KITAZIMA AND R. HARAKAWA, 1996: Partial characterization of citrus leprosis virus. p.179-188. *In: Proceeding of 13th Conference of International Organisation of Citrus Virologists, IOCV, Riverside*

LUTZ, A. AND J. MENGE, 1986: Citrus root health II: *Phytophthora* root rot. *Citrograph*,**72**: 33-35, Los Angeles.

MALI, V.R., K.G. CHAUDHARI AND S.D. RANE, 1976a: Leaf-curl virus disease of citrus in India. *Science and Culture*,**42**: 525-527,Calcutta.

MALI, V.R., K.G. CHAUDHARI AND S.D. RANE, 1976b: The vein enation virus disease of citrus in India. *Indian Phytopathology*, **92**: 43-45,New Delhi.

MALI, V.R., 1979: Indexing results of mosambi, sweet orange decline on rough lemon rootstocks in Maharashtra. *Indian Journal of Mycology and Plant Pathology*, **9**: 193-199,Udaipur.

MALI, V.R., K.G. CHAUDHARI AND S.D. RANE, 1975: Blastomania - a new bud transmissible disorder of citrus. *Current Science*, **44**: 627-628,Bangalore.

MANJUNATH, K.L., H.R. PAPPU, R.F. LEE, C.L. NIBLETT AND E.L. CIVEROLO, 1993: Studies on the coat protein genes of four isolates of citrus tristeza closterovirus from India:cloning,sequencing and expression.p.20-27 *In: Proceeding of 12th Conference of International Organisation of citrus Virologists, IOCV, Riverside.*

MARTELLI, G.P., G. MAJORAMA AND M. RUSSO, 1968: Investigations on the purification of citrus variegation virus. p. 267-273. *In: Proceeding of 4th Conference of International Organisation of citrus Virologists, IOCV, Riverside.*

MARAIS, L.J., R.F. LEE, J.H.J. BREYTENBACH, B.Q. MANICOM AND S.P. VUUREN, 1996: Association of a viroid with gum pocket disease of trifoliate orange. p. 236-244. *In : Proceeding of 13th Conference of International Organisation of citrus Virologists, IOCV, Riverside.*

MILNE, R.G., K. DJELOUAH, M.L. GARCIA, E. DAL BO AND O. GRAU, 1996: Structure of ringspot psorosis - associated virus particles : implications for diagnosis and taxonomy. p 78. *In : Proceeding of 13th Conference of International Organisation of Citrus Virologists, IOCV, Riverside* (Abstract).

MIYAKAWA, T., D. GONSALVES AND S.M. GARNSEY, 1977: Some recently discovered and recognized virus diseases and their potential threat to citrus production. p. 941-945. *In : Proceeding of Third International Society of Citriculture, Riverside.*

MIYAKAWA, T., 1980: Occurrence and varietal distribution of tatter leaf, citrange stunt virus and its effects on Japanese citrus. p. 220-224. *In: Proceeding of 8th Conference of International Organisation of Citrus Virologists, IOCV, Riverside.*

MISHRA, M.D., R.W. HAMMOND, R.A. OWENS, D.R. SMITH AND T.O. DIENER, 1991: Indian bunchy top disease of tomato plants is caused by a distinct strain of citrus exocortis viroid.*Journal of General Virology*, **72** : 1-5.London.

MOREIRA, S., 1942: Observacoes sobre a "tristeza" dos citrus ou "podridao dos radiacelas" *Biologica,* 8:269-272, Oulu, Finland.

MURTI, V.D. AND G.S. REDDY, 1975: Mosaic-a transmissible disorder of sweet oranges._*Indian Phytopathology,* **28:** 398-399,New Delhi.

NARAYANAPPA, M., (1990). Chemical control of powdery mildew on coorg mandarin (*Citrus reticulata*). *Indian Journal of Agricultural Sciences,* **60**: 363-364, New Delhi.

NAGPAL, R.L. 1959 : Tristeza and other diseases of citrus found in Bombay State. *Citrus Industry,* **40**(1) : 14-15, Barlow, Florida.

NARIANI, T.K., S.P. RAYCHAUDHARY, B.C. SHARMA, 1968: Exocortis in citrus in India, *Plant Disease Reporter,* **52**:834, Washington, D.C.

NAVARRO, L., J. JUAREZ, J.F. BALLESTER AND J.H. PINNA, 1980: Elimination of some citrus pathogen producing psorosis like symptoms by shoot tip grafting in vitro. p.162-166 *In: Proceeding of 8th Conference of International Organisation of Citrus Virologists, IOCV, Riverside.*

NAVAS-CASTILLO, J., P. MORENO, J. BALLESTER, J.A. PINA AND A. HERMOSA DE MENDOSA, 1991: Detection of a necrotic strain of citrus ringspot in star ruby grapefruit in Spain. p. 345-351. *In : Proceeding of 11th Conference of International Organisation of Citrus Virologists, IOCV, Riverside.*

NAVAS-CASTILLO, J., P. MORENO AND N. DURAN-VILLA, 1995. Citrus psorosis, ringspot, cristaçortis and concave gum pathogens are maintained in callus culture. *Plant Cell Tissue Organ Culture,* **40** : 133-137, Dordrecht, Netherlands.

NIRVAN, R.S., 1961: Citrus canker and its control. *Horticulture_Advances,* **5**: 171-175, Saharanpur, Uttar Pradesh.

NOUR-ELDIN, F., 1956: Phloem discoloration of sweet orange. *Phytopathology,* **46** : 238-239, St.Paul, Minnesota.

OBEBHOLZER, P.C.J., 1947 : The present status of citrus nutrition in South Africa. *South Africa Department of Agriculture, Citrus Nutrition Bulletin,* **271** : 14.

OHIRA,K.S., M. NAMBA, T. ROZANOVKUSUMI AND T. TSUCHIZAKI,, 1995: Complete nucleotide sequence of an infectious full length cDNA clone of citrus tatter leaf Ilarvirus. Comprative sequence analysis of capillovirus genome. *Journal of General Virology,* **76**:2305-2309, London.

ONELGE, N., A. CINAR, U. KERSTING AND J.S. SEMANCIK, 1996: Citrus viroids associated with gummy bark disease of sweet orange in Turkey. p. 245-248. *In : Proceeding of 13th Conference of International Organisation of Citrus Virologists, IOCV, Riverside.*

PANT, R.P. AND Y.S. AHLAWAT, 1997: Partial characterization of citrus mosaic virus. *Indian Phytopathology,* **50** (4): 557-564, New Delhi.

PANT, R.P., Y.S. AHLAWAT AND R.G. MILNE, 1997: Studies on citrus ringspot virus in India. National symposium on citriculture, Nov. 17-19, 1997 at NRCC, Nagpur : 82-83 (Abstract).

PARSAI, P.S., 1959: Citrus canker. p. 91-95. *In : Proceeding on Seminar on Diseases of Horticultural Plants,* Shimla.

PATEL, M.K. AND A.C. PADHYA, 1964: Sodium arsenite, copper sulphate spray for the control of citrus canker. *Current Science,* **33** : 87-88, Bangalore.

PATEL, R.S. AND M.V. DESAI, 1970: Control of citrus canker. *Indian Journal of Horticulture,* **27** : 93-98, Bangalore.

PATIL, B.P. AND D.C. WARKE, 1968: A note on existence of exocortis virus in India. *Current Scince,* **37** : 469-470, Bangalore.

PRASAD, M.B.N.V., R. SINGH AND A. REKHA, 1997: ALH-77: An interspecific hybrid lime with good fruit quality and resistant to citrus bacterial canker disease. National Symposium on Citriculture, held at NRCC, Nagpur 17-19 Nov., 1997. p. 1. (Abstract).

PUCHTA, H., K. RAMM, R. HADIDI, M. BAR-JOSEPH, R. LUCKINGER, K. FREIMULLER AND H. SANGER, 1989: Nucleotide sequencing of hop stunt viroid (HSVd) isolate from grapefruit. *Nucleid Acid Research,* **17**:1247, London.

PUJOL, A.R. AND H.N. BENATENA, 1965: A study of psorosis in concordia, Argentina. p. 179-184. *In: Proceeding of 3rd Conference of International Organisation of Citrus Virologists, IOCV, Riverside.*

RAM, B., R. NAIDU, N.N. RAO, B.A. ULLASA, H.S. SOHI AND D.G. RAO, 1977: Fungal diseases of mandarin in Malnad region of Karnataka and their control. *Proceeding of International Symposium on Citriculture. USA.*

RAM, G., R.S. NIRWAN, AND M.L. SAXENA, 1972: Control of Citrus canker. *Progressive Horticulture,* **12** : 240-243, Ranikhet, Uttar Pradesh.

RAMACHANDRAN, P., P.K. PANDEY, Y.S. AHLAWAT, A. VARMA AND S.P. KAPUR, 1993: Viroid Diseases of Citrus in India.p. 438-440. *In: Proceeding of 12th Conference of International Organisation of Citrus Virologists, IOCV, Riverside.*

RAMAKRISHNAN, T.S., 1954: Common diseases of citrus in Madras State, Govt. of Madras publication, Chennai.

RAMANJULU, V. AND M.R.S. REDDY, 1989: Pre-harvest stem end rot of Sathgudi sweet orange caused by *Gloeosporium limetticola. Indian Phytopathology,* **42** : 108-109, New Delhi.

RAO, M.R.K. AND B. NARASIMHAN, 1974: The effect of different citrus viruses on photosynthetic pigments. *Current Science,* **43**: 86-87.Bangalore.

RAO, N.N.R., 1985: Relative efficacy of a single premonsoon spray of some fungicides for the control of brown fruit rot of Citrus. *Indian Journal of Agricultural Sciences,* **55**: 189-192,New Delhi.

RAO, Y.P. AND M.K. HINGORANI, 1963: Survival of *Xanthomonas citri* (Hasse) Dowson in leaves and Soil. *Indian Phytopathology,* **16** : 362-364,New Delhi.

RAO, Y.P. AND R.C.K. SOUMINI, 1957: Diseases of Citrus canker in Madras State. *Indian Horticulture,* **5**: 50-57,New Delhi.

RAYCHAUDHURY,S.P., T.K. NARIANI AND Y.S. AHLAWAT, 1977: Dieback of citrus in India.p.914-918. *In: Proceeding of International Citrus Congress Orlando,* Florida, USA.

REDDY, B.V.B. 1997. Characterization of Citrus mosaic virus and to develop methods for its quick detection. *Ph.D. Thesis, IARI,* New Delhi.

REDDY, G.S. AND A.P. RAO, 1960: Control of canker in Citrus nurseries. *Andhra Agriculture Journal,* **7**: 11-13,Bapatla.

REDDY,G.S., V. DAKSHINAMURTI AND V.B.K. REDDY, 1974: Yellow corky vein: First report of a new graft-transmissible disorder of sathgudi in Andhra pradesh. *Indian Phytopathology,* **27**:82-84,New Delhi.

REICHART, I. AND E. HELLANGER, 1930: Internal decline physiological disease of citrus fruits new to palestine. *Hader* (Tel-Aviv) **3**: 220-224, Tel-Aviv.

REICHART, I. AND J. PERLBERGER, 1934: Xyloporosis, the new citrus disease. *Jewish Agency for Palestine Agriculture Experiment Station.[Rehovot] Bulletin,* **12** : 1-50.

ROSSETTI, V., M.J.G. BARETTA, A.R.R. TEIXEIRA, 1991: Experimental transmission of declinio by approach-root grafting in Sao Paulo State,

Brazil, p. 250-255. *In: Proceeding of 11th Conference of International Organisation of Citrus Virologists, IOCV, Riverside.*

ROISTACHER, C.N., 1963: Effect of light on symptoms expression of concave gum virus in certain mandarin. *Plant Disease Reporter*, **47**: 914-915, Washington,D.C.

ROISTACHER, C.N., E.C. CALAVAN, R.L. BLUE, L. NAVARRO AND R. GANZALEZ, 1977: A new more sensitive citron indicator for detection of mild isolates of citrus exocortis viroid (CEV). *Plant Disease Reporter*, **53**:33-336, Washington,D.C.

ROISTACHER, C.N., D.J. GUMPF, E.M. NAUUER AND R. GONZALES, 1983: Cachexia disease virus or viroid. *Citrograph*, **68**: 111-113,Berkeley.

ROISTACHER, C.N., J.E. PEHRSON AND J.S. SEMANCIK, 1991: Effect of citrus viroid and citrus exocortis viroid on performance of sweet orange on three root stock, p. 234-239. *In: Proceeding of 11th Conference of International Organisation of Citrus Virologists, IOCV, Riverside.*

ROISTACHER, C.N., 1993: Psorosis- A Review. p 139-154. *In: Proceeding of 12th Conference of International Organisation of Citrus Virologists, IOCV, Riverside.*

ROSSETTI, V., 1993: Citrus variegated chlorosis, a new severe disease in Brazil. A Review. p.449-452. *In: Proceeding of 13th Conference of International Organisation of Citrus Virologists, IOCV, Riverside.*

ROSSETTI,V., M. GARNIER, J.M. BOVE, M.J.G. BERETTA, A.R.R. TEIXEIRA, J.A. QUAGGIO AND J.D. DENEGRI, 1990: Presence de bacteries dans le xyleme d'orangers atteints de agrumes au Brazil. *Comptes Rendus des seances de L' Academie Paris. Sirie* III: 345-349.

RUGGIERI, G., 1955: Le arana impietrate. *Riv. agrumic*, **1**: 65-69.

RUGGIERI, G., 1961: Observations and research on impietratura, p. 182-186. *In : Proceeding of 2nd Conference of International Organisation of Citrus Virologists, IOCV, Riverside.*

RUSSO, F. AND L.J. KLOTZ, 1963: Tarocco pit. *California- Citrograph*, **49** : 221-222, Los Angeles.

RUSTEM ALI, F.A. 1998. Investigations on the citrus yellow corky vein disease to establish its etiology. *Ph.D. Thesis, IARI*, New Delhi.

SMITH, P.F. AND H.J. REITZ, 1977: A review of the nature and history of citrus blight in Florida. p. 881-884. *In : Proceeding of 3rd International Society of Citriculture.*

SALIBE, A.A., 1959: Leaf curl - a transmissible virus disease of citrus. *Plant Disease Reporter*, **43** : 1081-1083,Washington,D.C.

SEARLE, C.M., 1969: A preliminary report on off type trees. Mazoe Citrus Estates (Rhodesia), p. 15.

SAGLIO, P., D. LEFLECHE, C. BONISSOL AND J.M. BOVE, 1971a: Isolement et culture in vitro des mycoplasmes associes au 'stubborn' des agrumes et ieur observation au microscope electronique. *Comptes Rendus desscances de L'Academic d' Agriculture de France.* **272**: 1387-1390.

SAGLIO, P., M. L'HOSPITAL, D. LAFLECHE, G. DUPONT, J.M. BOVE, J.G. TULLY AND E.A. FREUNDT, 1972: *Spiroplasma citri* gen. and sp.n. a mycoplasma-like organism associated with "Stubborn" disease of citrus. *International Journal of Systematic Bacteriology*, **23**: 191-204, Washington, D.C.

SEMANCIK, J.S., C.N. ROISTACHER AND N. DURAN-VILA. 1988 : A new viroid is the causal agent of the citrus ca chexia. p. 125-135. *In: Proceeding of 10th Conference of International Organisation of Citrus Virologists, IOCV, Riverside.*

Semancik, J.S. and N. Duran-Vila, 1991: The grouping of citrus viroids: Additional physical and biological determinants and relationship with disease of citrus, p.178-188. *In: Proceeding of 11th Conference of International Organisation of Citrus Virologists, IOCV, Riverside.*

Semancik, J.S. and L.W. Weathers, 1972: Exocortis disease: evidence for a new species of infectious low molecular weight of RNA in plants. *Nature New Biology.* p. 242-244.

Sawant, I.S., S.D. Sawant and K.A. Nanaya, 1995: Biological control of *Phytophthora* root rot of Coorg mandarin (*Citrus reticulata) by Trichoderma sp.* grown on coffee waste. *Indian Journal of Agricultural Sciences,* **65** : 842-846,New Delhi.

Sawant, D.M., A.G. Ghawte, J.V. Jadhav and K.G. Chaudhari, 1985: Control of Citrus canker in acid lime. *Maharashtra Journal of Horticulture,* **2**: 55-58, Rahuri.

Sawant, S.D., K.L. Manjunath and I.S. Sawant, 1990: Chemical control of stump rot and gummosis of coorg mandarin in Kodagu. *Plant Disease Research,* **5** : 191-193,Washington,D.C.

Schneider, H., 1954: Anatomy of bark of bud union trunk and roots of quick decline affected sweet orange trees on sour orange root stock. *Hilgardia,* 22, No.16: 567-581, Berkeley, California.

Schwarz, R.E. and A.P.D. McClean, 1969: Gum-pocket, a new virus-like disease of *Poncirus trifoliata. Plant Disease Reporter,* **53** : 336-338, Washington,D.C.

Semancik, J.S., 1986: Separation of viroid RNAs by cellulose chromatography indicating conformational distinctions. *Virology,* **155** : 39-45, New York, N.Y.

Sharma, D.C. and P.K. Pandey, 1983 : Leaf yellow mid vein, a previously undescribed virus disease of Citrus. *International Journal of Tropical Plant Disease,* **1** : 145-152, New Delhi.

Sharma, Neeta., 1993: Post harvest biological control of Citrus fruit rot. *Journal of Biological Control,* **7**: 84-86, Bangalore.

Singh, Dhanbir and R.K. Agarwala, 1987: Differential reaction of fungicides to anthracnose of citrus (*Collectotrichum gloeosporioides*) in vitro and in vivo. *Indian Journal of Mycology and Plant Pathology,* **17** : 323-324,Udaipur.

Singh, R. 1994: Studies on bacterial diseases of fruit crops. *Annual_Report,* IIHR, Bangalore.

Singh, R.S. and R.V. Khanna, 1966: Black core rot of mandarin oranges caused by *Alternaria tenuis. Plant Disease Reporter,* **50** : 127-131, Washington, D.C.

Swingle, W.T. and H.J. Webber, 1896: The principle diseases of citrus fruits in Florida. *United States Deptartment of Agriculture Division of Vegetable Physiology and Pathology.Bulletin* **8** : 42.

Tanaka, H. and Imada, J., 1976: Purification of viruses of citrus mosaic and navel orange infectious mottling. p. 116-118. *In : Proceeding of 5th Conference of International Organisation of Citrus Virologists, IOCV, Riverside.*

Takahashi, T., N. Saito, A. Goto, A. Kawai, S. Namba and S. Yamashita, 1990:

Apple stem grooving virus isolated from Japanese apical (*Prunus mume*) imported from China. *Research Bulletin Plant Protection, Japan*, **26:** 15-21.

TANAKA, H., S. YAMADA AND NAKANISHI, 1971: Approach to eliminating tristeza virus from citrus trees by using trifoliate orange seedlings. *Bulletin Horticulture Research Station Japan, B.No.11*:157-163.

TIMMER, L.W. AND S.M. GARNSEY, 1980: Natural spread of citrus ringspot virus in Texas and its association with psorosis like diseases in Florida and Texas. p. 163-173. *In : Proceeding of 8th Conference of International Organisation of Citrus Virologists, IOCV, Riverside,*

TIMMER, L.W. AND S.E. ZITKO, 1992: Timing of fungicidal application for control of post bloom fruit drop of citrus in Florida. *Plant Disease*, **76** : 820-823, St.Paul,Minnesota.

TSAO, P.H., 1969: Studies on saprophytic behaviour of *Phytophthora parasitica* in soil. *Proceeding of First International Citrus Symposium*, **3**: 1221.

UYEDA, I. AND G.I. MINK, 1983: Relationship among ilarviruses: proposed revision of sub group A. *Phytopathology*, **73**:47-50, St.Paul,Minnesota.

USUGI, T. AND Y. SAITO, 1979: Satsuma dwarf virus. CMI/AAB Description of plant viruses, No.208 Commonwealth Mycological Institute and Association of Applied Biologists, Kew, Surry, England.

VARMA, A., Y.S. AHLAWAT, N.K. CHAKRABORTY, M. GARNIER AND J.M. BOVE, 1993: Geographical distribution of greening in India as determined by leaf mottle and detection of the greening BLO by electron microscopy, hybridization and ELISA. p. 280-285.*In: Proceeding of 12th Conference of International Organisation of Citrus Virologists, IOCV, Riverside.*

VASUDEVA, R.S. AND S.P. CAPOOR, 1958: Citrus diseases in Bombay state. *Plant Protection Bulletin*, **6**: 91, New Delhi.

VASUDEVA, R.S., P.M. VERMA AND D.G. RAO, 1959: Transmission of citrus decline virus by *Toxoptera citricidus* Kirk. in India. *Current Science*, **28:** 418, Bangalore.

VAUTERIN, L., B. HOSTE, K. KWRSTERS AND J. SWINGS, 1995: Reclassification of *Xanthomonas. International Journal of Systematic Bacteriology*, **45**: 472-489, Washington, D.C.

VERMA, P.M., D.G. RAO AND S.P. CAPOOR, 1965: Transmission of tristeza virus by *Aphis craccivora* (Koch) and *Dactynotus jaceae* L. *Indian Journal of Entomology*, **27**:67-71,New Delhi.

VERMA, P.M., D.G. RAO AND R.S. VASUDEVA, 1960: Additional vectors of tristeza disease of citrus in India.*Current Science*, **29**:359,Bangalore.

VISVADER, J.E., AND R.H. SYMONS, 1985: Eleven new sequence variants of citrus exocortis viroid and the correlation of sequence with pathogenicity. *Nucleic Acid Research*, **13**: 2907-2920, London.

VISVADER, J.E., A.R. GOULD, G.E. BRUENING AND R.H. SYMONS, 1982: Citrus exocortis viroid : nucleotide sequence and secondary structure of an Australian isolate.*FEBS Letters*, **137** : 288-292.

VOGEL, R. AND J.M. BOVE, 1964: Stem pitting sun bigarudier et sur organger tarocco en corse: une maladie a virus. *Friuts*, **19**: 269-274, Paris.

VOGEL, R. AND J.M. BOVE, 1972: Relationship of citrus cristacortis virus to other viruses. p 178-184. *In*: *Proceeding of 5th Conference of International Organisation of Citrus Virologists, IOCV, Riverside.*

VOGEL, R. AND BOVE, J.M., 1980: Citrus ringspot in corsica. p. 180-182. *In : Proceeding of 8th Conference of International Organisation of Citrus Virologists, IOCV, Riverside,*

VOGEL, R. AND J.M. BOVE, 1987: Transmission of some infectious diseases of citrus using leaf discs. *Fruits - Paris,* **42** : 231-234, Paris.

WALKER, G.E., 1988: *Phytophthora* root rot of container grown citrus as affected by foliar sprays and soil drenches of phosphorus and acetyl - salicylic acids. *Plant and Soil,* **107**: 107-112, The Hague.

WALLACE, J.M., 1945: Techniques for hastening foliage symptoms of psorosis of citrus. *Phytopathology,* **35**: 535-541, St.Paul,Minnesota.

WALLACE, J.M. AND R.J. DRAKE, 1960: Woody galls on citrus associated with vein-enation virus infection. *Plant Disease Reporter,* **44**:580-584,Washington D.C.

WALLACE, J.M. AND R.J. DRAKKE, 1962: Tatter leaf, a previously undescribed virus effect on citrus. *Plant Disease Reporter,* **46**:211-212, Washington, D.C.

WALLACE, J.M. AND R.J. DRAKE, 1953: A virus-induced vein enation in citrus. *Citrus leaves,* **33** : 22-23.

WALLACE, J.N. AND R.J. DRAKE, 1968: Citrus stunt and ringspot, two previously undescribed virus diseases of citrus. p. 177-183. *In : Proceeding of IVth Conference of International Organisation of Citrus Virologists, IOCV, Riverside.*

WEATHERS, L.G., 1957: A vein-yellowing disease of citrus caused by a graft-transmissible virus. *Plant Disease Reporter,* **41** : 741-742, Washington, D.C.

WEATHER, L.G., 1960: Yellow vein disease of citrus and studies of interaction between yellow vein and other viruses of citrus. *Virology,* **11**: 753-764, New York, N.Y.

WEBBER, H.J., 1943. The 'tristeza' disease of sour orange rootstock. *Proceeding of American Society of Horticulture Science,* **43**: 160-168.

WHITESIDE, J.O., 1972: Foot rot of citrus trees : the importance of high budding as a prevention measure. *Citrus Industry,* **53** : 14-16 Barlow, Florida.

WHITESIDE, J.O., S.M. GARNSEY AND L.W. TIMMER, 1988: Compendium of citrus diseases,p.80. APS press, 3340 Pilot Knob Road, St. Paul, Minnesota 55121, USA.

WUTSCHER, H.K., R.E. SCHWARZ, H.G. CAMPIGLIA, C.S. MOREIRA AND V. ROSSETTI, 1980: Blight like citrus tree decline in South America and South Africa. *Horticulture Science,* **15:** 588-590,Alexandria,Virginia.

YAMADA, S. AND K. SAWAMURA, 1952: Studies on the dwarf disease of satsuma orange. Citrus unshiv marcov. (Preliminary reports) *Horticulture Division Tokai Kinki Agricultural Experiment Station Bulletin 1: 61-67.*

YORA, K., Y. DOI, S.P. RAYCHAUDHARY AND Y.S. AHLAWAT, 1977: Infectious variegation in India. *Citrograph,* **62**:342, Los Angeles.

YOT-DAUTHY D. AND J.M. BOVE, 1968 : Purification of citrus crinkly leaf virus. p. 255-263 *In : Proceeding of 4th Conference of International Organisation of Citrus Virologists, IOCV, Riverside.*

4

NEMATODE PARASITES OF CITRUS

RENU SHARMA* AND S.B. SHARMA**

Plant parasitic nematodes, the microscopic worms, are very widespread in the developed as well as the developing countries. These parasites penetrate the roots, remove nutrients from plant system, hamper plant growth, and make the plants vulnerable to attack by other organisms. Roots being the major target of their attack, nematodes remain underground, out of sight of cultivators as well as scientists, and cause unhindered damage which is cumulative in nature. Due to these reasons, these subterranean organisms are known as "hidden enemies of farmers" as they are responsible for causing 12% annual yield loss (= US $ 77 billion) worldwide to the major life sustaining and economically important crops (Sasser and Freckman, 1987).

As no nematodes were found directly attacking and consuming the citrus fruits, the nematode parasites of citrus remained hidden for a very long period of time. Nevertheless, these parasites attacked the roots of citrus trees and were hidden beyond detection in the soil and embedded in the citrus roots. The first record of an association of a parasitic nematode on citrus roots was that of Neal (1889), who observed *Heterodera radicicola* (= *Meloidogyne* spp.) attacking the citrus roots in Florida, USA. Subsequently, citrus nematode (*Tylenchulus semipenetrans*) was discovered in California and Florida, USA (Cobb, 1913; Thomas, 1913) and importance of this nematode as a constraint to citrus cultivation was gradually realised. Till 1969, 189 species of

*. Directorate of Plant Protection Quarntine & Storage, NH IV, Faridabad-121001.

** Head, Division of Nematology, Indian Agricultural Research Institute, New Delhi-110012.

nematodes belonging to 39 genera were reported in association with citrus roots (DuCharme, 1969) and subsequently in less than 20 years the number of associated species increased to more than 200 within 44 genera worldwide (Anonymous, 1986). Between 1986 and 1997 several new species of nematodes have been reported on citrus. Some potentially important new species described after 1990 include a species of root-knot nematode, (*Meloidogyne jianyangensis*) on mandarin oranges (Yang-Baojun, *et al.*, 1990), *M. citri* on *Citrus unshui* in China (Shaosheng, *et al.*, 1990), *Xiphinema karachensis* (Nasira *et al.*, 1992), and *Paralongidorus lemoni* on citrus (Nasira *et al.*, 1993), another species of burrowing nematode, (*Radopholus citri*) in Indonesia (Machon and Bridge, 1996). Association of large number of species with citrus roots indicates that citrus rhizosphere provides a very congenial atmosphere for reproduction and development of nematodes. Though many species of nematodes have been associated with citrus but only a few are highly pathogenic and are of economic importance at global, regional or local levels (Table 1). The important species of nematode parasites of citrus are: citrus nematode (*Tylenchulus semipenetrans*) (Cobb, 1914), burrowing nematode (*Radopholus citrophilus* = *R. similis* citrus race) (Suit and DuCharme, 1953), sting nematode (*Belonolaimus longicaudatus*), stubby nematode (*Trichodorus christiei*) (Standifer and Perry, 1960), lesion nematode (*Pratylenchus brachyurus*) (Brooks and Perry, 1967), and sheath nematode (*Hemicycliophora arenaria*) (VanGundy and Rackman, 1961). In India, The nematode constraints to citrus production are well documented (Bindra, 1970; Rao and Thammiraju, 1975; Sharma and Swarup, 1982; Chandel, 1986; Chandel and Sharma, 1989; Ganguly, 1988; Mani, 1994; Sundaram and Vadivelu, 1994), *T. semipenetrans* (Siddiqi, 1961), *Pratylenchus coffeae* (Siddiqi, 1964), *Hoplolaimus indicus* (Gupta and Atwal, 1972) and *Meloidogyne javanica* (Mani, 1986) are pathogenic to citrus.

Surveys conducted particularly in last 10 years have also indicated high levels of infestation of the citrus nematode along with other important plant parasitic

Table 4.1. Important (and potentially important) nematode parasites of citrus.

Common name	Nematode species	Root symptoms	Distribution
ECTOPARASITES			
Dagger nematode	*Xiphinema basiri*	Browning	Sudan, India
	X. brevicolle	Browning	Israel
	X. index	Browning	Israel
Sheath nematode	*Hemicycliophora arenaria*	Galled tips	USA
	H. nudata	Galled tips	Australia
Sting nematode	*Belonolaimus longicaudatus*	Blinded root tips	USA
Ring nematode	*Criconemella australis*	Girdled roots	Australia
	C. citri	Girdled roots	USA
Pin nematode	*Paratylenchus hamatus*	Necrosis	USA
SEMI-ENDOPARASITES			
Citrus nematode	*Tylenchulus semipenetrans*	Dirty root	Worldwide
MIGRATORY ENDOPARASITES			
Burrowing nematode	*Radopholus citrophilus*	Burrows and lesions	USA
Lesion nematode	*Pratylenchus coffeae*	Lesions	India, Italy, Tropics
	P. brachyurus	Lesions	USA
Lance nematode	*Hoplolaimus indicus*	Lesions	India
Spiral nematode	*Helicotylenchus multicinctus*	Lesions	Canary Islands
SEDENTARY ENDOPARASITES			
Root-knot nematode	*Meloidogyne javanica*	Root-galls	India, Israel

nematodes in the different citrus growing areas of the country. The nematode is present in many citrus growing tracts of Assam, Punjab, Delhi, Uttar Pradesh, Rajasthan, Maharashtra, Bihar, West Bengal, Kerala, Tamil Nadu, Himachal Pradesh, Haryana, Gujarat, Orrisa, Andhra Pradesh, and Karnataka (Reddy and Singh, 1979; Mani and Murthi, 1986). High frequencies of citrus nematode are recorded in the Nilgiri hills and Shevroys hills in Tamil Nadu, in the north-eastern part of the country and parts of Andhra Pradesh (Sinha and Rahman, 1994; Sundaram and Vadivelu, 1994; Sundaram *et al.*, 1990; Mani, 1994). Some other nematodes recorded on citrus are *Helicotylenchus spp.*, *Rotylenchulus reniformis, Xiphinema basiri, Hemicricone-moides spp.*, *Pratylenchus coffeae*, *Pratylenchus spp.*, and *Tylenchorhynchus spp.* Fortunately, some of the very damaging nematode species such as *Radopholus citrophilus* and *Belonolaimus longicaudatus* are not present in the country.

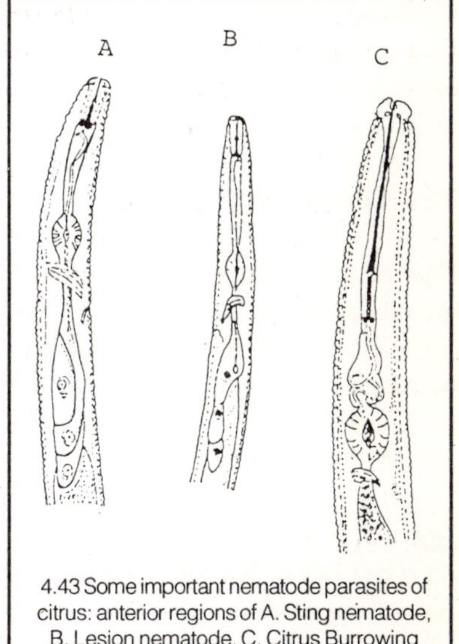

4.43 Some important nematode parasites of citrus: anterior regions of A. Sting nematode, B. Lesion nematode, C. Citrus Burrowing nematode.

Nematode parasites of citrus differ in their mode of parasitism. They are ectoparasites (*e.g.*, *B. longicaudatus*), .semi-

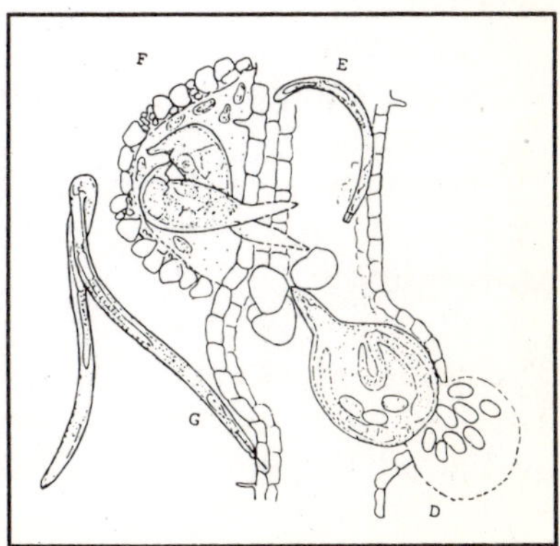

4.44 Parasitism of citrus root by different types of nematodes: D. Female of a rootknot nematode (sedentary endoparasite) with egg sac protruding out of the root. E. Burrowing nematode (migratory endoparasite), F. Females of citrus nematode (semi-endoparasite) with their posterior part hanging out of root and it is covered by soil particles attached to the mucilaginous martrix. G. Female of sting nematode (ectoparasite).

endoparasites (*e.g., T. semipenetrans*) and endoparasites (*e.g., M. javanica*). The endoparasites are either sedentary (*e.g., M. javanica*) or migratory (*e.g., R. citrophilus*) based on their ability to move (4.43 and 4.44).

The Citrus Nematode

Tylenchulus semipenetrans, the citrus nematode, causes 'slow decline' of citrus. Within less than 100 years of discovery of this nematode in the USA, it has been reported in citrus orchards from virtually all over the world. The far and wide distribution of this nematode, along with other nematodes associated with citrus crop, has been described by DuCharme (1969). He has aptly mentioned that nematode parasites of citrus should erect a monument and dedicate it to man because with co-operation of man, nematodes have been carefully, perhaps even tenderly, transported and distributed throughout most tropical countries where citrus is cultivated. The human assistance has been so efficient that there is virtually no country in the world, in which citrus is grown and *T. semipenetrans* is not found. Man assisted fast movement of citrus nematode has increased due to the enhanced trade relations all over the world. This movement is in addition to the slow and cumulative natural movement of the nematode and passive spread along with the irrigation water, floods, and soil erosion.

Taxonomic position

Order	Tylenchida
Suborder	Tylenchina
Super family	Tylenchuloidea
Family	Tylenchulidae
Sub-family	Tylenchulinae

Morphology

The nematode is sexually dimorphic. The female is parasitic. The juveniles, males and pre-adult females are migratory, and adult females are sedentary.

Adult female: The anterior region of the body is embedded in the root as the nematode is a semiendoparasite. The embedded neck region is irregular, thin with fine cuticle, weak cephalic sclerotisation, stylet and oesophagous. Body expands behind the median bulb region, mostly on the dorsal side, ventrally arcuate with a thick cuticle. Well developed stylet with prominent knobs. Well developed excretory system with excretory pore slightly anterior to vulva. The excretory system secretes a gelatinous matrix that covers the body of the mature female and protects the eggs. Ovary single, prodelphic, convoluted with several eggs, and extends up to oesophageal region. Vulva in the posterior region of body, distinct with vulval lips. Post-vulval part of body is elongate and tapering. Anus is non-functional. Tail tapering.

Young female: Free living in soil, vermiform body, curved ventrally in the posterior region, less than 0.5 mm in length. Cuticle clearly striated. Weak cephalic sclerotisation, head region continuous with body. Medium stylet with rounded basal knobs. Median bulb developed but not separated from procorpus. Excretory pore and vulva in the posterior region of body, single genital tract with few oocytes.

Male: Vermiform body, cephalic sclerotisation, stylet and oesophagus reduced but not completely degenerated. Excretory pore near mid-body region. Testis outstretched tail elongated and pointed, bursa absent.

Biology

The body of adult female is covered with gelatinous matrix and eggs are laid in this matrix. The embryonic development and first moult are completed within the eggshell. The second-stage juveniles hatch out of the egg, start feeding on the cortical cells, and subsequently moult to form the third and fourth stage young females. The elongated neck region facilitates the young female to penetrate deeper into the cortex, just short of pericycle. The posterior portion of the body projects and hangs outside the root. The anterior portion remains irregular and slender whereas posterior region swells, with a pointed tail region. As the mature female feeds and develops, the reproductive organs develop and eggs are deposited in the gelatinous matrix. Male may fertilise a female, but unfertilised females also produce eggs. The male second-stage juveniles moult three times without feeding and may fertilise the adult females. One life cycle is completed in 4-8 weeks. The soil particles adhere to the gelatinous matrix and form a protective covering for the developing eggs from natural predators (Van Gundy, 1958; Maggenti, 1962; Cohn, 1965). This covering remains even after washing the infected roots. Several generations can be completed per year (Prasad and Chawla, 1965; Bello *et al.*, 1986; Bhagel and Bhatti, 1982; Salem, 1980; Duncan and Noling, 1988). A single female on *Citrus jambiri* lays on an average 95 eggs, and embryonic development takes 16 days after egg laying. The first stage is completed in the egg itself. The second-stage juveniles are of two types; larger and slender develop into females and shorter and broader ones develop into males. The duration of second, third and fourth stage juveniles, young, and mature females is 8, 7, 12, 7 and 14 days, respectively. All female stages develop in the roots. The duration of male stages is much shorter 24 hours (second stage), 29 (third stage), 64 (fourth stage), and 48 hours (adult). The duration of life cycle for male nematode is 16 days after hatching, and 48-55 days for female nematode (Jagdale *et al.*, 1984). The initial population level also effcts the final nematode population buildup (Das and Mukhopadhyay, 1983).

Ecology

The population biology of the citrus nematode is greatly influenced by the abiotic factors such as soil structure, texture, temperature, moisture, and pH. Population densities of the citrus nematode in India are often very high during late autumn and spring, low during winter, and reach their lowest level during mid-summer. The enhancement in nematode population density generally coincides with flushes of new feeder roots (Chawla and Sharma, 1984b; Hamid *et al.*, 1988). In Florida, USA, the nematode population density increases in late autumn (July-November) following a large flush of root growth (O'Bannon *et al.*, 1972; Duncan and Noling, 1987). This is also the period of high female fecundity (Duncan and Noling, 1988). The seasonal and annual trends of population densities are essentially the same for juveniles and males in soil and females in roots. Fluctuation in population density is bimodal; high density during late autumn and spring and low density during mid-winter and summer (Cohn, 1966; Duncan *et al.*, 1993). During low Winter temperatures the nematode populations decline and once again increase in Spring season when soil temperatures are favourable. Soil temperature and moisture stresses reduce the nematode population in Summer (Cohn, 1964). Lignin concentration in roots increases in May and June and it results in decline in susceptibility to the nematode (Cohn, 1964; Duncan *et al.*, 1993). Numbers of soil and root inhabiting *T. semipenetrans* positively correlate with fibrous root mass and root carbohydrates and negatively correlate with soil moisture content and root lignin concentration. Young roots are readily invaded than the old roots (Cohn, 1964). The citrus nematode populations are also affected by harvesting of citrus fruits, due to implication of carbohydrate competition. The feeder root mass density of defruited trees markedly increases and these roots have less insoluble starch as compared with fruited trees. Between 9 and 15 months following fruit removal, the nematode fecundity and population density on defruited trees increase at a very fast rate. It indicates that carbohydrate competition between developing citrus fruit and *T. semipenetrans* influences seasonal fluctuation in nematode population densities (Duncan and Eissenstat, 1993).

The nematode does not have much capacity to enter anhydrobiosis (Tsai and Van Gundy, 1988; 1989) and population development is favoured by rather dry, than wet conditions in soils with moderately high clay content, whereas in sandy soil wet conditions are favorable (Van Gundy *et al.*, 1964). The nematode prefers reduced water potentials, but has a poor ability to survive severe drought. Roots in dry soil maintain high population of *T. semipenetrans* when water is available from other portions of root (Duncan and Morshedy, 1996).

The vertical distribution and seasonal fluctuations in nematode populations are influenced by the age of plant and condition of the roots. Plants of intermediate age (15-20 years) show moderate decline symptoms but support heavy populations whereas old plants (8 years) show severe decline but support poor population. Young plants in the age of 3-10 years also have low population. The best suited pH for citrus nematode is 7.4-7.8 (Husain *et al.*, 1981). As the infection level increases, the plant tends to produce greater number of fresh feeder roots to compensate for the damaged roots whereas the affected roots die premature, leading to no substantial increase in root biomass (Hamid *et al.*, 1985). Thus the energy resources of infected trees are diverted to production of more roots, as compared to the healthy trees. Such qualitative differences in healthy and infested trees in terms of food quality and rhizosphere changes may influence the nematode population(Duncan and Noling, 1987).

Soil salinity increases the population density of the nematode (Machmer, 1958). Greater population densities of the nematode are found in orchards with known salinity problems (Cohn, 1976; Cohn *et al.*, 1965). Discontinuous salt treatment which is similar to irrigation with poor quality water under field conditions, supports highest mean population densities of *T. semipenetrans* on Rangpur lime as well as on sweet lime (Mashla *et al.*, 1992). High soil salinity predisposes citrus to nematode infection through accumulated salt stress, and nematode females under this treatment have highest fecundity. The continuous salt stress, however, inhibits nematode mobility (Kirkpatrick and Van Gundy, 1966; Lee and Atkinson, 1977). It is apparent that the citrus nematode problem may enhance in areas with known salinity problem (O'Bannon and Esser, 1985; Oster, 1984).

Soil type as well as organic amendments influence development, reproduction, and hatching of the citrus nematode (O'Bannon, 1968). Clay soil is most conducive for root invasion and nematode development. Calcareous and loamy soils rank second where considerable number of larvae invade the roots and develop into adult females (Youssef *et al.*, 1989). Irrigation systems also affect the citrus nematode population. On *Citrus sinensis* var. Valencia, high population density of the nematode has been found in drip irrigation system than in surface irrigation system (Mahros *et al.*, 1985). The nematode populations grow well at a temperature range between 21-30°C and 25°C is optimum (Van Gundy and Martin, 1961; Van Gundy *et al.*, 1964; O'Bannon *et al.*, 1966). The population growth of the nematode is slow on young trees until canopies are developed sufficiently to shade the soil and result in optimum temperatures (Reynolds and O'Bannon, 1963).

Citrus is a mycorrhiza-dependent plant and the non mycorrhizal seedling may be severely stunted in soils low in phosphorus. Mycorrhizae and the citrus nematode are found in most citrus soils. Mycorrhiza (*Glomus mossae*) enhances the seedling growth and the citrus nematode (*T. semipenetrans*) suppresses the growth of seedlings. The nematode infection in roots prevents the formation of fungal vesicles (O'Bannon and Nemec, 1978).

Symptoms

The nematode infection causes "slow decline" in yield and quality of citrus. However, when young seedlings are planted at the sites of old citrus trees, serious and rapid destruction occurs. The young seedlings may fail to establish in heavily infested soil or they may develop slowly. This problem is referred to as "Citrus replant problem" and heavy losses up to 50% may occur (Anonymous, 1986).

The direct damage caused by nematodes to the plants is mainly on the roots, hidden under the soil surface The infected roots are shortened and malformed as compared to healthy roots. They appear to be thicker and dirtier than normal roots due to deposition of soil particles on the gelatinous matrix secreted by the adult female for laying eggs. In fact 'dirty root' is the most important symptom of citrus nematode infection on host roots (B-4.17). The dirty roots are difficult to clean as the soil clings to the sticky gelatin, whereas in healthy roots the soil particles can be removed by shaking or washing the roots. These symptoms

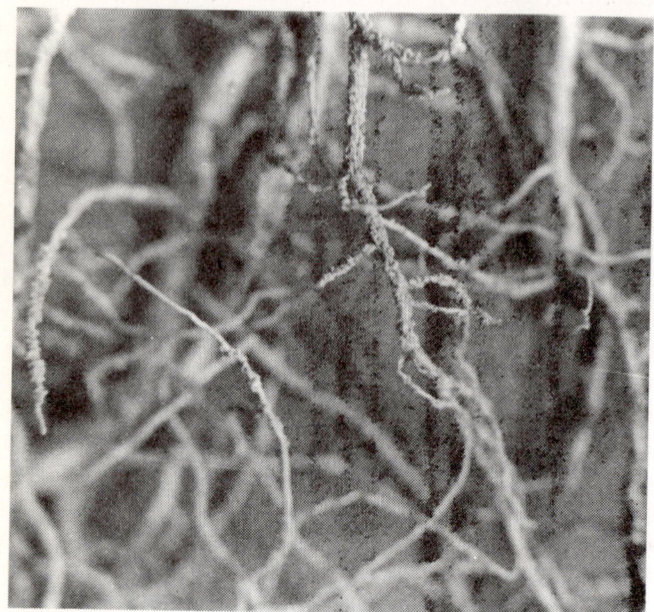

B-4.17 "Dirty root" of a citrus plant caused by the infection of *Tylenchulus semipenetrans*. The soil particles adhere to the mucilaginous martrix which covers the females and eggs on the root surface. (Courtesy: Dr. A.K. Sinha, Assam)

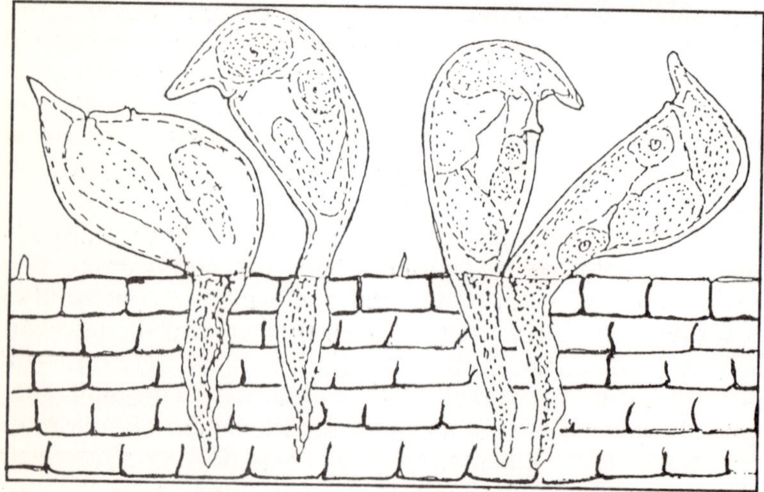

4.45 Diagrammatic presentation of citrus nematode females (semiendoparasites) feeding in the cortical zone of citrus root.

are clearly visible on heavily infected roots due to larger number of females and more gelatin secreted on the root surface. When the gelatinous matrix is removed, the females can be seen attached to the roots (4.45). However, less infected roots may go undetected because of their apparent healthy appearance. It is one of the major reasons for dissemination of the nematode with the young stock. The nematode being a semiendoparasite damages the epidermis and the cortical layers of the affected roots, providing easy access to the various secondary organisms like bacteria, virus and fungi (Schneider and Bains, 1964; Cohn, 1965a, Hamid *et al.*, 1985; Anonymous, 1986). The feeder roots decay very quickly due to the infection of nematodes. Root necrosis turns the roots dark and lateral roots are reduced and slightly thickened. The infested fibrous roots may show irregular thickening (Swarup *et al.*, 1964). In severe cases when the roots are gently rubbed between fingers the root cortex may come out in the form of sleeve leaving the inner vascular stele intact. The nematode infection may lead to inward curling of leaves and some twigs become partially or totally defoliated giving the upper most portion of the tree a naked appearance (O'Bannon and Esser, 1985). However, the above-ground manifestation of damage is not spectacular and needs special efforts to relate the stunted growth and reduced yield to presence of nematodes in roots and rhizosphere. The symptoms of nematode-caused damage are easily confused with nutritional disorders and diseases caused by other soilborne pathogens like viruses, bacteria and fungi. The sure way to assess the presence of nematodes is to test the soil and root samples in the laboratory and conduct microscopic examination of roots. The orchard conditions also play a vital role in expression or suppression of symptoms. Optimum conditions for plant growth may help the plant to stay healthy whereas adverse or less suitable conditions may enhance the symptom expression on nematode-infected plants (van Gundy *et al.*, 1964; Heald and O'Bannon, 1987). Heavily infected plants are stunted with defoliated twigs. Leaves are small and yellow. The extent of damage depends on the level of nematode population build up in soil and roots, tolerance of the host plant to nematode infection and environmental conditions. Infected trees give out more roots than healthy trees.

Management

The citrus tree harbours a large number of nematodes in its rhizosphere. It is not easy to control the nematode population, once the nematode gets established in a citrus orchard. Use of various combinations of rootstock and scions that differ in their reaction to nematode attack may further complicate the nematode management options in an orchard. Judicious combination of different management tactics is useful in suppressing the nematode population densities and plant damage caused by them.

Prevention

Prevention of nematode infestation in an orchard is a safe strategy for maintaining crop health and productivity. It is possible both through man-made means as well as by the already existing natural physical barriers. Natural water bodies and uncultivated areas surrounding the citrus producing areas are the natural barriers which check further spread of nematode populations. Strips of land act as barriers for citrus nematode; a strip of land about 3-8 m wide which surrounds the infested citrus area is treated with chemical nematicides on a half yearly basis to check the nematode movement from infested to uninfested area (Poucher *et al.*, 1967). Buffer land 5-18 m wide between the infested and uninfested areas also checks the spread. Destruction of roots of citrus in the buffer zone with nematicides also reduces the threat of spread (Suit and Brooks, 1957, Poucher *et al.*, 1967).

Another important means of preventing the nematode spread is through the use of clean and healthy planting material. Regular inspection and certification of nursery plants and soil is required (Milne, 1982). Though majority of the nursery plants in commercial nurseries in the state of Andhra Pradesh, India, (and possibly in other citrus growing states) are found to be heavily infected with the citrus nematode and the root-knot nematode (Mani, 1986, Mani *et al.*, 1988). There is as yet no law ensuring certification of nurseries for freedom from nematodes in the country. There is a tremendous scope to curb rate of spread of the citrus nematode through movement of infected planting material (Tarjan, 1956; Esser, 1984; Van Gundy and Meagher, 1977). Planting stocks must be certified as pest free to ensure that the nurseries are free of nematodes, regular samples should be drawn while inspecting the nurseries. Movement of the equipment used in the infested orchards should be limited to within the orchard itself to avoid any adhering infested soil or roots to move to fresh sites. It is also important to restrict the soil movement and a certificate for freedom from pests at the site of origin should be obligatory for any movement of soil in India. Farm machinery also needs to be cleaned thoroughly before moving to another area (Meagher, 1969). Biological barriers in the form of a band of resistant citrus trees or nonhost plants may be useful (DuCharme and Suit, 1955; Ford, 1967).

The citrus nematode spreads very slowly (Meagher, 1967; Tarjan, 1971 and Baines, 1974). However, irrigation water percolating through the nematode infested orchard is an effective source of nematode spread and it leads to fast spread of infection from one orchard to another (Cohn *et al.*, 1976). It is necessary to ensure regular monitoring of commercial nurseries, restrict movement of contaminated equipment and farm machinery, separate sources of irrigation for infested and nematode-free nurseries would greatly restrict nematode spread in future.

Cultural practices

Healthy citrus plants growing under optimum conditions have greater ability to tolerate the nematode-caused damage as compared with those plants that are under some kind of biotic or abiotic stress. It is important to carefully study the condition of the trees and other factors in an orchard to prescribe a holistic crop health management strategy that enhances the plant growth and suppresses the crop growth stress factors such as the nematodes. One of such practices that reduces the nematode populations as well as enhances the plant growth is the application of organic amendments in the vicinity of tree basin in the form of oil cakes. Addition of oil cakes enhances the microbial activity in soil and phenol content in the roots (Alam *et al.*, 1977) and causes reduction in the nematode population. The oilcakes of neem, castor, olive, cotton seed and mustard are some of the common organic amendments used to reduce the nematode population and improve the growth of citrus (Ahmad, 1985). Addition of 20 kg of oilcake at the tree basin once in four months substantially reduces the population of *T. semipenetrans*. Aqueous extracts of leaves of marigold (*Tagetes erecta* L.) are toxic to the second-stage juveniles of citrus nematode (Mani *et al.*, 1986). Leaf extracts of *Azadirachta indica, Datura metel, Nerium oleander* and *Calotropis procere* cause juvenile mortality (Awan *et al.*, 1992).

Intercropping with nematode antagonistic plants is a cultural practice which generally enhances health of citrus plants and reduces soil population of nematodes (Bindra, 1970). Interculture of marigold and mustard with acid lime reduces the rate of multiplication of the citrus nematode (Mani, 1988); in some regions, marigold (*T. patula*) does not affect the population of citrus nematode and inhibits the growth of sour orange (Cohn, 1975). However, recommendation of plant species should be made with caution because intercrops sometimes lead to establishment of other highly pathogenic nematodes such as the root-knot nematodes. Infection of root-knot nematode is greatly enhanced on citrus when it is intercropped with vegetables and tobacco (Tirumala, 1956).

In addition to the organic amendments of plant origin, the animal products such as cow manure diluted with water are also useful in reducing the population of the citrus nematode. Two applications reduce the nematode population within eight weeks and enhance the tree growth (Tichinovo, 1957). Addition of refuse compost enhances seedling vigour and reduces nematode population. Irrigation with sewage water reduces nematode population (Cohn *et al.*, 1965). Clandosan®, a Chitin-protein based material derived from crab shells, is effective in the management of *T. semipenetrans*. However, the dosage (4-8 t/ha) required are very high and may not be economical (Speigel *et al.*, 1989). There is a need to design economically viable organic amendment-based crop health management strategies to manage losses caused by the citrus nematode.

Soil solarisation is useful in reducing the citrus nematode infestation along with other soilborne pathogens and weeds. The disinfestation is obtained not only due to high temperatures, exceeding the thermal death points, but also due to the sublethal effect of fluctuating temperatures and population increase of antagonists in solarised soil. Solarisation may enhance the population of natural parasites in soil.

Another effective means of disinfestation of rootstocks before planting is by the use of hot water treatment for 5 min at 51°C. The roots immediately after the treatment are submerged in cold water to minimise any damage to the roots (Meagher, 1960).

Use of resistant genetic resources

Availability of nematode resistant and (or) tolerant high yielding cultivars is one of the best and economic means of combating nematode damage. Citrus nematode resistant root stocks, grafted to the scions with desirable fruit quality and yield, are valuable in the management of citrus nematode. The scion affects the rate of reproduction of the nematode, it does not affect the penetration and feeding (Scotto La Massese *et al.*, 1975, Van Gundy and Kirkpatrick, 1964). *Poncirus trifoliata*, a native of central Southern China, is a source of resistance to *T. semipenetrans* and several hybrids of *P. trifoliata* with citrus have exhibited resistance to the nematode (Reddy and Singh, 1978; Jagdale *et al.*, 1984; Ibrahim *et al.*, 1989; Mani and Reddy, 1986). Resistant rootstocks reduce the population of nematode and thereby enhance plant growth. As compared to susceptible citrus plants, only 14-20% of population of citrus nematode are recorded on *P. trifoliata*. The time taken to complete one generation on *P. trifoliata* is also nearly double when compared with that on susceptible plants; the life cycle is completed in 14 weeks on *P. trifoliata* and in 6-8 weeks on *C. aurantium* (Cohn, 1965). The nematode resistance genes in *P. trifoliata* are dominant and are inherited in all the hybrids with citrus. However, presence of biotypes of citrus nematode creates problems in resistance breeding as *P. trifoliata* has been found susceptible to the *Poncirus* biotype occurring in USA, Yugoslavia, and Japan (Vucinic and Tiodorovic, 1988). The extent of spread of this virulent strain needs to be ascertained prior to undertaking resistance breeding programmes, because planting *Poncirus* rootstocks in areas where this biotype exists may lead to extensive damage. Three biotypes of citrus nematode are known to exist: citrus, mediterranean and *Poncirus* (Inserra *et al.*, 1994). The citrus biotype infects genera in Rutaceae family, including *Citrus* spp., Troyer and Carrizo citrange, olive grape and persimmon, but reproduces poorly on *Poncirus trifoliata*. Mediterranean biotype has similar host range with the exception of olive. The *Poncirus* biotype reproduces on *P. trifoliata*, *Citrus* spp., and grape but not on olive (Verdejo-Lucas *et al.*, 1997).

Highly resistant citrus plants exhibit hypersensitivity to early feeding, moderately resistant plants produce wound phelloderm and toxins (Van Gundy and Kirkpatrick, 1964). Some citrus varieties exhibit tolerance to the nematode. Hybrids of *P. trifoliata* such as Troyer and Carrizo citrange rootstocks are resistant to the nematode; swingle citrumelo, a commercially useful rootstock is also resistant (Mani and Reddy, 1986; Reddy and Singh, 1978; Duncan and Cohn, 1990). Resistance to biotypes of *T. semipenetrans* is available in several selections, and some of these selections are resistant to *Phytophthora citrophthora* and *Tristeza* (Gottlieb et al., 1986). In India, utility of trifoliate orange rootstocks in India is limited due to several factors of plant growth and fruit quality. Rangpur lime (*Citrus lemonica* L.) and rough lemon (*C. jambhiri* Luch) are commercially used as rootstocks for sweet orange (*C. senensis* (L) Osbeck). The strains of Rangpur lime and rough lemon tested by Mani (1989) were not resistant to citrus nematode, however, five rough lemon strains were moderately resistant.

Screening citrus rootstocks for nematode resistance is

time consuming and tedious. As the roots produced by the leaves in citrus species and hybrids are functionally similar to the natural roots in their reaction to parasitism by the nematode, they can serve as useful tool in rapid evaluation of resistance to the citrus nematode (Gottlieb *et al.* ,1987).

Use of bioagents

Use of bioagents for pest management is an important environment-friendly option. Bioagents, naturally occurring or released artificially in the soil, normally do not lead to environmental pollution. Many naturally occurring bacteria and fungi have been identified as potential biological control agents for the citrus nematode (Tandingan,1985; Jatala,1985; Walter and Kaplan,1990; Osman *et al.*, 1988; Osman and Salem, 1995). Nematode trapping fungi and predatory mites have been reported from the citrus rhizosphere and some of them probably prey on the nematodes (Mankau, 1972; Gaspard and Mankau, 1986; Walter and Kaplan, 1990). A fungus, *Paecilomyces lilacinus* has been found to reduce populations of *T. semipenetrans* (Jatala, 1985). It is more effective in reducing the number of nematodes in roots and soil around citrus plants than the soil application of nematicides such as aldicarb, vydate and mocap. Integration of *P. lilacinus* with application of oil cakes of castor, *karanj* and *neem* under field conditions is useful in reducing the nematode population both in soil and roots. Addition of neem cake increases the incidence of fungus-infected eggs. Similarly another fungus, *Verticillium lecanii*, in combination with oilcakes of *karanj*, castor and *neem* reduces the nematode population. The oilcakes help in enhancing the fungal growth. Large number of nematode eggs are infected and greater number of fungal spores are found in soil (Anonymous, 1995). *Glomus fasciculatum* and *G. mossae* also reduce the nematode population (Anon., 1995). Sincocin AGTM at 200 ppm as a biocontrol agent reduces population of citrus nematode both in soil and roots of *C. sinensis* and enhances the growth of orange trees and soil mites by 243% in two months time (Osman and Salem, 1995).

Populations of a gram-positive bacterium *Pasteuria*

penetrans have been isolated from *T. semipenetrans*. The isolates from citrus nematode have spores which are morphologically similar to the *P. penetrans* spores isolated from the root-knot nematode, but they complete the life cycle in the second-stage juvenile and not in the adult female (Fattah *et al.*, 1989; Davies *et al.*, 1990). Predacious nematodes and micro-arthropods do not have any significant effect on populations of *T. semipenetrans*. Species of *Thornia sp.*, *Prionchulus punctatus*, and *Mylonchulus sigmaturus* do not seem to have any suppressive effect on the citrus nematode population (Boosalis and Mankau, 1965; Small, 1979).

Fumigants

These chemicals are used as a pre-plant treatment and incur high cost of application(O'Bannon and Tarjan, 1973). Young citrus trees replanted in old infested orchard soils are highly vulnerable to severe damage by the nematode. Multipurpose fumigants like methyl bromide, metham sodium, chloropicrin and methyl isothiocyanate (MIT) and true nematicides like dichloropropane-dichloropropene (DD), 1,3-dichloropropene (1,3D), ethylene dibromide (EDB) and Dibromochloropropane (DBCP) are used for pre-plant treatment of soil (Baines *et al.*, 1956; 1958; 1962 and 1965). These are contact poisons and directly act on the nematodes. Several soil factors influence the efficacy of these fumigants in soil. Well drained soils at field capacity with less than 20% clay and 5% organic matter are normally porous enough to allow volatile fumigants to percolate. Sandy soils are easier to fumigate than the clay and peats. Soil moisture content is an important factor in determining effectiveness of fumigation. Efficacy of fumigation enhances with the rise in temperature due to increase in vapour pressure. Application of fumigants in the field is cumbersome and toxic nature of these chemicals requires careful handling. The poisonous, odourless gas methyl bromide is almost always applied under airtight polyethyhene cover. Though the fumigants provide effective control of nematodes, but improper application can affect the young trees (Cohn *et al.*, 1968; Milne, 1974). It is important to maintain proper gap between application and planting to avoid phytotoxicity. Liquid formulations are applied with the help of hand held or tractor mounted injectors. The fumigants, as general biocides also kill the beneficial soil flora and fauna. Mycorrhizae can easily be eliminated by this treatment leading to phosphorus deficiency and need to be re-inoculated in the nurseries (O'Bannon and Nemec, 1978; Timmer and Leyden, 1978; Graham, 1986).

DBCP has been widely used for postplant treatment of living trees but is not used any more due to its pollution hazards (Reynolds and O'Bannon, 1958; Tarjan and O'Bannon, 1974). Control of citrus nematode on living trees was first practised in USA (Baines *et al.*, 1965; 1963; 1958; O'Bannon and Reynolds, 1967). Many reports are available on effective control of citrus nematode on living trees by DBCP (Cohn and Minz, 1965; Meagher, 1969; Mukhopadhyaya and Dalal, 1971; Oteifa *et al.*, 1965; Philis, 1969, Reynolds and O'Bannon, 1958; Toung, 1969, Vilardebo, 1963; Yokoo, 1964). Populations of *T. semipenetrans* are killed by initial contact with DBCP, but at least 50 days are required for maximum kill of the nematode (Baines *et al.*, 1958; Carter and Reynolds, 1969). The reduction in nematode populations results in improved tree growth and fruit production; the beneficial effect lasts for 3-4 years. Chisel injection and flood irrigation have been found to be very suitable application methods. In USA, greater fruit size and yield were obtained by applying DBCP at 38-50 kg/ha, whereas best nematode control was obtained at a rate of 77 kg/ha. Application of chemical emulsion with sprinkler irrigation is one of the best methods (Tarjan and O'Bannon, 1974).

Non-volatile compounds like organophosphates and oxim carbamates are often applied as post plant granules. These chemicals have high toxicity to mammals, but when applied at recommended dosages they show little or no phytotoxicity. These are also formulated as emulsifiable concentrates and for tree crops can be applied along with drip irrigation system. The commonly available non-volatile compounds are carbofuran, aldicarb, diazinon, oxamyl, phenamiphos, fensulfothion, and prophos (Davis and Wilhite,1985; Childers *et al.*, 1987; Meher, 1995; Thapa and Mukhopadhyay, 1994) and new product like cadusaphos has also been found effective in reducing citrus nematode population (McClure and Schmitt, 1996). Most of these post-plant nematicides are systemic in nature and are translocated in the tree system. Two important factors to be considered in the post plant chemical treatment include the time of application (chemical should be applied, wherever possible prior to active invasion of nematodes in the root system), and the area of chemical application. Nematicides in large commercial orchards are applied in bands down the tree rows or through low volume irrigation system. Application in this manner ensures that nematicide reaches the area of greatest root mass and nematode density. Nematicide bands are most useful when applied within the tree canopy because nematode density and feeder root mass in the uppermost soil horizon decrease with distance from tree trunk (Nigh, 1981; Duncan, 1986).

Citrus Burrowing Nematode

Radopholus citrophilus is popularly known as the Citrus Burrowing Nematode or CBN. It causes "spreading decline" of citrus. The average yearly rate of spread in Florida (USA) groves has been recorded as 1.6 trees or 15.2 m per year. The down hill spread of nematode may reach 66 m per year, whereas uphill spread is restricted to 8 m per year (Ducharme, 1954; 1955). The spreading decline has been described as group of trees that show the same degree of decline and the area increases in size each year (Poucher *et al.*, 1967). The spreading decline was first observed in Florida (USA) in 1928 and later in 1953 the burrowing nematode, *R. similis* was identified as the causal agent of spreading decline and for a long time it was known as the citrus race of the burrowing nematode. Huettel *et al.*(1984) described this race as a new species (*R. citrophilus*) because of its differences from banana race in chromosome number, isozyme pattern, mating behaviour, and host preferences. The citrus burrowing nematode adversely affects the mycorrhizal stimulation of seedling growth. The nematode causes a significant reduction in vesicle formation. Disruption of cortex by *R. citrophilus* leaves very less space for *G. etunicatus* and severely limits its development on root system. While mycorrhizal development retards the nematode infection (O'Bannon and Nemec, 1978).

Taxonomic position

Order	Tylenchida
Suborder	Tylenchina
Super family	Hoplolaimoidea
Family	Pratylenchidae
Sub-family	Radopholinae

Morphology

Sexual dimorphism in anterior region well marked. Males have higher rounded and more offset cephalic region than the females. The female is parasitic not males. The juveniles and adult stages are migratory endoparasites.

Adult female: Vermiform body, cephalic region low, continuous to slightly offset and strongly sclerotised. Stylet well developed; conus about as long as shaft. Oesophageal glands elongated, dorsal to intestine. Excretory pore near oesophago-intestinal junction. Vulva at 50-70% of body length. Didelphic and amphidelphic.

Male: Vermiform body, cephalic sclerotisation, stylet and oesophagus markedly reduced. Tail generally more tapering than that of female. Bursa subterminal, spicule cephalated and slightly arcuate.

Biology

It is a migratory endoparasite. All the juveniles and adult stages possess the ability to penetrate the roots. Juveniles and adult females enter the root tips and feed on the cortical parenchyma. The female lays eggs in the roots singly, the juveniles hatch within 3-7 days and invade the roots. The life cycle, from egg to egg, is completed in three weeks at 24-26 ^{o}C (DuCharme and Price, 1966). The nematode reproduces parthenogenetically as well as sexually (Brooks and Perry, 1962; Huettel *et al.*, 1982). The short duration of life cycle ensures rapid build up of nematode population under favourable conditions, enhancing the disease severity. From a single female, 11,000 individuals can be obtained in three months under gnotobiotic conditions (Duncan and Cohn, 1990). The nematode migrates from the decaying roots to healthy roots. The roots turn foul due to accumulation of nematode waste product and wound gum deposition by the plant (DuCharme, 1968). Males do not invade the roots.

Ecology

The range of optimum temperature for nematode reproduction is 20-32 ^{o}C with 24 ^{o}C as the optimum temperature (DuCharme, 1969). The nematode populations are usually concentrated in deeper layers of soil and greatest densities occur 30 cm to 180 cm deep. Due to the migratory nature of the nematode, population levels vary at any given site and greater number of nematodes are found in newly infected roots than in the old roots that have been infected for a longer time (DuCharme and Price, 1966). Low population level of the nematode in top layer of soil is probably due to the high soil temperature in Florida (Tarjan, 1961). Therefore, about 25-30% of the feeder roots are affected that are 25-75 cm deep, and 90% of roots 75 cm deep are affected (Ford, 1953). Sandy soils are more conducive for spread of this nematode and rate of movement through fine sand is 100 mm in four days.

Fortunately, this highly destructive nematode is very restricted in its distribution. It is present only in parts of USA. This species does not seem to have good survival abilities and the nematode populations do not survive in soil for more than six months in absence of hosts. However, some individual nematodes may survive in root debris scattered in soil. Soil encumbered on the vehicles is a good medium of nematode dissemination (Esser, 1984).

Symptoms

The nematode infected trees show poor growth, die back in the upper canopy, reduction in the size of fruit, leaf, and tree, and poor response to the addition of fertilisers. The affected trees may have 50% fewer functional roots in well drained soil than the uninfected trees. The disease initially appears in a group of trees within a grove and the decline areas increase annually in size. The tissue most commonly invaded is cortex, but phloem, cambium, apical meristem, xylem parenchyma and pericycle are also invaded. It also causes hyperplasia in pericycle and endodermis (DuCharme, 1959). Available food supply and competition with other organisms limits population growth of the nematode (DuCharme and Price, 1966). The infected root has lesions due to nematode feeding and large number of nematodes inhabit these lesions. Poucher *et al.*, (1967) isolated 1739 nematodes from a single lesion. The decline symptoms appear after about a year of the initial infection.

Management

The nematode resistant rootstocks are available (Ford and Feder, 1969; O'Bannon and Ford, 1976). Milam, Ridge Pineapple, and Algerian Naval are resistant and Estes Rough Lemon is tolerant. (Kaplan and O'Bannon, 1985). *Radopholus citrophilus* has two morphologically and karyotypically identical populations which differ in their ability to damage and reproduce in the roots of citrus rootstocks previously identified as resistant or tolerant. These biotypes are responsible for causing damage to the resistant rootstock such as Milam lemon, Carrizo citrange, and sweet orange. Wide use in replanting and also in buffer zone may prove dangerous. Vegetative cloning of rootstocks that suppress the nematode population may be useful (Kaplan, 1986) . Phenamiphos and DBCP are very effective in nematode management (O'Bannon and Tarjan, 1979).

Colonisation of citrus feeder roots by the VAM fungi enhances the uptake of phosphorus and minor elements and lowers the nematode densities (Smith and Kaplan, 1988).

Sting Nematode

Belonolaimus longicaudatus is a migratory ectoparasite of

roots. The nematode attacks the roots from outside, causing damage with its long stylet. It is reported from 64% citrus orchards in Florida (Esser *et al.*, 1993; Duncan *et al.*, 1996) and is widely distributed on a large number of cultivated and non cultivated plants in south Eastern United States. In the central ridge region of Florida it was detected in 50% of the samples and in 64% of the orchards. The infected trees were 23% smaller, had 25% less root density, and 57% reduction in fruit yield as compared to the healthy trees. It has a wide host range and many weeds and grasses are good hosts. It is a pathogen of grapefruit seedlings (Abu-Gharbieh and Perry, 1970). The nematode is not reported outside of USA but has the potential of spread along with the infected plant roots, even without any adhering soil particles, despite being an ectoparasite (Kaplan, 1985).

Morphology

It is a large sized (2-3 mm), slender (a =50-80) nematode, with a single incisure marking the lateral field from head to tail. The cephalic region is rounded, offset and separated into four sectors by the lateral incisors, stylet is very long, rounded median bulb with a short isthmus. Female with paired ovaries.

Ecology

The sting nematode appears to be injurious only in soils with a minimum of 80% sand content and a maximum of 10% clay (Robbins and Barker, 1974). It reproduces best in moderately dry soil with 7% moisture at 25-30 ^{o}C. It can be extracted from soil with 65% extraction efficiency by centrifugation method and with 41% efficiency by combining semiautomatic elutriation with centrifugation (Byrd *et al.*, 1976). Maximum population in the rhizosphere is distributed in the upper 45 cm soil layer. The nematode can not be dislodged easily from roots, because it is firmly attached to the roots with the help of its long stylet.

Symptoms

The citrus trees infected with sting nematode are stunted and possess small stem and chlorotic leaves. The infected rootstocks have coarse roots with few lateral roots and swollen root terminals (Kaplan, 1985). Even a low nematode density of 40 individuals per litre of soil causes damage and symptoms on the above-ground parts of the plant. Damage symptoms caused by the nematode are similar to those caused due to copper toxicity. The nematode infection results in enlargement of cortical cells, root terminals with damaged apical meristems, and cells without any cytoplasm lining the cavities. Young trees replanted in fumigated soil yield twice as much fruit in the first five years as compared with trees in the non fumigated soil (Bistline *et al.*, 1967).

Management

Hot water treatment of seedling roots at 49 ^{o}C for 5 min. is effective in reducing the nematode population (Kaplan, 1985).

Lesion Nematodes

Several species of the lesion nematodes (*Pratylenchus spp.*) are parasites of citrus (Duncan and Cohn, 1990). Association of a *Pratylenchus* sp. with citrus was first reported by Suit and DuCharme (1953) and later parasitism of citrus roots by the lesion nematodes was studied (Brooks and Perry, 1967). *Pratylenchus coffeae* is one of the important species on citrus and it is reported to cause severe damage to citrus plants in warmer regions of Japan (Yokoo and Ikegemi, 1966), USA (O'Bannon *et al.*, 1972), Taiwan (Huang and Chang, 1976), South Africa (Milne, 1982), and India (Siddiqi, 1964). The infected plants exhibit poor growth, general die back and rotting of roots (Siddiqi, 1964; O'Bannon and Tommerlin, 1969). The nematode reduce, plant growth by 22% within a year and the damage is severe on young trees (O'Bannon and Tommerlin, 1973). *Pratylenchus brachyurus* and *P. vulnus* also affect citrus crop; the former is widely distributed and is potentially important constraint to citrus production.

Biology and Ecology

Pratylenchus coffeae migrates readily from infected to uninfected roots. However, movement in soil is very slow, about 100 cm in a year (O'Bannon, 1980). Soil type does not have any marked effect on the pathogenecity of *P. coffeae* on rough lemon roots (O'Bannon *et al.*, 1976) but fine textured soils tend to favour nematode activity on citrus roots. It is obligately amphimictic species. Males, like females, feed in cortical region of the roots and constitute 30-40% of the total population (Radewald *et al.*, 1971). The nematode completes a life cycle is less than a month at optimum temperature of 26-30 ^{o}C and sometimes nematode population may reach up to 10,000 nematodes/g root (O'Bannon and Tommerlin, 1969; Radewald *et al.*, 1971). The nematode normally invades the root cortex and feeds there but it damages endodermis when in very high number.

Symptoms and Management

The symptoms of lesion nematode infection are similar to the spreading decline caused by the citrus burrowing nematode. Under conditions favourable to *R. citrophilus* and *P. coffeae*, the latter is more damaging to citrus. When present together, these nematodes mutually inhibit each other due to competition for common feeding sites and antagonism. The population of *P. coffeae* increases at a much faster rate than *R. citrophilus* (Kaplan and Timmer, 1982).

The rootstock suitability for nematode development determines the extent of damage and fruit yield. Commercial rootstocks resistant to this nematode are not yet available.

The Reniform Nematode

Rotylenchulus reniformis, a semiendoparasite, has been reported in association with citrus in several tropical and subtropical areas (Ayala and Ramirez, 1964). Investigations on reniform nematode on citrus are mainly confined to population dynamics in soil and not on the pathogenicity. A slight infection of reniform nematode on rough lemon roots was reported from Ghana. Ganguly (1988) has reported reniform nematode infection on the roots of rough lemon and mandarin. However, there are not many reports indicating that the reniform nematode is an important parasite of citrus (Inserra *et al.*, 1989). Inserra and Duncan (1996) found no evidence of nematode infection and reproduction on rough lemon roots from plants grown alone or in combination with cowpea, whereas cowpea roots were heavily infected. The nematode density declined on rough lemon within five months. Sweet orange is a poor host of reniform nematode (Singh, 1974) and lemon is a non host (Goulard and Monterio,1995).

The Root-knot Nematodes

Meloidogyne spp. (*M. exigua, M. incognita, M. indica* and *M. javanica*) attack citrus roots worldwide (Chitwood and Toung, 1960) particularly *M. exigua* in Surinam (Den Ouden, 1965) and Guadeloupe (Scott La Massese, 1969), *M. incognita* in Australia (Colbran, 1958), and *M. indica* in India (Whitehead, 1968). "Asiatic Pyroid citrus nematode" is the common name of a root-knot nematode species (*M. indica*) which is pathogenic to citrus in Taiwan and India. In China, two other species of the root-knot nematodes, *Meloidogyne fujianesis* and *M.oteifae*, have been found on *C. reticulata*, and *M. fujianesis* appears to be widespread (Pan, 1984;1985).

Of all these species, *M. javanica* is most common on citrus; some populations of *M. javanica* fail to complete the life cycle generally due to lack of syncytial formation (Orion and Cohn, 1975; Inserra *et al.*, 1978). In Andhra Pradesh, India, the root-knot infection particularly of *M. javanica* on citrus is widespread (Thirumala, 1956; Rao and Thammiraju, 1976; Singh *et al.*, 1979; Ahmad, 1985) probably due to frequent intercropping of okra, brinjal, cucurbits, and tobacco in citrus orchards (Mani *et al.*, 1986). The nematode is pathogenic to acid lime and sweet orange in the coastal region. Presence of this nematode in the commercial citrus nurseries is a potential threat to citrus cultivation as the nematode is likely to spread via the nursery plants to non-infested orchards (Mani *et al.*, 1988).

Symptoms

The nematode infected trees grow poorly, with unthrifty appearance. The roots have nematode-caused galls. Symptoms on roots and foliage advance with time due to prolific enhancement of infective nematode population.

Management

Trap crops have been used as cover crops to manage the nematode population and crop damage. *Crotolaria* spp., strawberry, groundnut and soybean are poor hosts (Chitwood and Toung, 1960).Troyer citrange, rough lemon, lime, grapefruit and trifoliate orange have been identified as highly resistant (Ibrahim *et al.*, 1989). However, *P. trifoliata* and Troyer citrange (*Citrus sinensis* and *P. trifoliate*) rootstocks are parasitised by populations of *M. javanica* and roots show swellings and malformed tips. The root-knot populations also parasitise a common weed of citrus groves, *Oxalis pes-caprae*. Nematode-caused damage is enhanced when citrus plants are grown along with the weed (Ciancio *et al.*, 1992). Application of organic amendments such as castor and *neem* (20 Kg/ tree) reduces *M. javanica* population by 25% after three months of application (Mani, 1986). Bare root dip in 1000 ppm of monocrotophos, chlorpyriphos and carbofuran causes 85-88% reduction in nematode reproduction (Mani, 1989).

Concluding Remarks

It is apparent that the major nematode pests of citrus have been identified and management options for their effective control have been developed. There is still a wide gap between the knowledge base available to the nematologists on these deceitful parasites and to the citrus growers. It is unfortunate that many growers in India and in many other developing countries are not even aware that plant parasitic nematodes busy feeding on the roots of citrus plants in their orchards. One of the crucial reasons for this lack of transfer of knowledge and technology to the growers is that there are very few extension nematologists in the country to educate the growers about these organisms and assist them in applying management strategies. It is, therefore, imperative that the extension nematology must be strengthened to check this depredation. The extension nematologists must have a high priority public awareness agenda to educate the growers in particular. The recent advances in media technology would be handy in quick dissemination of information to large number of growers, policy makers and general public.

The research nematologists have an important mission to devise simple decision making tools to assist the extension nematologists and growers to determine whether or not to impose nematode management options in a given orchard. Understanding of nematode population densities and their distribution patterns in soil and root is essential for developing nematode management decision options (Goodell and Ferris, 1980; McSorely, 1982). Accurate estimation of nematode population densities further relies heavily on the sampling and extraction procedures (McSorely and Perrado, 1982). The problem of precise sampling is complicated in case of citrus because of vertical as well as horizontal spread of nematodes along with the roots. The density of nematodes is generally greatest in the zone of active feeder roots. The feeder root mass under the tree canopy declines at dripline. The extension nematologist may be able to reduce the sample size and the cost of sampling by sampling during the season in the highest feeder root mass zone (Nigh, 1981, Duncan, 1986). For proper estimation of nematode populations in soil and root, the samples may be taken at the drip line, under the canopy, and combination of the two may be used for assessing the nematode density (Duncan, 1986). The nematode extraction methods vary in their efficiency and may influence the judgment of decision makers if the methods are not standardized (Chawla and Sharma, 1984 a and b).

Root mass of nematode-infected plants is affected by many other factors. If root initiation and growth is reduced, the nematode population would be less abundant on such trees as compared to young growing trees or on trees that compensate for root damage with higher root growth rates. Stress factors such as water logging, low temperatures trigger root initiation (Reitz and Long, 1955; Ford, 1963, 1965). and regrowth of roots following such events is suitable for infection and damage by the nematodes. Moderately declined trees support maximum population of different plant-parasitic nematodes and population peaks coincided with root growth maxima (Mukherjee and Dasgupta, 1993). Thus stratification of orchards into areas of healthy and unhealthy trees would also improve sample precision (Scotto La Massese, 1980). Thirty-two soil cores taken at the dripline of randomly selected trees provide an estimate within 50% of the mean of soil-stage population density (McSorely and Parrado, 1982). Five composite samples of 12 cores each may be sufficient to estimate population density within 20% of the mean in a grapefruit orchard (Davis, 1984), and 20-60 samples are required to estimate population density within 40% of the mean in Egyptian orchards (Abd-Elgawad, 1992). In India, there is a need for developing such decision making tools for each location. Once the decision is made to control the nematodes, a combination of inexpensive, locally available options may be suggested and used; if such options are not locally available they must be developed on priority.

It is evident that some highly virulent nematode parasites of citrus are not present in the country. Any planting material originating from regions where these virulent nematodes are present must be examined very carefully at the port of entry. Presently, the number of nematologists in the quarantine directorate is inadequate to check the entry of these nematodes. The policy makers must strengthen the quarantine nematology to prevent entry of these insidious organisms along with the planting material. Enhancing citrus nematology in India seems to be a good investment!

REFERENCES

ABU-GHARBIEH, W.I. AND V.G. PERRY, 1970: Host differences among Florida populations of *Belonolaimus longicaudatus* Rau. *Journal of Nematology*, **2**: 209-216, Lake Alfred Florida.

ABD-ELGAWAD, M.M. 1992: Spatial distribution of phytonematode community in Egyptian citrus groves. *Fundamental and Applied Nematology*, **15**:367-373, Rue Linois, Paris.

AHMAD, N.S. 1985: Studies on citrus nematode, *Tylenchulus semipenetrans* Cobb with reference to distribution, biology and control. *M.Sc. (Ag) Thesis, A.P. Agricultural University*, Hyderabad, 84pp.

ALAM, M.M., S.A. SIDDIQUI AND A.M. KHAN, 1977: Use of organic amendments in nematode management. *Indian Journal of Nematology*, **7**: 27-31, New Delhi.

ANONYMOUS, 1986: Plant parasitic nematodes of bananas, citrus, coffee, grapes and tobacco. *Union Carbide Agricultural Products Company*, N.C. 71p. U.S.A.

ANONYMOUS, 1995: Annual Report, 1994-95: Indian Inst. of Horticultural Research, Hasargatta, Bangalore. pp. 63-65.

AWAN, M.N., N. JAVED, R. AHMAD AND M. INAM-UL-HAQ, 1992: Effect of leaf extract of four plant species on larval mortality of citrus nematode (*Tylenchulus semipenetrans* Cobb.) and citrus plant growth. *Pakistan Journal of Phytopathology*, **4**:41-45, Pakistan.

AYALA, A. AND C.T. RAMIREZ, 1964: Host range, distribution and bibliography of the reniform nematode, *Rotylenchulus reniformis*, with special reference to Puerto Rico. *Journal of Agriculture of the University of Puerto Rico*, **48**:140-161, Puerto Rico.

BAGHEL, P.P.S. AND D.S. BHATTI, 1982: Vertical and horizontal distribution of phytonematodes associated with citrus. *Indian Journal of Nematology*, **12**:339-344, New Delhi.

BAINES, R.C. 1974 : Effect of soil type on movement and infection rate of larvae of *Tylenchulus semipenetrans*. *Journal of Nematology*, **6**:60-62, Lake Alfred, Florida.

BAINES, R.C., F.J. FOOTE AND J.P. MARTIN, 1956: Fumigate soil before replanting citrus for control of the citrus nematode. *Citrus Leaves*, **36**:6-8, 24, 27, U.S.A.

BAINES, R.C., L.J. KLOTZ, T.A. DE WOLFE, R.H. SMALL AND G.O. TURNER, 1966: Nematocidal and fungicidal properties of some soil fumigants. *Phytopathology*, **56**:691-698, St Paul, Minnesota.

BAINES, R.C., J.P. MARTIN, T.A. DE WOLFE., S.B. BOSWELL AND M.J. GARBER, 1962: Effect of high doses of D-D on soil organisms and the growth and yield of lemon trees. *Phytopathology*, **52**:723, St. Paul, Minnesota.

BAINES, R.C., R.H. SMALL AND L.H. STOLZY, 1965: DBCP recommended for control of citrus nematode on bearing trees. *California Citrograph*, **50**:342-346, Los Angeles.

BAINES, R.C., L.H. STOLZY, O.C. TAYLOR, R. H. SMALL AND G.E. GOODALL, 1958. Nematode control on bearing trees. *California Citrograph*, **43**:328-329, Los Angeles.

BELLO, A., A. NAVAS AND C. BELART, 1986: Nematodes of citrus groves in Spanish Levante Ecological study focused to their control. p. 217-226.

In: Cavalloro, R., and Di Martino, E. (Eds). Integrated Pest Control in Citrus Groves. A.A. Balkema Publ. Cp., Rotterdam, Boston.

BINDRA, O.S. 1970: Nematodes. p. 53-56. *In* : Citrus Decline in India: Chadha, K.L. (Ed), Punjab Agriculture University, Ludhiana.

BISTLINE, F.W., B.L. COLLIER AND C.E. DIETER, 1967: Tree and yield response to control of a nematode complex including *Belonolaimus longicaudatus* in replanted citrus. *Nematologica*, **13**:137-138, Leiden, Netherlands.

BOOSALIS, M.G. AND R. MANKAU, 1965: Parasitism and predation of soil microorganisms. p. 374-391. *In*: Barker, K.F. and Snyder, W.C. (Eds.) Ecology of Soil-borne Plant Pathogens, 374-91. University of California Press, Los Angeles.

BROOKS, T. L. AND V.J. PERRY, 1962: Apparent parthenogenetic reproduction of the burrowing nematode *Radopholus similis*. *Proceedings of Soil Crop Science Society of Florida*, **22**:160-162, Florida.

BROOKS, T. L. AND V.J. PERRY, 1967: Pathogenicity of *Pratylenchus brachyurus* to citrus. *Plant Disease Reporter*, **51**:569-573, Washington, D.C.

BYRD, D.W. JR., K.R. BARKER, H. FERRIS, C.J. NUSBAUM, W.E. GRIFFIN, R.H. SMALL AND C.A. STONE, 1976: Two semiautomatic elutriators for extracting nematodes and certain fungi from soil. *Journal of Nematology*, **8**:206-212, Lake Alfred, Florida.

CARTER, W.W. AND H.W. REYNOLDS, 1969: Mortality rate of the citrus nematode after DBCP treatment. *Plant Disease Reporter*, **53**:690, Washington, D.C.

CHANDEL, Y.S., 1986: Studies on the nematodes associated with citrus in Himachal Pradesh. *PhD Thesis, DR. Y.S. Parmar University of Horticultural Science and Forestry*, Solan, H.P., India.

CHANDEL, Y.S. AND N.K. SHARMA, 1989: Interaction of *Tylenchulus semipenetrans* with *Fusarium solani* on citrus. *Indian Journal of Nematology*, **19**: 21-24, New Delhi.

CHAWLA, M.L. AND S.B. SHARMA, 1984: Efficacy of some nematode extraction methods in the recovery of citrus nematode *Tylenchulus semipenetrans* larvae from soil. *Indian Journal of. Nematology*, **14**:63-64, New Delhi.

CHAWLA, M.L. AND S.B. SHARMA, 1984a: Effect of temperature on the recovery of second stage larvae of citrus nematode *Tylenchulus semipenetrans* through modified Baermann funnel technique. *Indian Journal of Nematology*, **14**:64-65, New Delhi.

CHAWLA, M.L. AND S.B. SHARMA, 1984b: Horizontal and vertical distribution of citrus nematode, *Tylenchulus semipenetrans*. *Indian Journal of. Nematology*, **14**:193-195, New Delhi.

CHILDERS, C.C., L.W. DUNCAN, T.A. WHEATON AND L.W. TIMMER, 1987: Arthropod and nematode control with Aldicarb on Florida citrus. *Journal of Economic Entomology*, **80**:1064-1071, Lanham, Maryland.

CHITWOOD, B.G. AND M.C. TOUNG, 1960: Host-parasite interaction of the Asiatic pyroid citrus nema. *Plant Disease Reporter*, **44**:848-854, Washington, D.C.

CIANCIO, A., V.L. GIVDICE, R. BONSIGNORE AND G. ROCCUZZO, 1992: [Root-knot nematodes attacking weeds in southern Italy]. *Informatore Fitopatologico*, **42**:55-57, Bologna, Italy.

COBB, N.A. 1913: Notes of *Mononchus* and *Tylenchulus*. *Journal of*

Washington.Acadamy of Sciences, **3**:287-288, Washington, D.C.

COBB, N.A. 1914: Cirtus-root nematode. *Journal of Agricultural Research*, **2**:217-230, London.

COHN, E. AND M. MORDECHAI, 1974: Experiments in suppressing citrus nematode populations by use of a marigold and a predacious nematode. *Nematologia Mediterranea* **2**:43-53, Amendola, Bari, Italy.

COHN, E. 1964: Penetration of the citrus nematode in relation to root development. *Nematologica*, **10**:594-600, Leiden, Netherlands.

COHN, E. 1965: On the feeding and histopathology of the citrus nematode. *Nematologica*, **11**:47-54, Leiden, Netherlands.

COHN, E. 1965a: The development of citrus nematode on some of its hosts. *Nematologica*, **11**:593-600, Leiden, Netherlands.

COHN, E. 1966: Observations on the survival of free- living stages of citrus nematode. *Nematologica*, **12**:321-327, Leiden, Netherlands.

COHN, E. 1976: Report of investigations on nematodes of citrus and subtropical fruit crops in South Africa. Citrus and Subtropical Fruit Res Inst. Nelspruit. 41p, South Africa.

COHN, E., W.A. FEDER AND M. MORDECHAI, 1968: The growth response of citrus to nematicide treatments. Isreal *Journal of Agricultural Research*, **18**:19-24.

COHN, E., A. HOUGH AND N. MULDER, 1976: Elimination of nematodes and other plant pathogens from irrigation water by filtration techniques. *13 International nematology Symposium*, Sept.5-11 Dublin, Ireland,:16-17, Ireland.

COHN, E. AND G. MINZ, 1965: Application of nematicides in established orchards for connrolling the citrus nematode, *Tylenchulus semipenetrans* Cobb. *Phytopathologia Mediterrania*, **4**:17-20, Bologna, Italy.

COHN, E., G. MINZ AND S.P. MONSELISE, 1965: The distribution, ecology and pathogenicity of the citrus nematode in Israel. *Israel Journal of Agricultural Research* **15**:187-200.

COLBRAN, R.C. 1958: Studies of plant and soil nematodes. 2. Queensland host records of root-knot nematodes (*Meloidogyne species*). *Queensland Journal of Agricultural Sciences*, **15**:101-135, Brisbane, Queensland.

DAS, T.K. AND M.C. MUKHOPADHYAY, 1983: Influence of varying levels of initial populations of *Tylenchulus semipenetrans* on Citrus paradisi and on final nematode population. *Indian Journal of Nematology*, **13**:128-130, New Delhi.

DAVIES, K.G., V. LAIRD AND B.R. KERRY, 1990: The life cycle, population dynamics and host specificity of a parasite of *Heterodera avenae*, similar to *Pasteuria penetrans*. *Revue de nematologie*, **13**:303-309, Paris.

DAVIS, R.M., 1984: Distribution of *Tylenchulus semipenetrans* in a Texas grapefruit orchard. *Journal of Nematology*, **16**: 313-317, Lake Alfred, Florida.

DAVIS, R.M. AND H.S. WILHITE, 1985: Control of *Tylenchulus semipenetrans* on citrus with Fenamiphos and Oxamyl. *Plant Disease* **69**:974-976, St. Paul, Minnesota.

DEN OUDEN, H. 1965: An infection on citrus in Surinam caused by *Meloidogyne exigua*. *Surinam Landbouw*, 13:34, Pas a dena.

DUCHARME, E.P. 1954: Cause and nature of spreading decline of citrus. *Proceedings of the Florida State Horticultural Society*, **67**:75-81, Deland.

DUCHARME, E.P. 1955: Sub-soil drainage as a factor in the spread of the burrowing nematode. *Proceedings of the Florida State Horticultural Society*, **68**: 29-31, Deland.

DUCHARME, E.P. 1959: Morphogenesis and histopathology of lesions induced on citrus roots by *Radopholus similis. Phytopathology*, **49**:388-395, St. Paul, Minnesota.

DUCHARME, E.P. 1968. Burrowing nematode decline of citrus, a review. *In*: Smart, G.C. and Perry, V.G. (Eds), *Tropical Nematology*, 20-37, University of Florida Press, Gainesville.

DUCHARME, E.P. 1969: Nematode problems of citrus. p. 225-237. *In*: Peachey, J.E. (Ed), Nematodes of Tropical Crops, Common Wealth Agricultural Bureau, England.

DUCHARME, E.P. AND W.C. PRICE, 1966: Dynamics of multiplication of *Radopholus similis. Nematologica*, **12**:113-121, Leiden, Netherlands.

DUCHARME, E.P. AND R.F. SUIT, 1955: Immunity of the lychee from burrowing nematode. *Proceedings of the Florida Station Horticulture Society*, **68**:270-272, Deland.

DUNCAN, L.W. 1986: The spatial distribution of citrus feeder roots and of the citrus nematode, *Tylenchulus semipenetrans. Revue de Nematologie*, **9**:233-240, Paris.

DUNCAN, L.W. AND E. COHN, 1990: Nematode parasites of citrus. p. 321-346. *In*: Plant Parasitic Nematodes in Subtropical and Tropical Agriculture: Luc. M, Sikora, R.A. and Bridge, J. (Eds.), CAB International.

DUNCAN, L.W. AND D.M. EISSENSTAT, 1993: Responses of *Tylenchulus semipenetrans* to citrus fruit removal: implications of carbohydrate competetion. *Journal of Nematology*, **25**:7-14, Lake Alfred, Florida.

DUNCAN, L.W. AND M.M. EL-MORSHEDY, 1996: Population changes of *Tylenchulus semipenetrans* under localized versus uniform drought in citrus root zone. *Journal of Nematology*, **28**:360-368, Lake Alfred, Florida.

DUNCAN, L.W. AND J.W. NOLING, 1987: The relationship between development of the citrus root system and infestation by *Tylenchulus semipenetrans. Revue de Nematologie*, **10**:61-66, Paris.

DUNCAN, L.W. AND J.W. NOLING, 1988: Computer simulated management of *Tylenchulus semipenetrans* on citrus. p. 262. *In* : *International Citrus Congress Middle East*, Book of abstracts.

DUNCAN, L.W., J.H. GRAHAM AND L.W. TIMMER, 1993: Seasonal patterns associated with *Tylenchulus semipenetrans* and *Phytophthora parasitica* in the citrus rhizosphere. *Ecology and Epidemiology*, **83**:573-581.

DUNCAN, L.W., J.W. NOLING, R.N. INSERRA AND D. DUNN, 1996: Spatial patterns of *Belonolaimus* spp. among and within citrus orchards on Florida's central ridge. *Journal of Nematology*, **28**:352-359, Lake Alfred, Florida.

ESSER, R.P. 1984: How nematodes enter and dispers in Florida nurseries via vehicles. *Florida Department of Agriculture and Consumer Services, Division of Plant Industry, Nematology Circular No.109.2p*

ESSER, R.P. AND S.E. SIMPSON, 1984: Sting nematode on citrus. *Florida Department of Agriculture and Consumer Services, Division of Plant*

Industry, Nematology Circular No.106, 2p.

ESSER, R.P., G.T. SMITH AND J.H. O'BANNON, 1993: An eleven year phytoparasitic nematode survey of Florida citrus groves and their environs. *Florida Department of Agriculture and Consumer Services, Gainsville. Bulletin No. 15.*

FATTAH, F.A., H.M. SALEH AND H.M. ABOUD, 1989: Parasitism of the citrus nematode, *Tylenchulus semipenetrans* by *Pasteuria penetrans* in Iraq. *Journal of Nematology,* **21**:431-433, Lake Alfred, Florida.

FORD, H.W. 1967: Burrowing nematode resistant rootstocks as biological barriers in citrus groves. *Proceedings of the Florida State Horticultural Society,* **80**: 56-59, Deland.

FORD, H.W. AND W.A. FEDER, 1964: Three citrus root stocks recommended for trial in spreading decline areas. *Florida Agriculture Experiment Station Circular,* 5-151, Deland.

GANGULY, S. 1988: Phytonematodes associated with betavine and citrus in Nagpur, Maharashtra, India. *Indian Journal of Nematology,* **18**:142-144, New Delhi.

GASPARD, J.T. AND R. MANKAU, 1986: Nematophagous fungi associated with *Tylenchulus semipenetrans* and the citrus rhizosphere. *Nematologica,* **32**:359-63, Leiden, Netherlands.

GOTTLIEB, Y., E. COHN AND P. SPEIGEL-ROY, 1986: Biotypes of the citrus nematode (*Tylenchulus semipenetrans* Cobb) in Israel. *Phytoparasitica,* **14**:193-198, Rehovot, Isreal.

GOODELL, P. AND H. FERRIS, 1980: Plant parasitic nematode distribution in an alfalfa field. *Journal of Nematology,* **12**: 136-141, Lake Alfred, Florida.

GOULART, A.M.C. AND A.R. MONTEIRO, 1995: [Resistance of citrus limonica to *Rotylenchulus reniformis*]. *Nematologia Brasileira,* **19**:86-88, Brazil.

GRAHAM, J.H. 1986: Citrus mycorrhiza: potential benefits and interactions with pathogens. *Horticultural Science,* **21**:1302-1306, Alexandria, Virginia.

GUPTA, J.C. AND A.S. ATWAL, 1972: Biology and ecology of *Hoplolaimus indicus* (Hoplolaiminae : Nematoda) 3. Influence of population density on tomato and citrus plants. p. 26. *In : Proceedings of 11th International Symposium on Nematology,* Reading, UK.

HAMID, G.A., S.D. VAN GUNDY AND C.J. LOVATT, 1985: Citrus nematode alters carbohydrate partiotioning in the "Washington" naval orange. *Journal of the American Society for Horticultural Science,* **110**:642-646.

HAMID, G.A., S.D. VAN GUNDY AND C.J. LOVATT, 1988: Phenologies of the citrus nematode and citrus roots treated with oxamyl. p. 993-1004. *In*: Diseases and Nematodes, R. Goren and K. Mendel, (Eds), Vol. 2. Margraf Publ., Weikersheim, Germany.

HANNON, C.I. 1962: The occurrence and distribution of the citrus-root nematode, *Tylenchulus semipenetrans* Cobb. In Florida. *Plant Disease Reporter,* **46**:451-455, Washington, D.C.

HANNON, C.I. 1963: Longevity of *Rodopholus similis* under field conditions. *Plant Disease Reporter,* **47**:812-816, Washington, D.C.

HEALD, C.M. 1970: Distribution and control of the citrus nematode in the lower Rio Grande Valley of Texas. *Journal of Rio Grande Valley Horticulture Society,* **24**:32-35, Weslaco.

HEALD, C.M. AND J.H. O'BANNON, 1987: Citrus decline caused by nematodes. V. Slow decline. *Florida Department of Agriculture and Consumer Services, Division of Plant Industry, Nematology Circular No. 143. 4p.*

HUANG, C.S. AND Y.C. CHANG, 1976: Pathogenicity of *Pratylenchus coffeae* sunki orange. *Plant Disease Reporter*, **60**:957-960, Washington, D.C.

HUSAIN, S.I., H.Y. MOHAMMAD AND A.J. AL-ZARARI, 1981: Studies on vertical distribution and seasonal fluctuation of the citrus nematode (*Tylenchulus semipenetrans*) in Iran. *Nematologia Mediterranea*, **9**:7-19, Amendola, Bari, Italy.

HUETTEL R.N., D.W. DICKSON AND D.T. KAPLAN, 1982: Sex attractants and behaviour in the two races of *Radopholous similis*. *Nematologica*, **28**:360-369, Leiden, Netherlands.

HUETTEL, R.N., D.W. DICKSON AND D.T. KAPLAN, 1984: *Radopholous citrophilus* n. sp. (Nematoda), a sibling species of *Radopholous similis*. *Proceedings of Helminthological Society of Washington*, **51**:32-35, Washington, D.C.

IBRAHIM, I.K.A., M.W. TAHA AND M.W.A. HASAN, 1989: Resistance of some citrus rootstocks and grape cultivars to *Tylenchulus semipenetrans* and *Meloidogyne* spp. *International Nematology Network Newsletter*, **6**:3-7.

INSERRA, R.N. AND L.W. DUNCAN, 1996: Host status of rough lemon to *Rotylenchulus reniformis*. *Nematropica*, **26**:87-90, Auburn, Alabama.

INSERRA, R.N., L.W. DUNCAN, J.H. O'BANNON AND S.A. FULLER, 1994: Citrus nematode biotypes and resistance of citrus rootstocks in Florida. *Florida Department of Agriculture and Consumer Services, Division of Plant Industry, Nematology Circular*, **205**:4, Gainesville, Florida.

INSERRA, R.N., R.A. DUNN, R. MCSORLEY, K.R. LANGDON AND A.V. RICHMER, 1989: Weed hosts of *Rotylenchulus reniformis* in ornamental nurseries of southern Florida. *Nematology circular* No. 171. Gainesville, Florida.

INSERRA, R.N., G. PERROTTA, N. VOVLAS AND A. CATARA, 1978: Reaction of citrus root stocks to *Meloidogyne javanica*. *Journal of Nematology*, **10**:181-184, Lake Alfred, Florida.

JAGDALE, G.B., R.N. POKHARKAR AND K.S. DAREKAR, 1984: Relative susceptibility of citrus rootstocks to citrus nematode (*Tylenchulus semipenetrans* Cobb). *Biovigyanam*, **10**:149-156, Pune.

JATALA, P. 1985: Biological control of nematodes. p.303-308. *In*: *An Advanced Treatise on Meloidogyne* Vol. I. *Biology and Control* : Sasser,J.N. and Carter, C.C. (Eds.) North Carolina State University Graphics, Raleigh, USA.

KAPLAN, D.T. 1981: Characterization of citrus rootstock responses to *Tylenchulus semipenetrans*. *Journal of Nematology*, **13**:492, Lake Alfred, Florida.

KAPLAN, D.T. 1985: Influence of the sting nematode *Belonolaimus longicaudatus* on young citrus trees. *Journal of Nematology*, **17**:408-414, Lake Alfred, Florida.

KAPLAN, D.T. 1986: Variation in *Radopholus citrpohilus* population densities in the citrus rootstock Carrizo citrange. *Journal of Nematology*, **18**:31-34, Lake Alfred, Florida.

KAPLAN, D.T. AND L.W. TIMMER, 1982: Effects of *Pratylenchus coffeae-Tylenchulus semipenetrans* interactions on nematode population dynamics in citrus. *Journal of Nematology*, **14**:368-373, Lake Alfred,

Florida.

KAPLAN, D.T. AND J.H. O'BANNON, 1985: Occurrence of biotypes in *Radopholus citrophilus*. *Journal of Nematology*, **17**:158-162, Lake Alfred, Florida.

KIRKPATRICK, J.C. AND S.D. VAN GUNDY, 1966: Soil salinity and citrus nematode survival. *Nematologica*, **12**:93-94, Leiden, Netherlands.

LEE, D.L. AND H.J. ATKINSON, 1977. *Physiology of nematodes*. New York: Columbia University Press.

MACHMER, J.H. 1958. Effect of soil salinity on nematodes in citrus and papaya plantings. *Journal of the Rio Grande Valley Horticultural Society*, **12**:57-60, Weslaco.

MACHON, J.E. AND J. BRIDGE, 1996: *Radopholus citri* n.sp. (Tylenchida : Pratylenchidae) and its pathogenicity on citrus. *Fundamental and Applied Nematology*, **19**(2):127-133, Rue Linois, Paris.

MAGGENTI, A.R. 1962: The production of the gelatinous matrix and its taxonomic significance in *Tylenchulus*. *Proceedings of Helminthological Society of Washington*, **29**:139-144, Washington, D.C.

MAHROS, M.E., M.M. ABOU-EL-NAGA, S.A. MONTASSER AND H.E. LOKMA, 1985: Plant parasitic nematodes associated with certain crops under different irrigation systems at Khattara project, newly reclaimed area, Egypt. *Zagazig Journal of Agricultural Research* **12**(1):707-723.

MANI, A. 1986: Occurrence of *Meloidogyne javanica* on citrus in Andhra Pradesh (India). *International Nematology Network Newsletter*, **3**: 9-10.

MANI, A. 1988: Effect of interculture of marigold and mustard with acid lime on citrus nematode, *Tylenchulus semipenetrans*. *International Nematology Network Newsletter*, **5**:14-15.

MANI, A. 1989: Control of citrus nematode, *Tylenchulus semipenetrans* by bare root dip treatment of acid lime seedlings in pesticides. *International Nematology Network Newsletter*, **6**:20-21.

MANI, A. 1994: Occurrence and distribution of *Tylenchulus semipenetrans* in Andhra Pradesh. *Indian Journal of Nematology*, **24**:106-111, New Delhi.

MANI, A., B.N. AHMED, P. KAMESHWARA RAO AND V. DAKSHINAMURTI, 1986: Plant products toxic to the citrus nematode, *Tylenchulus semipenetrans*. Cobb. *International Nematology Network Newsletter*, **3**:14-15.

MANI, A. AND V. DAKSHINAMURTHI, 1986: Citrus nematode and root-knot nematodes : distribution, crop losses, symptoms and management. p. 246-260. *In*: Plant Parasitic Nematodes of India: Problems and Progress: Swarup, G and Dasgupta, D.R. (Eds), Indian Agricultural Research Institue, New Delhi.

MANI, A., V. DAKSHINAMURTHI AND G.S. REDDY, 1988: Distribution of *Tylenchulus semipenetrans* and *Meloidogyne javanica* in commercial citrus nurseries in Andhra Pradesh. *Indian Journal of Nematology*, **18**: 38-339, New Delhi.

MANI, A. AND G.S. REDDY, 1986: Studies on the varietal reaction of some species of citrus and *Poncirus* to citrus nematode. *Indian Journal of Nematology*, **16**:267-268, New Delhi.

MANKAU, R., 1972: Utilization of parasites and predators in nematode pest management ecology. *Proceedings of Annual Tall Timbers Conference on Ecological Animal Control by Habitat Management* **4**:129-143.

MANKAU,R. AND R.J. AND MINTEER, 1962: Reduction of soil populations of the citrus nematode by the addition of organic materials. *Plant Disease Reporter*, **46**:375-378, Washington, D.C.

MASHELA, P., L. W. DUNCAN AND R. MCSORELY, 1992: Salinity reduces resistance to *Tylenchulus semipenetrans* in citrus rootstocks. *Nematropica*, **22**:7-12, Auburn, Alabama.

MCCLURE, M.A. AND M.E. SCHMITT, 1996: Control of citrus nematode, *Tylenchulus semipenetrans*, with Cadusafos. *Supplement to Journal of Nematology*, **28**:624-628, Lake Alfred, Florida.

MCSORLEY, R. 1982: Simulated sampling strategies for nematodes distributed accordng to a negative binomial model. *Journal of Nematology*, **14**:517-522, Lake Alfred, Florida.

MCSORLEY, R. AND J. L. PARRADO, 1982 : Plans for the collection of nematode soil samples from fruit groves. *Nematropica*, **12**:257-267, Auburn, Alabama.

MEAGHER, J.W., 1960: Root-knot nematode of the grapevine control in rooted stocks by hot water treatment. *J. Agric. Vict. Dept. Agric.*, **58**:419-423.

MEAGHER, J.W. 1967: Observations on the transport of nematodes in subsoil drainage and irrigation water. *Australian Journal of Experimental Agricultural Animal Husbandry*, **7**:577-579, East Melbourne, Victoria.

MEAGHER, J.W. 1969: Nematodes as a factor in citrus production in Australia. *Proceeding of First International Citrus Symposium, Riverside, California* **2**:999-1006, Riverside, USA.

MEHER, H.C. 1995: Persistence and nematicidal efficacy of phorate in soil and residues in fruit of *Citrus reticulata* (L.). *Indian Journal of Nematology*, **25**:38-42, New Delhi.

MILNE, D.L. 1974: *Citrus seedbeds: methyl bromide* and *mychorrizae*. Citrus and Subtropical Fruit Research Institute, Nelspruit:9-11.

MILNE, D.L. 1982: Nematode pests of citrus. p. 12-18. *In*:. Nematology in Southern Africa.: Keeth, D.P. and Heyns,J. (Eds) Republic of South Africa, Department of Agriculture and Fisheries Science Bulletin No.400.

MUKHERJEE, B. AND M. K. DASGUPTA, 1993: Population dynamics and association of plant parasitic nematodes in the decline of *Citrus limettoides* L. in West Bengal. *Indian Journal of Nematology*, **23**: 69-70, New Delhi.

MUKHOPADHYAYA, M.C. AND M.R. DALAL, 1971: Effects of two nematicides on *Tylenchulus semipenetrans* and on sweetlime yield. *Indian Journal of Nematology*, **11**:195-97, New Delhi.

NASIRA, K., K. FIROZA AND M.A. MAQBOOL, 1992: Description of *Xiphinema karachiense* sp.n. and morphometric data on *Enchodelus macrodorus* (de Man, 1880), Thorne, 1939 (Nematoda: Dorylaimida) from Pakistan. *Fundamental and Applied Nematology*, **15**:421-426, Rue Linois.

NASIRA, K., F. SHAHINA, K. FIROZA, M.A. MAQBOOL, 1993: Description of *Paralongidorus lemoni n.*sp. and redescription of *P. major* Verma, 1973 (Nematoda: Longidoridae) from Pakistan with SEM observation. *Pakistan Journal of Nematology*, **11**:79-91, Pakistan.

NEAL, J.C. 1889: The root-knot disease of the peach, orange and other plants in Florida, due to the work of *Angullila. Bulletin of Bureau of Entomology US Department of Agriculture*, No.20:31pp, USA.

NIGH, E.L. JR. 1981: Relation of citrus nematode to root distribution in flood irrigated citrus. *Journal of Nematology*, **13**:451-452, Lake Alfred, Florida, USA.

O'BANNON, J.H. 1968: The influence of an organic soil amendment on infectivity and reproduction of *Tylenchulus semipenetrans* on two citrus rootstocks. *Phytopathology*, **58**:597-601, St. Paul, Minnesota.

O'BANNON, J.H. 1980: Migration of *Pratylenchus coffeae* and *P. brachyurus* on citrus and in soil. *Nematropica*, **10**:70, Auburn, Alabama.

O'BANNON, J.H. AND H.W. FORD, 1976: An evaluation of several *Radopholus similis* - resistant or tolerant root stocks. *Plant Disease Reporter*, **60**:620-624, Washington, D.C.

O'BANNON, J.H. AND R.P. ESSER, 1985: Citrus declines caused by nematodes in Florida. I. Soil factors. *Nematology circular. Florida Dept. Of Agriculture and Consumer Services, Division of Plant Industry, No. 114*, Gainesville, Florida.

O'BANNON, J.H. AND S. NEMEC, 1978: Influence of soil pesticides on vescicular arbuscular mycorrizae in citrus soil. *Nematropica*, **8**:56-61, Auburn, Alabama.

O'BANNON, J.H., J.D. RADEWALD AND A.T. TOMMERLIN, 1972: Population fluctuation of three parasitic nematodes in Florida citrus. *Journal of Nematology*, **4**:194-199, Lake Alfred, Florida.

O'BANNON, J.H., J.D. RADEWALD, A.T. TOMMERLIN AND R.N. INSERRA, 1976: Comparative influence of *Radopholus similis* and *Pratylenchus coffeae* on citrus. *Journal of Nematology*, **8**: 58-63, Lake Alfred, Florida.

O'BANNON, J.H., AND H.W. REYNOLDS, 1967: The effect of chemical treatment on *Tylenchulus semipenetrans* and citrus tree response during eight years.. *Nematologica*, **13**:131-136, Leiden, Netherlands.

O'BANNON, J.H., H.W. REYNOLDS AND C.R LEATHERS, 1966: Effects of temperature on penetration, development, and reproduction of *Tylenchulus semipenetrans*. *Nematologica*, **12**: 483-487, Leiden, Netherlands.

O'BANNON, J.H. AND A.C. TARJAN, 1973: Preplant fumigation for citrus nematode control in Florida. *Journal of Nematology*, **5**: 88-95, Lake Alfred, Florida.

O'BANNON, J.H. AND A.C. TARJAN, 1979: Management of *Radopholus similis* infecting citrus with DBCP or Phenamiphos. *Plant Disease Reporter*, **63**:456-460, Washington, D.C.

O'BANNON J.H. AND A.T. TOMMERLIN, 1969: Population studies on two species of *Pratylenchus* on citrus. *Journal of Nematology*, **1**:299-300, Lake Alfred, Florida.

O'BANNON J.H. AND A.T. TOMMERLIN, 1973. Citrus tree decline caused by *Pratylenchus coffeae*. *Journal of Nematology*, **5**: 311-316, Lake Alfred, Florida.

ORION, D. AND E. COHN, 1975: A resistant response of citrus roots to the root-knot nematode, *Meloidogyne javanica*. *Marcellia*, **38**:327-328, Marcellia, Strasbourg, France.

OSMAN, G.Y. 1990: Preliminary study on induction of sex reversal in *Meloidogyne javanica* (Chitwood) and *Tylenchulus semipenetrans* (Cobb) by testosterone. *Anzeiyer fuer schaed lingskunde Pflan zenschutz Umweltschutz*, **63**:14-15.

OSMAN, G.Y. AND SALEM, F.M. 1995: Bio-efficacy of Sincocin AGTM to control

Tylenchulus semipenetrans (Tylenchida, Nematoda) in citrus orchard. *Anzeiger fuer Schadlingskunde Pflanzenschutz Umweltschutz*, **68**:179-181, Hamburg, Germany.

OSMAN, G.Y., A.M. ZAKI, F.M. SALEM AND E.T.E DARWISH, 1988: Biological control study on *Tylenchulus semipenetrans* (Cobb) by certain soil Mesofauna. *Anzeiger fuer Schaedlingskunde Pflanzenschutz Umneltschutz*, **61**:116-118, Hamburg, Germany.

OSTER, J.D. 1984: Leaching for salinity control. p. 175-189. *In*: Ecological studies Vol. **51**: W.D. Bilings, F. Golley, O.L. Laange, J.S. Ollson, and H. Remert (Eds), New York: Springer Verlag.

OTEIFA, B.A., F.A. SHAFIEE AND F.M. EISSA, 1965: Efficacy of DBCP flood irrigation in established citrus. *Plant Disease Reporter*, **49**:598-599, Washington, D.C.

PAN, C. 1984: [Studies on plant parasitic nematodes on economically important crops in Fujian. 1. Species of root-knot nematodes (*Meloidogyne* species) and their host-plants.] *Acta Zoologica Sinica*, **30**:159-167.

PAN, C. 1985. [Studies on plant parasitic nematodes on economically important crops in Fujian. 111. Description of *Meloidogyne fujianensis* n. sp.(Nematoda:Meloidogynidae) infesting citrus in Najing county.] *Acta Zoologica Sinica*, **31**:263-268.

PEOPLES, S.A., K.T. MADDY, W. CUSIK, T. JACKSON, C. COOPER AND A.D. FREDERICKSON, 1980: A study of samples of well water collected from selected areas in California to determine the presence of DBCP and certain other pesticide residues. *Bulletin of Environmental Contamination and Toxicology*, **24**:611-618, Berlin.

PHILIPS, J. 1969: Control of citrus nematode. *Tylenchulus semipenetrans* with DBCP in established Cyprus citrus groves. *Plant Disease Reporter*, **53**:804-806, Washington, D.C.

POUCHER, C., H.W. FORD, R.F. SUIT AND E.P. DUCHARME, 1967: Burrowing nematode in citrus. Florida *Department of Agriculture, Division of Plant Industry Bulletin No.* **7**. 63p., Florida.

PRASAD, S.K. AND M.L. CHAWLA, 1965: Observations on the population fluctuations of citrus nematode, *Tylenchulus semipenetrans* Cobb, 1913. *Indian Journal of Entomology*, **27**: 450-454, New Delhi.

QUENEHERRE, P., F. DROB AND P. TOPART, 1995: Host status of some weeds to *Meloidogyne* spp., *Pratylenchus* spp. *Helicotylenchus* spp. And *Rotylenchulus reniformis* associated with vegetables cultivated in polytennels in Martinique. *Nematropica*, **25**:149-157, Auburn, Alabama.

RADEWALD, J.D., J.H. O'BANNON AND A.T. TOMMERLIN, 1971: Temperature effects on reproduction and pathogenecity of *Pratylenchus coffeae* and *P. brachyurus* and survival of *P. coffeae* in roots of *Citrus jambhiri*. *Journal of Nematology*, **3**: 390-394, Lake Alfred, Florida.

RADEWALD, J.D., J.H. O'BANNON AND A.T. TOMMERLIN, 1971: Anatomical studies of *Citrus jambhiri* roots infected by *Pratylenchus coffeae*. *Journal of Nematology*, **3** : 409-416, Lake Alfred, Florida.

RAO, K.B.H. AND N.B. THAMMIRAJU, 1975 : Nematodes associated with citrus. p. 85. *In*: Citrus Decline in Andhra Pradesh: Reddy, G.S. and V.D. Murthy (Eds), A.P.A.U., Hyderabad, India.

REDDY, P.P. AND D.B. SINGH, 1978: Evaluating the reaction of some of the species and varieties of *Citrus* and *Poncirus* to the citrus nematode. *Indian Journal of Nematology*, **8**:82-84, New Delhi.

Reddy, P.P. and D.B. Singh, 1979: Nematode problems of citrus in India. *PANS*, **25**:409-419, London.

Reitz, H.J. and W.T. Long, 1955: Water table fluctuation and depth of rooting of citrus trees in the Indian river area. *Proceedings of Florida Station Horticulture Society*,**68**: 24-29., Deland.

Reynolds H.W. and J.H. O'Bannon, 1958: The citrus nematode and its control on living citrus in Arizona. *Plant Disease Reporter*, **42**:I288-1292, Washington, D.C.

Reynolds H.W. and J.H. O'Bannon, 1963: Factors influencing the citrus replant in Arizona. *Nematologica*, **9**:337-340, Leiden, Netherlands.

Robbins, R.T. and K.R. Barker, 1974: Effect of soil type, particle size, temperature and moisture on reproduction of *Belonolaimus longicaudatus*. *Journal of Nematology*, **6**:1-6, Lake Alfred, Florida.

Salem, Ahmed Abdul-Magid, 1980: Observations on the population dynamics of the citrus nematode, *Tylenchulus semipenetrans* in Sharkia Governorate, Egypt. *Journal of Phytopathology*, **12**: 31-34.

Sasser, J.N. and D.W. Freckman, 1987: A world perspective on nematology: The role of society. p. 7-14. *In*: Vistas on Nematology:Veech, J.A. and Dickson, D.W. (Eds.), Society of Nematologists, Hyattsville, USA.

Schneider, H. and R.C. Bains, 1964: *Tylenchulus semipenetrans*: parasitism and injury to orange tree roots. *Phytopathology*, **54**: 1202-1206, St. Paul, Minnesota.

Scotto La Massese, C. 1969. The principal plant nematodes of crops in the French West Indies. p. 168-169. *In*: *Nematodes of tropical crops*: J.E. Peechy (Ed), Commonwealth Bureau of Helminthology (Great Britain) Technical Communication: 40.

Shaosheng, L., G. Rixia, W. Ziming, 1990:[*Meloidogyne citri* n.sp. (Meloidogynidae), a new root-knot nematode parasitizing citrus in China]. *Journal of Fujian Agricultural College*, **19**:305-311.

Sharma, S.B. and G. Swarup, 1982: The citrus nematode *Tylenchulus semipenetrans*. p. 76-85. *In*: Problems of Citrus Diseases in India : Raychaudhury, S.P. and Y.S. Ahlawat (Eds), Surabhi Printers and Publishers, New Delhi

Smith, G.S. and D.T. Kaplan, 1988: Influence of mycorrhizal fungus, phosphorus and burrowing nematode interactions on growth of rough lemon citrus seedlings. *Journal of Nematology*, **20**:539-544, Lake Alfred, Florida.

Siddiqi, M.R. 1961: Occurrence of the citrus nematode, *Tylenchulus semipenetrans* Cobb, 1913 and the reniform nematode, *Rotylenchulus reniformisin* India. *Proceedings of 48th Indian Science Congress* Part **111**: 504.

Siddiqi, M.R. 1964: Studies on nematode root rot of citrus in Uttar Pradesh, India. *Proceedings of the Zoological Society* (Calcutta), **17**: 67-65, Calcutta.

Siddiqi, M.R. 1986: Tylenchida : Parasites of Plants and Insects, 645pp. *Common Wealth Institute of Parasitology*,UK.

Singh, D.B., V.R. Rao and P.P. Reddy, 1979: Plant parasitic nematodes associated with horticultural crops in South India. *Indian Journal of*

Nematology, **9**:183-186, New Delhi.

SINGH, N.D. 1974: Some host plants of the reniform nematode in Trinidad, 1974. p. 119-125. *In*: Crop Protection in the Caribbean: Brathwaite, C.W.D., Phelps, R.H., Bennett, F.D.

SINGHAL, V., 1995: Handbook of Indian Agriculture, 147- 149pp. Vikas Publ. New Delhi.

SINHA, A.K. AND M.F. RAHMAN, 1994: Occurrence of citrus nematode, *Tylenchulus semipenetrans* Cobb, 1913 in Assam. *Indian Journal of Nematology*, **24**:246-247, New Delhi.

SMALL, R.W., 1979: The effects of predatory nematodes on populations of plant parasitic nematodes in pots. *Nematologica*, **25**:94-103, Leiden, Netherlands.

SPEIGEL, Y., E. COHN AND I. CHET, 1989: Use of chitin for controlling *Heterodera avenae* and *Tylenchulus semipenetrans. Journal of Nematology*, **21**:419-422, Lake Alfred, Florida.

STANDIFER, M.S. AND V.G. PERRY, 1960: Some effects of sting and stubby root nematodes on grape fruit roots. *Phytopathology*, **50**:132-156, St. Paul, Minnesota.

SUIT, R.F. AND T.L. BROOKS, 1957:Current information relating to barriers for the burrowing nematode. *Proceedings of the Florida State Horticutural society*,**70**: 55-57, Deland.

SUIT, R.F. AND E.P. DUCHARME, 1953:The burrowing nematode and other parasitic nematodes in relation to spreading decline.*Citrus Leaves*, **33**:8-9.

SUNDARAM, R., S. THAMBURAJ AND S. VADIVELU, 1990: Studies on citrus nematode in Shevroy hills of Tamil Nadu. *Journal of Hill Research*, **3**:212-215.

SUNDARAM, R., AND S. VADIVELU, 1994: Intensity of nematode infestation in manderin orange crop in the Nilgiri hills region. *Indian Journal of Nematology*, **24**:185-188, New Delhi.

SWARUP, G., C.L. SETHI AND J.S. GILL, 1964: Some records of plant parasitic nematodes in India. *Current Science*, **33**:593, Bangalore.

TANDINGAN, I.C., 1985: Biological control of nematodes *Tylenchulus semipenetrans* Cobb on citrus and *Radopholus similis* (Cobb) Thorne on banana using soil fungi *Paecilomyces lilacinus* (Thom) Samson and *Penicillium anatoicum* Stolk. *University of Philippines* circular, 89p

TARJAN, A.C., 1956: The possibility of mechanical transmission of nematodes in citrus groves. *Proceedings of the Florida State Horticultural Society*, **69**: 34-37, Deland.

TARJAN, A.C. 1961: Longivity of *Radopholus similis* (Cobb) in host -free soil. *Nematologica*, **61**:170-175, Leiden, Netherlands.

TARJAN, A.C. 1971: Migration of three pathogenic citrus nematodes through two Florida soils. *Proceedings of the Soil Crop Science Society of Florida*, **31**:253-255.

TARJAN, A.C. AND J.H. O'BANNON, 1974: Postplant fumigation with DBCP for citrus nematode control in Florida. *Journal of Nematology*, **6**:41-48, Lake Alfred, Florida.

THAPA, A. AND S. MUKHOPADHYAY, 1994: Studies on the problems of citrus nematode in orchard and nurseries of manderine orange in Darjeeling hills and their management. *Indian Journal of Nematology*, **24**:214-220, New Delhi.

THIRUMALA RAO, V. 1956: Stray notes on some crop pest outbreaks of south India. *Indian Journal of Entomology*, **18**:123-126, New Delhi.

THOMAS, E.E. 1913 : A preliminary report of a nematode observed on citrus roots and its possible relation with the mottled appearance of citrus trees. *Circular of California Agricultural Experimental Station*, No. **85**:14.

THOMPSON, I.J. 1987: Challenges facing nematology: environmental risks with nematicides and the need for new approaches. p. 469-476. *In*: Vistas on Nematology: Veech, J.A. and Dickson, D.W. (Eds), E.O. Painter Printing Company, Deleon Springs, Florida.

THORNE, G. 1961: Principles of Nematology, 553 pp. McGraw-Hill, New York and London.

TICHINOVA, L.V. 1957: Eelworms of agriculture crops in the Vzebk. SSR" Tashkent. *Akad Nauk, Vzebksoi SSR*: 101-131

TIMMER, L.W. AND R.F. LEYDEN, 1978: Relationship of seedbed fertilization and Fumigation to infection of sour orange seedlings by mychorrizal fungi and *Phytophthora parasitica. Journal of the American Society For Horticultural Science*, **103**:537-541., Alexandra Virginia.

TOUNG, M. 1969: Control of the citrus nematode. *Tylenchulus semipenetrans* with 1,2, dibromo-3-chloropropane in Taiwan. *Plant Disease Reporter*, **53**:300-301, Washington, D.C.

TSAI, B.Y. AND S.D. VAN GUNDY, 1988: Comparison of anhydrobiotic ability of the citrus nematode with other plant parasitic nematodes. *Proceedings of Sixth International Citrus Congress*, **2**:983-992

TSAI, B.Y. AND S.D. VAN GUNDY, 1988: Tolerance of proto- anhydrobiotic citrus nematodes to advers conditions. *Revue de nematologie*, **12**:107-112, Paris.

VAN GUNDY, S.D. 1958: The life history of the citrus nematode, *Tylenchulus semipenetrans* Cobb. *Nematologica*, **3**:283-294, Leiden, Netherlands.

VAN GUNDY, S.D., A.F. BIRD AND H.R. WALLACE, 1967: Aging and starvation in larvae of *Meloidogyne javanica* and *Tylenchulus semipenetrans. Phytopathology*, **57**:599-571, St. Paul, Minnesota.

VAN GUNDY, S.D. AND S. GARABEDIAN, 1982: Alternatives to DBCP for citrus nematode control. *Proceedings of the International Society of Citriculture*. November, 1981:387-390.

VAN GUNDY, S.D. AND J.D. KIRCKPATRICK, 1964: Nature of resistance in certain rootstocks to citrus nematodes. *Phytopathology*, **54**: 419-427, St. Paul Minnesota.

VAN GUNDY, S.D. AND J.P. MARTIN, 1961: Influence of *Tylenchulus semipenetrans* on the growth and chemical composition of sweet orange seedlings in soils of various exchangeable cation ratios. *Phytopathology*, **51**:146-151, St. Paul Minnesota.

VAN GUNDY, S.D., J.P. MARTIN AND P.H. TSAO, 1964: Some soil factors influencing reproduction of citrus nematode and growth reduction of sweet orange seedlings. *Phytopathology*, **54**:294-299, St. Paul, Minnesota.

VAN GUNDY, S.D. AND J.W. MEAGHER, 1977: Citrus nematode (*Tylenchulus semipenetrans*) problems worldwide.1977 *International Citrus Congress, Orlando, Florida*:7, Orlando.

VAN GUNDY, S.D. AND R.L. RACKHMAN, 1961: Studies on the biology and pathogenicity of *Hemicycliophora arenaria. Phytopathology*, **51**:393-

397, St. Paul Minnesota.

VERDEJO-LUCAS, S., F.J. SORRIBAS, J. PONS, J.B. FORNER AND A. ALCAIDE, 1997: The Mediterranean biotypes of *Tylenchulus semipenetrans* in Spanish citrus orchards. *Fundamental and Applied Nematology*, **20**:399-404, Rue Linois, Paris.

VILARDEBO, A. 1963: Etude sur les nematode parasites des agrumes au Maroc. *Al Awamia*, **7**:57-69, Rabat, Morocco.

VUCINIC, L., J. TIODOROVIC, 1988: [Occurrence of nematode, *Tylenchulus semipenetrans* Cobb, 1913 on the coastal part of Montenegro (Yugoslavia)]. *Poljoprivreda* I *Sumarstvo* (Yugoslavia) :33-42.

WALTAR, D.E. AND D.T. KAPLAN, 1990: Antogonists of plant parasitic nematodes in Florida citrus. *Journal of Nematology*, **22**:567-573, Lake Alfred, Florida.

WHITEHEAD, A.G. 1968: Taxonomy of *Meloidogyne* (Nematodea: Heteroderidae) with description of four new species. *Trans Zoological Society of London*, **31**:263-401, London.

WRIGHT, D.J. 1981: Nematodes: Mode of action and new approaches to chemical control. p. 421-499. *In*: Plant Parasitic Nematodes Vol. III:, B.M. Zuckerman and R.A. Rohde., (Eds), Academic Press, London and New York.

YANG-BAOJUN, HU-KAIJI, CHEN-HUI, ZHU WEISHENG, 1990: A new species of root-knot nematode *Meloidogyne jianyangensis* n.sp. parasitizing mandarin orange. Acta *Phytopathologica - Sinica* (China), **20**:259-264, Beijing, Chino.

YOKOO, T. 1964. Studies on the citrus nematode (*Tylenchulus semipenetrans* Cobb. 1913) in Japan. *Agricultural Bulletin, Saga University*, **20**:71-109.

YOKOO, T. AND Y. IKEGAMI, 1966: Some observations on growth of the new host plant, snapdragon (*Antirrhinum majus* L.) attacked by root lesion nematode, *Pratylenchus coffeae* and control effect of some nematicides. *Agricutural Bulletin, Saga University*, **22**:83-92..

YOUSSEF, G.M., A.A. FARAHAT, M.Y. YASSIN AND H.I. ELNAGAR, 1989: Reproduction of *Tylenchulus semipenetrans* on sour orange seedlings cultivated in different soil types (Egypt). *Bulletin of Faculty of Agriculture*, **40**:505-514.

APPENDIX

INSECT PEST OF CITRUS

CLASSIFIED LIST

ORTHOPTERA Latreille, 1793

ACRIDIDÆ Brunner, 1990 — Grasshoppers

Patanga succincta (Linnæus), Bombay Locust

Poecilocerus pictus (Fabricius), Ak grasshopper

GRYLLDÆ Saussure, 1894 — Crickets

Brachtrypes portentosus (Lichtenstein) 1796, Large brown cricket.

ISOPTERA Brulle, 1832

TERMITIDÆ (Westwood, 1940) Light, 1921 — Termites

Nasutitermes spp. — South America

Odontotermes obesus (Rambur), 1842

HETEROPTERA Laterille, 1810

COREIDÆ Leach, 1815 — Coreid bugs

Anoplocnemis phasiana (Fabricius), 1781

Dasynus antennatus (Kirby), 1891

Leptoglossus australis (Fabricius, 1781), Leaf-footed bug

L. membranaceus (Fabricius, 1781)

PENTATOMIDÆ (Leach 1815) — Stink bugs

Antestia cruciata (Fabricius, 1775)

Cappoea taprobanensis (Dallas, 1851)

Chrysochoris grandis (Thunberg, 1783)

Eurostus grossipes Dallas

Nezara viridula (Linnæus, 1758), Green stink bug

Rhynchocoris humeralis (Thunberg, 1783) (Citrus green bug

Scutellara perplexa (Westwood)
Vitellus orientalis (Distant, 1900)

HOMOPTERA

Homoptera Leach, 1815

CERCOPIDÆ Leach, 1815 — Spittle bugs
Cosmoscarta funeralis Butler, 1874

MEMBRACIDÆ Germar, 1821 — Treehoppers
Leptocentrus tanrus (Fabricius, 1775)
Oxyrhachis tarandrus (Fabricius, 1798)
Tricentrus bicolor, Distant, 1907

CICADELLIDÆ Laterille, 1802 — Leafhoppers
Idioscopus nigroclypeatus (Melichar, 1903), Mango-hopper
I. nivesoparsus (Lethierry, 1889), Mango-hopper
Tettigoniella illustris Distant, 1907

PSYLLIDÆ Latreille, 1807 — Jumping plantlice
**Diaphorina auberii* Hollis — Comoro Island (Indian ocean)
D. citri (Kuwayama, 1907), Citrus psyllid
D. communis Mathur, 1935
Leuronota minuta (Crawford, 1912)
**Psylla citrionga* — China
**P. citricola* — China
**P. murraji.* Mathur — Malaysia
**Trioza erytreae* del Guerico — Citrus psyllid. China, and S. Africa

ALEYRODIDÆ Westwood, 1840 — Whiteflies
Aleurocanthus citriperdus Quaintance & Baker, 1916
A.husaini Corbett, 1935, Citrus blackfly
A.spiniferus Quaintance, 1903, spiny whitefly
A.woglumi Ashby, 1915 (= *punjabensis*, Corbett, 1935), Citrus blackfly
Aleurocybotus setiferus (Quaintance & Baker, 1917)

---

* Not yet intercepted from India.

Aleurolobus citrifolli (Corbett, 1935)
A.marlatti (Quaintance, 1903)
.*Aleurothrixus floccosus* (Maskell, 1895) (= *howardi* Quaintance, 1907), Wooly whitefly
Aleurotuberculatus citrifolii (Corbett, 1935)
A.murrayae (Singh, 1931)
Bemisia giffardi (Kotinsky, 1907)
Dialeurodes citri (Ashmead, 1885), Citrus whitefly
D.citrifolii (Morgan, 1893), Cloudy winged whitefly
Dialeurolonga elongata (Dozier, 1928), Litchi whitefly

APHIDIDÆ (Buckton, 1881) — Aphids/plantlice
Aphis citricola van der Goot, 1912 (=*pseudopomi* Blanchard 1933; *spiraecola* Patch, 1914).
A.gossypii Glover, 1877, Cotton aphid
A.pomi de Geer 1773, Green apple aphid
Brachycaudus helichrysi (Kaltenbach, 1843) (=*xanthii* del Guercio, 1913)
Myzus persicae (Sulzer, 1776), Green peach aphid
Toxoptera auranti (Boyer de Fonscolombe, 1841)
T. citricidus (Kirkaldy, 1907) (=*argentinensis* Blanchard, 1944; *tavaresi* del Guercio, 1980)
T. odinae van der Goot, 1917

MONOPHLEBIDÆ (MARGARODIDÆ Newstead, 1910) — Margarodid scales
Drosicha contrahens (Walker, 1858)
D.dalbergiae (Green, 1903)
D.mangiferae (Green, 1903)
D.stebbingi (Green, 1902), Giant mealybug
Hemiaspidoproctus (Walkeriana) *cinerae* (Green, 1908)
Icerya aegyptiaca (Douglas, 1890)
I.minor Green, 1908
I.purchasi Maskell, 1878, Cottony cushion scale
I.seychellarum (Westwood, 1855), Seychelles fluted scale
Labinoproctus poleii (Green, 1896)
Periossopneumon tamarindus (Green, 1908)
Steatococcus assamensis Rao, 1950

ORTHEZIDÆ (Green, 1896) — Ensign scales

Orthezia insignis Browne, 1887 (nec. Douglas), Lantana bug

LACCIFERIDÆ Chamberlin, 1925 — Lac-Insects
Keria lacca (Kerreman, 1782)
Paratachardina lobata (Green)

PSEUDOCOCCIDÆ Heymons, 1915 — Mealybugs
Ferrisia virgata (Cockerell, 1893), Striped mealybug
Maconellicocus hirsutus (Green, 1908)
Nipaeococcus vastator (Maskell, 1895)
N.viridis (Newstead, 1895)
Planococcus citri (Risso, 1823), Root mealybug
P.lilacinus (Cockerell, 1905)
Pseudococcus adonidum (Linnæus), Long-tailed mealybug
P.citriculus, (Green, 1922), Citrus mealybug
P.comstocki (Kuwana, 1902)
Rastrococcus iceryoides (Green, 1908)

COCCIDÆ Stephens, 1829 — Soft Scales
Ceroplastes ceriferus (Anderson, 1791)
C.floridensis Comstock, 1880
C.rubens Maskell, 1892, Pink waxy scale
C.sinesis del Guercio, Chinese waxy scale
Chloropulvinaria psidii (Maskell, 1892) Guava scale
Coccus alphbius De Lotto, Soft green scale—E. Africa
C.bicruciatus (Green)
C.discrepans (Green, 1904)
C.hesperideum Linnæus, Brown scale
C.viridis (Green), Soft green scale
Gascardia brevicauda (Hall—East Africa)
G.destructor (Newstead), White waxy scale—Africa, Australasia, Florida, Mexico
Paracoccus species
Parasaissetia nigra (Neither, 1861)
Pulvinaria cellulosa Green, 1909
P.maxima Green, 1900
Saissetia coffeae (Walker), Helmet scale
S.hemiphaerica (Targioni, 1867)

S.oleae (Bernard), Black scale
Vinsonia stellifera (Westwood, 1871)

DIASPIDIDÆ Maskell, 1878 — Armoured Scales
Aonidiella auranti (Maskell, 1878), California red scale
A.citrina (Coquillett, 1891)
A.comperei Mc Kenzie, 1937
A.orientalis (Newstead, 1894) Oriental yellow scale
Aspidiotus destructor Signoret, 1869, Coconut scale
A.excisus Green
A.tamarindi Green, 1919
**Chionaspis pusa* Rao, 1953
Chrysomphalus dictyospermi (Morgan, 1889)
C.ficus Ashmead 1880 (=*aonidum* (Linnæus, 1758) Citrus pimple scale
Cornuaspis beckii (Newman)
Fiorinia proboscidaria Green, 1900
F.theae Green, 1900
Hemiberlesia lataniae (Signoret, 1869)
H.palmae (Cockerell)
H.rapax (Comstock, 1881)
Howardia biclavis (Comstock, 1883)
Insulaspis citrina Borchsenius, 1964)
I.gloverii (Packard, 1896), Glover scale
I.lasianthi (Green, 1900)
**I.longirostris* (Signoret), Black line scale — Pantropical
I.pallidula Williams, 1969
Lindingaspis rossi (Maskell, 1891)
Parlatoria camelliae Comstock, 1883
P.cinerea Hadden, 1909
P.oleae (Colvee, 1880)
P.pergandii Comstock, 1881, Chaff scale
P.zizyphus (Lucas, 1853)
Pinnaspis aspidistrae (Signoret, 1869), Fern scale
P.strachani (Cooley)
P.minor (Maskell, 1884)
Pseudaonidia trilobitiformis (Green, 1896) — Trilobate

* *Chionaspis* pusa Green mentioned by Ayyar (1930) under manuscript name is *nomen nundum* species.

scale

Quadraspidiotus perniciosus (Comstock, 1881), San Jose scale

Selenaspidus articulatus (Morgenson, 1889)

Unaspis articolor (Green, 1922)

U.citri Comstock, Citrus snow scale.

THYSANOPTERA Haliday, 1836

THRIPIDÆ Stephens, 1829 — Thrips*

Frankliniella dampfi Priesner, 1923 — Flower thrips

Heliothrips haemorrhoidalis (Bouché, 1833), Greenhouse thrips

Ramaswamiahiella subnudula Karny, 1926

Scirtothrips aurantii Faure, Citrus thrips

S.citri (Moulton), Citrus thrips

S.dorsalis Hood, 1919 — Chilli thrips

Thrips flawus Schrank, 1776

T.florum Schmutz, 1913

T.nilgiriensis (Ramakrishna, 1928)

T.pandu Ramakrishna, 1928

T.tabaci Lindeman, 1888, Onion, tobacco thrips

COLEOPTERA Linnæus, 1758

CETONIIDÆ MacLeay, 1819 — Flower-beetles

Heterorrhina elegans (Fabricius)

Oxycetonia albopunctata (Fabricius, 1798)

O.histro (Oliver, 1789)

O.jucunda (Faldermann, 1835)

MELOLONTHIDÆ Leach, 1817, MacLeay, 1819 — Cockchafers

Aserica nilgiriensis Sharp

Autoserica insanabilis Brenske

Ectinophoplia nitidriventris Arrow

Lachnosterna (*Hototrichia*) *consanguinea* (Blanchard)

H.repetita Sharp

H. rufoflava Brenske

Schizonycha repetita (Fabricius)

---

* Erroneously called thrip — plural thrips, thripses.

S.ruficollis (Fabricious)

RUTELIDÆ MacLeay, 1819 — Leaf-chafers

Adoretosoma citricola Ohaus

Anomala (Singhala) tenella (Blanchard, 1850), Flower beetles

CERAMBYCIDÆ Leach, 1815 — Long-horned beetles

Aesopida malasiaca Thomson

Anoplophora versteegi (Ritsema)

Blepephæus succincter (Chevrolat)

Chelidonium (Caltichroma) cinctum (Guerin-Meneville) 1844

Chloridolum alcmene Thomas, 1865

C.demene Thomas

Demonax balyi (Pascoe, 1859)

Gnatholea eburifera (Thomson, 1861)

Monohammus spp., Citrus longhorns beetle

**Niphonoclea capita* (Newman), Mango twig broer — Philippines

Peltotrachelus pubes Foerster

Oberea mangalorensis Foerster

Sthenias grisator Fabricius, 1787

Stromatium (Callidium) barbatum (Fabricius, 1775)

Xoanodera trigona Pascoe, 1859

CHRYSOMELIDÆ (Leach 1815, 1819) — Leaf eating beetles

Aspidomorpha santaecracis (Fabricius)

*Argopistes spp. Citrus flea beetles — South China

Cassida spp.

Colasposoma coerulecatum Baly, 1879

C.semicostatum Jacoby, 1908

C.splendidum (Fabricius, 1792)

**Prodagricomela nigricollis*, Citrus flea beetles — South China

Throscoryssa citri Maulik

BOSTRYCHIDÆ (Leach 1815, 1817) — False powderpost beetles

Apote monachus Fabricius, Black borer — Africa, West Indies

Sinoxylon crassum Lesne, 1897

BUPRESTIDÆ Leach, 1815 — Flat-headed Wood-borers
Agrilus grisator Kerremans, 1893
A.medicris Kerremans, 1900
Belionota prasina Thunberg, 1789, Mango buprestid
Chrysochroe fulminans (Fabricious), Jewel beetle — Philippines

MELOIDÆ Thomson, 1857 — Blister beetes
Epicauta horticornis (Hagg-Rutenberg)

SCOLYTIDÆ (Latreille, 1907), Kirby 1836 (*Ipidæ*) (Ganglbaur, 1903) — Bark beetles
Xyleborus affinis Eichhoff, 1867
X.testaceus (Walker, 1829)

ANTHRIBIDÆ (Sharp, 1873) — Fungus weevils
Araecerus fasciculatus de Geer, 1775

CURCULIONIDÆ (Leach, 1815) Stephens, 1829 — Weevils/snout beetles
Amblyrrhinum poricollis Bohemen
Astucus chrysochlorus Wiedemann
A.immunis (Walker)
Hypomeces (Cuculo) squamosus (Fabricius, 1792), Gold-dust weevil
Meleuterpes dentipes Heller
Mylocerus dentifer (Fabricius, 1792)
M.discolor Boheman, 1834
M.evasus Marshall, 1916
M.laetivirens Marshall, 1916
**Pachnaeus litus*, Citrus weevil — Cuba
Peltotrachelus pubes Faust
Phytoscaphus indicus Boheman
Sternochetus decipiens Marshall, 1916

LEPIDOPTERA Linnæus, 1758

GRACILLARIDÆ Rebel, 1901 — Leaf-bloch miners

Norsoa propolia Hampson, 1900

PHYLLOCNISTIDÆ — Small delicate moths
Phyllocnistis citrella Stainton, 1856

LIMACODIDÆ (Walker, 1855, Hübner, 1822, *Eucleidæ* Comstock, 1895) — Slug caterpillars
Latoia lepida (Cramer, 1779)

YPONEMENTIDÆ Stephens, 1927 (HYPNOMEUTIDÆ) — Ermine moths
Prays citri (Milliere, 1873)
P.endocarpa Meyrick, 1919

OECOPHORIDÆ (= DEPRESSARIIDÆ)
Agonopteryx species
Psorosticha zizyphi (Stainton, 1859)

METARBELODÆ (ARBELLIDÆ Moore, 1879) — Bark-eating caterpillars
Indarbela quadrinotata (Westwood, 1856)
I.tetraonis (Moore, 1879)

TORTRICIDÆ (Stephens, 1822) — Leaf-rollers moths
Cryptoiphlebia illepida (Butler, 1822)
Enarmonia promonella Linnæus, 1758
Cryptophlebia (Argyroploce) leucotreta (Meyrick)
Cacoecia epicyrta Meyrick
C.isocyrta Meyrick
C.pensilis Meyrick
Archips micaceana (Walker, 1863)
Ulodemis trigraphia Meyrick, 1907

PYRAUSTIDÆ (Smith, 1891) Meyrick 1896 — Pyraustid moths
Dichocrocis punctiferalis (Guenee, 1854)

GEOMETRIDÆ Stephens, 1829 — Geometrid moths
Grammodes geometrica (Fabricius)

PSYCHIDÆ Boisduval, 1829 — Bagworm moths
Euneta crameri (Westwood, 1854)

Psychid species

COSSIDÆ Leach, 1815 — Carpenterworm moths
Zeuzera coffeae Neitner, 1861

ARCTIIDÆ Stephens, 1829 — Tiger moths
Creatonotus transiens (Walker, 1855)
Estigmene perrottetic Guenee
Pelochyta astrea Drury

NOCTUIDÆ Stephens, 1829 — Underwings/owlet moths
Achaea janata (Linnæus, 1766)
Adris sikkimensis Butter
Anomis coronata
A.fulvida (Guenee, 1852)
A.trihaca Cramer, 1780
Anua coronata (Fabricius, 1794)
A.mejanesi Guenee, 1852
A.tirchoaca Cramer
Calpe emarginanta (Fabricius, 1775)
C.fasciata Moore, 1882
C.ophideroides Guenee, 1852
Calyptera bicolor (Moore, 1883)
Ercheia diversipennis Walker
E.cyllaria (Cramer, 1882)
Helicoverpa (*Heliothis*) *armigera* (Hübner, 1807)
Lagoptera (Ophiusa) dotata (Fabricius, 1775)
L.honesta (Hübner, 1780)
L.submira (Walker)
Maruca testualis Geyer, 1832
Mocis frugalis (Fabricius, 1832)
Nyctipao (Palamia) hierogluphica (Drury, 1770)
Ophiusa coronata (Fabricius)
O.tirhaca (Cramer)
Othreis ancila (Cramer, 1780), Fruit sucking moth
O.aurantia Moore, 1877
O.bicolor
O.cajeta (Cramer, 1783)
O.cocalus (Cramer, 1780)
O.discrepans Walker, 1857

O.fullonia (Clerck)
O.hypermnestra (Cramer, 1783)
O.materna (Linnæus, 1766)
O.multiscipta Walker
O.salaminia (Fabricius, 1775)
O.sikkimensis Susainathan
O.tyrannus (Guenee, 1852)
Polyresma quenavadi Guenee, 1852
Parallelia algira (Linnæus, 1776)
P.palumba (Guenee)
Pericyma glaucinaus (Westwood, 1848)
Phyllodes consobrina (Guenee)
Simplicia robustalis G
Sphingomorpha chlosea Cramer, 1780
Spodoptera littoralis (Fabricius, 1775)
**Tiracola plagiata* (Walker) Banana fruit caterpillar — South-East Asia

LYMENTRIIDÆ Hampson, 1892 — Tussock moths
Dasychira mendosa (Hübner, 1823)
D.mendosa fusiformis (Walker, 1855)
Euproctis fraterna (Moore, 1883)
Orygia postica Walker, 1885

HYPSIDÆ
Argina cribaria (Clerck, 1759)

SPHINGIDÆ Leach, 1819 — Sphinx moths
Macroglossum bellis (Linnæus, 1758)

LASIOCAMPIDÆ Harris, 1841 — Tent caterpillar moths
Paralebeda plagifera (Walker, 1855)
Suana concolor (Walker, 1855)

LYCÆNIDÆ Leach 1815 — Blues/coppers/hair streaks
Chilades laius Cramer, 1782
Tarucus theophrastus Fabricius, 1793
Virachola isocrates Fabricius, 1793

NOTODONTIDÆ Stephens 1829 — Notodontied moths

Thiacides postica Walker, 1855

PAPILIONIDÆ Leach, 1819 — Swallow tail butterflies
Papilio demoleus demoleus (Linnæus, 1758)
P.helenus Linnæus, 1758
P.helenus daksha, Moore, 1888
P.machaon Linnæus, 1758
P.machaon asiatica Menestries, 1855
P.menmon Linnæus, 1758
P.polyctor Boisduval, 1836
P.polymnestor Cramer, 1775
P.polytes alphenor Cramer
P.polytes pammon Linnæus
P.porinda Moore
P.protenor Cramer, 1775
P.rumanzovia Eschscholz
P.polytes polytes Linnæus, 1758

DIPTERA Linnæus, 1758

CECIDOMYIDÆ Macquart, 1838 — Gallmidges
Dasyneura citri Gover & Prasad, Citrus blossom midge

DROSOPHILIDÆ Loew, 1862 — Pomace/vinegar-flies
**Zaprionus multistriata,* Citrus fly

TRYPETIDÆ Loew, 1862 — Fruit flies
**Anastrepha ludens* (Loew), Mexican fruit-fly — Mexico, Central America.
Callantra icariiformis (Enderlein, 1920), Citrus fruit-fly
C.mimax (Enderlein, 1920), Citrus fruit-fly.
**Ceratitis capitata* (Wiedemann), Mediterranean fruit-fly — Mediterranean region
**C.rosa* Karsch, Natal fruit-fly — Africa
Dacus ciliatus (Loew, 1862), Ethiopian, fruit-fly
D.corrects (Bezzi, 1916), Sapota fruit-fly
D.cucurbitae (Coquillett, 1899), Melon fruit-fly
D.diversus (Coquillett, 1904), Guava fruit-fly
D.dorsalis (Hendel, 1912), Oriental Fruit-fly
D.scutellaris (Bezzi, 1913)

D.tau (Walker, 1849) (= *caudatus* Fabricius, *hageni* de Meijere)

D.zonatus (Saunders, 1841), Peach fruit-fly

**Paradalaspis quinaria* Bezzi, Rhodesian fruit-fly — South and North-west Africa

**Tetradacus isuneonis* (Miyake), Japanese orange fruit-fly — China, Japan.

HYMENOPTERA Linnæus, 1758

FORMICIDÆ Laterille, 1802 — Ants

**Atta* spp., Leaf cutting ants

Camponotus Species

Oecophylla maragdina (Fabricius, 1775), red ants

Solenopis geminata (Fabricius, 1804)

VESPIDÆ Leach, 1815 — Wasps/hornets

Polistes olivaceius (Fabricius)

P.herbæus (Fabricius)

CHALCIDIDÆ Walker, 1833 — Chalcids

Myrumicaria brunnea (Saunders, 1841).

AUTHOR INDEX

SPECIES INDEX

SUBJECT INDEX